D0712265

ENVIRONMENT AND CHEMICALS IN AGRICULTURE

Symposium jointly organised by

—the Commission of the European Communities, Brussels
—the National Institute for Physical Planning and Construction Research, Dublin
—the National Board for Science and Technology of Ireland, Dublin

under the patronage of

—Dr Karl-Heinz Narjes, Member of the Commission of the European Communities
and
—Liam Kavanagh, Minister for the Environment, Ireland

SCIENTIFIC COMMITTEE

P. G. Balayannis	Agricultural College of Athens, Greece
A. F. H. Besemer	Landbouwhoogschool, Wageningen, The Netherlands
Ph. Bourdeau	Commission of the European Communities
D. Calamari	University of Milan, Italy
A. Fairclough	Commission of the European Communities
N. J. King	Department of the Environment, London, UK
K. Mortensen	Landbrugministeriet, Copenhagen, Denmark
A. Noirfalise	Faculté des Sciences Agronomiques de l'Etat, Gembloux, Belgium
V. O'Gorman	National Board for Science and Technology, Dublin, Ireland
J. Thiault	Ministère de l'Agriculture, Paris, France
G. Weinschenck	Universität Stuttgart-Hohenheim, Federal Republic of Germany
F. P. W. Winteringham	Formerly Head of Joint FAO/IAEA Chemical Residues Programme, Vienna

ENVIRONMENT AND CHEMICALS IN AGRICULTURE

Proceedings of a symposium held in Dublin, 15–17 October 1984

Edited by

F. P. W. WINTERINGHAM
Formerly Head,
Joint FAO/IAEA Chemical Residues Programme,
Vienna, Austria

ELSEVIER APPLIED SCIENCE PUBLISHERS
LONDON and NEW YORK

ELSEVIER APPLIED SCIENCE PUBLISHERS LTD
Crown House, Linton Road, Barking, Essex IG11 8JU, England

Sole distributor in the USA and Canada
ELSEVIER SCIENCE PUBLISHING CO., INC.
52 Vanderbilt Avenue, New York, NY 10017, USA

WITH 98 TABLES AND 59 ILLUSTRATIONS

British Library Cataloguing in Publication Data
Winteringham, F. P. W.
Environment and chemicals in agriculture: proceedings of a symposium held in Dublin, 15–17 October 1984.
1. Agricultural chemicals—Environmental aspects
I. Title
631 S585

Library of Congress Cataloging in Publication Data

Environment and chemicals in agriculture.

Papers presented at the Symposium 'Environment and Chemicals in Agriculture'.
Bibliography: p.
Includes index.
1. Agricultural chemicals—Congresses. 2. Agricultural chemicals—Environmental aspects—Congresses.
I. Winteringham, F. P. W. II. Symposium 'Environment and Chemicals in Agriculture' (1984: Dublin)
S583.2.E58 1985 630 95-20735

ISBN 0-85334-404-3

Publication arrangements by Commission of the European Communities, Directorate-General Information Market and Innovation, Luxembourg

EUR 10050

Printed in Great Britain by Galliard (Printers) Ltd, Great Yarmouth

Foreword

Global and regional trends in the use of land and water resources, in population and habitat, in food and energy demands *per capita*, have led to a dramatic intensification of agriculture and forestry practices in recent decades. This intensification has involved a greatly increased use of chemicals, especially of fertilisers and pesticides.

The contribution made by chemicals to modern agriculture in terms of higher harvest yields, improved product quality, post-harvest protection and the maintenance and improvement of soil fertility is well recognised. The more complex questions of their undesirable side-effects and overall impact on the environment are less well understood and quantified.

To address these questions authoritative guidelines and adequate information are both needed. To this end objective studies of both the risks and benefits of agrochemical usage are essential, whether to allay unjustified fears or to provide the basis for imposing constraints on unnecessary or undesirable chemical use.

These issues are of special importance for the European Community (EC) because of its relatively high population densities, effectively limited land and water resources, and growing concern with environmental quality protection.

It was against this background that the Commission of the European Communities and the Irish Government jointly organised the Symposium 'Environment and Chemicals in Agriculture' which was held in Dublin, 15–17 October, 1984.

The Symposium was attended by more than two hundred delegates from thirteen EC and other countries of Western Europe (Annex I), and was held under the patronage of Commissioner Dr Karl-Heinz Narjes (for the EC) and Mr Liam Kavanagh, Minister for the Environment (for Ireland).

The proceedings, summing-up and recommendations which follow provide a basis on which possible regulatory actions over agrochemical use, improved guidelines for their safe distribution, management and effective utilisation and related research and development activities can all be considered.

Contents

Umwelt und Chemie in der Landwirtschaft: Rede anlässlich der Eröffnung des Symposiums

DR KARL-HEINZ NARJES
Mitglied der Kommission der Europäischen Gemeinschaften

Es ist eine grosse Freude und Ehre für mich, dieses Symposium über Chemie und Landwirtschaft in der Hauptstadt der Republik Irland zu eröffnen. Ich darf gleich einleitend den Dank der Kommission gegenüber der Regierung der Irischen Republik ausdrücken, dass Sie die vor etwa Jahresfrist entstandene Initiative der EG-Kommission zur Durchführung dieses Symposiums aufgegriffen und sehr tatkräftig unterstützt hat.

Mein herzliches Willkommen gilt den Vertretern des Europaïschen Parlaments sowie aus den Parlamenten der Mitgliedstaaten, es gilt den zahlreichen Repräsentanten aus den berufsständigen Organisationen der Landwirtschaft, der Industrie sowie aus den Umweltschutzorganisationen, den Verbrauchervertretungen und—last but not least—den Vertretern der Forschung.

Die Kommission ist der Auffassung, dass ein derartiges Symposium, auf dem alle Beteiligten in sachlicher Atmosphäre die Ergebnisse ihrer Arbeiten und Überlegungen vortragen können, die geeignete Atmosphäre bietet um Probleme in ihre richtigen Dimensionen zu rücken und Lösungsansätze zu finden. Bereits in der Antike war diese Form der Auseinandersetzung gewählt worden, nämlich die eines Gastmahls mit ernsten und heiteren Gesprächen, wie uns durch Plato überliefert wurde. Mögen diese beiden Facetten auch hier zu Tage treten.

Das Thema—Chemie und Landwirtschaft—ist von hoher Aktualität.

61 % der Landoberfläche der Gemeinschaft wird für agrarische Zwecke genutzt. 3·9 % des Bruttoinlandsprodukts der Gemeinschaft werden in der Landwirtschaft geschaffen. Auf Erzeugnisse der Agrikulturchemie entfallen 10 % der gesamten Erzeugung der chemischen Industrie in der Gemeinschaft. Soweit die Erzeugungsseite.

Was die Finanzierungsseite betrifft, so entfällt der grösste Teil der Finanzierung des Management der Märkte auf den Haushalt der Europäischen Gemeinschaft, in dem auf die Ausgaben der Landwirtschaft mit rund 18 Mrd. ECU rund 70% der gesamten Ausgaben des Gemeinschaftshaushalts entfallen. Im Vergleich dazu, für die seit 1972 entstandene Umweltpolitik der Gemeinschaft stehen nur 0·1% des Gemeinschaftshaushalts zur Verfügung, insbesondere für die Schaffung von Pilotvorhaben und die Durchführung von Studien.

Dennoch ist die Umweltpolitik in den Mitgliedstaaten der Europäischen Gemeinschaft von hoher Priorität. Nach einer im Herbst 1983 durchgeführten Umfrage wäre die Mehrzahl der Bürger in der Europäischen Gemeinschaft bereit, Stagnation oder verlangsamten Anstieg der Einkommen hinzunehmen, wenn dadurch eine Verbesserung der Umwelt erreicht wird. Damit soll nicht gesagt werden, dass dies notwendig ist. Es verdeutlicht jedoch das Gewicht, das der Umweltpolitik in den Ländern der Europäischen Gemeinschaft zukommt. Es ist wohl nicht übertrieben, davon auszugehen, dass die Umweltpolitik in den meisten Mitgliedstaaten die zweite Priorität nach der Beschäftigungspolitik einnimmt.

Eine der Besorgnisse unserer Mitbürger ist die Entwicklung des *Bodens*.

Hier muss verantwortliches Handeln darauf gerichtet sein, die Funktionsfähigkeit und Regenerationsfähigkeit des Naturhaushalts auf lange Sicht zu schonen und zu erhalten. So sichert die Landwirtschaft in manchen Regionen durch bestimmte Bewirtschaftungsformen die Lebensgrundlage und damit die Existenz für solche wildlebenden Pflanzen- und Tierarten deren Lebensvoraussetzungen an diese Bewirtschaftungsformen gebunden sind.

Dennoch ist auch unbestreitbar, dass die moderne Landwirtschaft auch Umwelt- und Naturbelastungen verursacht.

Mir ist völlig klar, dass der einzelne Landwirt selbst kein Interesse daran hat, seine eigene Wirtschafts- und Existenzgrundlage zu zerstören. Er ist es wohl auch, der die Gesetze der Natur deshalb am stärksten beachtet, und dies trotz all der chemischen, technischen und sonstigen Möglichkeiten.

Ein wichtiger Aspekt zur Erhaltung und Entwicklung des Bodens ist das Zusammenwirken von Chemie und Landwirtschaft. Dies ist das Thema dieses Symposiums. Darauf möchte ich im folgenden näher eingehen.

Die gemeinsame Agrapolitik wurde durch die Römischen Verträge im Jahre 1957 begründet. Ihr sind (in Artikel 39) fünf Zielsetzungen gestellt:

Steigerung der Produktivität,
Erhöhung des Pro-Kopf-Einkommens der in der Landwirtschaft tätigen

Personen,
Stabilisierung der Märkte,
Sicherstellung der Versorgung,
Belieferung der Verbraucher zu angemessenen Preisen.

Die Aspekte der Umweltpolitik sind nur angedeutet mit dem Hinweis in Absatz 2 dieses Artikels, nämlich dass bei der Gestaltung der Gemeinsamen Agrarpolitik die besonderen Eigenarten der landwirtschaftlichen Tätigkeit zu berücksichtigen sind auch im Hinblick auf die naturbedingten Unterschiede der verschiedenen landwirtschaftlichen Gebiete.

Die Umweltpolitik der Gemeinschaft entstand erst 15 Jahre nach der gemeinsamen Agrarpolitik, nämlich durch eine Konferenz der Staats- und Regierungschefs im Oktober 1972 in Paris.

In der Erklärung zur Begründung der Umweltpolitik wurde ausdrücklich darauf Wert gelegt, dass die Umweltpolitik dazu beitragen soll, eine schrittweise Annäherung der Wirtschaftspolitik der Mitgliedstaaten und eine harmonische Entwicklung des Wirtschaftslebens innerhalb der Gemeinschaft zu erreichen. Dies sind elementare Ziele des Vertrages von Rom.

Der Europäischen Umweltpolitik wurde mit dem Dritten Aktionsprogramm über die Umweltpolitik vom 7. Februar 1983 eine Dimension gegeben, die richtungsweisend für das Thema dieses Symposiums ist, und zwar wurden drei Grundsätze festgeschrieben:

1. Der Grundsatz der *Vorsorge*, d.h. keine Repression, sondern Prävention. Vorsorgen ist nicht nur wirksamer, es ist auch billiger als nachträgliches Heilen.
2. Der Grundsatz der *Subsidiarität*, d.h. es sollen auf der geeigneten Aktionsebene diejenigen Aktionsebenen vorgenommen werden, die dort am besten durchgeführt werden können. Für die Frage des Umweltschutzes in der Gemeinsamen Agrarpolitik beantwortet sich diese Frage von selbst.
3. Der Grundsatz der *Verursacherhaftung*, d.h. der Verursacher haftet für die entstandenen Umweltschäden. Ihm werden die Kosten zugerechnet. Damit wird der Umweltschutz in einer Marktwirtschaft überhaupt erst effizient. Damit wird auch sichergestellt, dass der Umweltschutz Bestandteil des Marktprozesses und damit einer optimalen Ressourcenverteilung wird.

In dem Dritten Aktionsprogramm für die Umweltpolitik ist darüber hinaus ausdrücklich festgestellt worden, dass die Unweltpolitik in allen übrigen Politiken Berücksichtigung finden muss, insbesondere gilt es, ‘die positiven Umweltauswirkungen der Landwirtschaft, ... zu verstärken und die negativen Wirkungen abzuschwächen'.

Dazu möchte ich ausdrücklich feststellen und anerkennen, dass die Landwirtschaft in der Vergangenheit die Umwelt geschaffen und bewahrt hat, die uns heute soviel wert ist. Die Landwirtschaft ist auch der Erwerbszweig, der mehr als jeder andere in Zukunft vor der Aufgabe steht, im eigenen Interesse und im Interesse der gesamten Bevölkerung die natürliche Umwelt zu erhalten und zu entwickeln. Mithin ist ein fortgesetzter Schutz der Umwelt und eine anhaltende Entwicklung der Landwirtschaft als ein kompatibles und voneinander abhängiges Paar von Zielen zu sehen, das in der gesamten Gemeinschaft verfolgt werden muss.

Dazu ist auch festzustellen, dass viele der modernen Bewirtschaftungsmethoden der Landwirtschaft, so zum Beispiel der integrierte Pflanzenschutz, sowohl dem Agrar- als auch dem umweltpolitischen Ziel dienen. Dieses Symposium wird zeigen, in welchem Ausmass dieses Zielpaar auf anderen Wegen erreicht werden kann.

Die moderne Landwirtschaft ist ohne den Einsatz chemischer Produkte nicht vorstellbar. Pflanzenschutzmittel haben die Reduzierung oder Vermeidung von Verlusten durch Pflanzenschädlinge ermöglicht. Mineraldünger verschiedener Ausprägung haben überhaupt Erträge und Pflanzenqualitäten ermöglicht, die diesem Kontinent die Selbstversorgung gebracht haben. Ich kann mir auch nicht vorstellen, dass ohne die Agrikulturchemie eine ausreichende Versorgung der Bevölkerung in den entwickelten Ländern überhaupt möglich gewesen wäre.

So wichtig diese chemischen Produkte in der Landwirtschaft sind, so oft wird indessen die Frage nach ihren möglichen unerwünschten Nebenwirkungen gestellt. Deshalb geht es um die ganz zentrale Frage: Wie kann die Landwirtschaft weiterhin Vorteile aus der Anwendung chemischer Erzeugnisse ziehen, ohne die Umwelt in unzumutbarer Weise zu belasten?

Hier geht es ganz eindeutig um ein Optimierungsproblem, für dessen Bewältigung die Politik auf den Rat der Wissenschaft angewiesen ist.

Nun ist die Wissenschaft, soweit wir dies bis zum jetzigen Zeitpunkt wissen, zwar in der Lage, eine Reihe von Einzelwirkungen verschiedener chemischer Stoffe zu analysieren, eine mittel- oder langfristige Wirkungsanalyse der Auswirkungen von Pflanzenschutzmitteln und von Mineraldüngern auf die Umwelt ist indessen erst im Werden. Nach neueren Veröffentlichungen aus den Vereinigten Staaten gibt es für 38 % der auf den Märkten angebotenen Pflanzenschutzmittel keine Informationen über die toxischen Wirkungen, die für Menschen und Umwelt auftreten können. Für 26 % sind die verfügbaren Informationen nicht ausreichend, um eine

Abschätzung möglicher Gefahren zu ermöglichen. Mit anderen Worten, für mehr als die Hälfte besteht ein Grad der Unsicherheit, aus dem sich ganz klar ein dringender Forschungsbedarf nur für den Bereich des Pflanzenschutzes ergibt.

Ein weiteres Problem, auf das unsere Wissenschaftler hingewiesen haben, stellt sich dadurch, dass viele neu in die Umwelt eingebrachten Pflanzenschutzmittel dort auf chemische Stoffe treffen, deren synergistische Wirkungen äussert schwierig abzuschätzen sind.

Hinzu kommt das Problem der Anpassung von Mikroorganismen an bekannte Pflanzenschutzmittel und die darauf folgende Resistenz, aber auch die damit einhergehende Gefahr einer Störung des agro-ökologischen Gleichgewichts. Ein besonders bekannter Störungsfaktor liegt in der Tatsache, dass eine Reihe von Pflanzenschutzmitteln nicht nur die Schädlinge sondern auch die natürlichen Feinde der Schädlinge bekämpfen, wodurch der gesamte Bodenkreislauf beeinträchtigt werden kann.

Soweit aus den vorliegenden Unterlagen ersichtlich, geht die Tendenz dahin, diesen Problemen durch integrierte Pflanzenschutzkontrolle Rechnung zu tragen. Mit Interesse sehe ich den Überlegungen der Teilnehmer dieses Symposiums zu diesem Fragenkreis entgegen.

Einige Anmerkungen zur Problematik der Düngemittel. Aus der Sicht des interessierten Beobachters der Landwirtschaft ist festzustellen, dass mit dem Streben nach immer höheren Erträgen und immer besserer Qualität landwirtschaftlicher Produkte eine ständige Zunahme der Mineraldüngung einhergegangen ist. Eine ernste Serie von Konsequenzen ist für die Sauberhaltung des Grundwassers und den Schutz der Gewässer entstanden. Regional kann die Überdüngung mit Stickstoff oder Phospaten Probleme aufwerfen, wenn diese Stoffe durch Abschwemmung in die Gewässer gelangen. Dabei können gerade Stickstoffauswaschungen zu erheblichen Nitratanreicherungen im Grundwasser führen, ein Problem, dem wir uns wohl im nächsten Jahr verstärkt werden stellen müssen, wenn die Richtlinie der Gemeinschaft über die Qualität von Wasser für den menschlichen Gebrauch vom 15. Juli 1980 für diesen Bereich der Chemikalien in Kraft tritt.

In der Bewertung der Wirkungen von Pflanzenschutzmitteln und Düngemitteln kann die Bewertung der Umweltprobleme nicht das alleinige Kriterium sein. Zu berücksichtigen sind gleichermassen die Vorteile, die sich für die Landwirtschaft selbst, für die wirtschaftliche Tätigkeit und für die Beschäftigung in den Regionen und in den betroffenen Wirtschaftszweigen ergeben. Auch hier gilt es, die mittel- und langfristigen Perspektiven soweit wie möglich aufzuzeigen.

Lassen Sie mich für Ihre Gespräche noch einige Hinweise geben und Bitten äussern in der Absicht, soviel wie möglich von Ihren Diskussionen zu profitieren:

Zunächst darf ich betonen, dass in der Gemeinschaft von 270 Millionen Menschen, die sich von den Inseln Nordschottlands bis nach Kreta erstreckt, die unterschiedlichsten Landschaften und die verschiedensten Formen der Landbewirtschaftung anzutreffen sind. Diese Unterschiede spiegeln sich wider in einem grossen Gefälle der Landbautechniken, in grossen Unterschieden in der Ausbildung der Landwirte und dem Niveau der landwirtschaftlichen und technischen Forschung. Die Bitte, dies zu berücksichtigen, soll nicht die Resignation fördern, sondern vielmehr darauf abheben, die Perspektive zu erweitern.

Alle Forschung, jede Erkenntnis und jede Problemlösung bleiben taubes Korn, wenn sie nicht auf einen fruchtbaren Boden fallen, der eine Verbreitung der Erkenntnisse ermöglicht. Ich würde es sehr begrüssen, wenn die Probleme der Bildung und Ausbildung der Bevölkerung in den ländlichen Gebieten ein Punkt sein könnte, den Sie in Ihre Überlegungen einbeziehen.

Zum dritten soll sich diese Verbreitung der Kenntnisse nicht nur auf die von der Landwirtschaft unmittelbar betroffene Bevölkerung beschränken. Die Verbraucher allgemein sollten soweit wie möglich über Merkmale hochwertiger Agrarerzeugnisse unterrichtet werden, und zwar von Erzeugnissen, die aus gesunden Böden kommen, die unbedenklich für die Gesundheit sind und die auch den Kriterien einer gesunden Ernährung entsprechen.

Dies sind die drei Bitten, die ich an Sie herantragen möchte.

Abschliessend lassen Sie mich auf ein Wort des französischen Historikers Fernand Braudel zurückkommen, der uns bereits vor mehr als einem Jahrhundert auf die Bedeutung unserer Thematik hinwies (freie Übersetzung):

'Letztlich, und davon bin ich überzeugt, ist der Wohlstand und die Stabilität einer Gesellschaft nicht von den menschlichen Aktionen abhängig, sondern vielmehr von der Produktivität seines Bodens, der Verfügbarkeit und der Nutzung von Trinkwasser, der Fruchtbarkeit der Natur und ihres Kreislaufs.'

Dem möchte ich nichts hinzufügen. Ihnen möchte ich die Erweiterung und Vertiefung dieses Themas überlassen.

Opening Address

LIAM KAVANAGH, T.D.
Minister for the Environment, Dublin, Ireland

It is a great pleasure to extend a warm welcome to you all to Dublin. I am particularly pleased that this important Symposium is taking place here during my term as President of the EEC Council of Environment Ministers. I am happy and honoured to act as co-patron of the Symposium with Dr Narjes.

The general theme of your symposium—the environmental impact of chemicals used in agriculture—is of paramount importance to all Member States of the Community. I hope that your deliberations and your conclusions will contribute, at one and the same time, to the development of agricultural practice and to the preservation and enhancement of the quality of the natural environment. We have a particular interest in this topic in Ireland as agriculture is still our foremost industry.

Agriculture has been revolutionised in the past few decades and, to a great extent, this has been due to the use of chemicals. Pesticides and fertilisers are among the most useful tools available to modern man. Many crops could not be grown or harvested without them. As a result of their use, food quality has improved and the cost of food is less than it would otherwise be. Without pesticides and fertilisers, the problem of world hunger would be much more serious than it is; indeed, attempts to deal with that problem involve increasing the use of fertilisers and pesticides and putting them to more effective use.

Questions arise, however, about the undesirable effects, direct and indirect, of new agricultural practices on the environment and on food. It is obvious that the continued use of chemicals in agriculture must be based on a comparison of costs and benefits and it is of paramount importance that risk-benefit ratios are based on reliable data and full information. Decisions cannot be made on the basis of emotive, unsubstantiated and exaggerated claims or reports which too often receive wide publicity. The scientific community has a major role to play in counteracting such reports and in educating the public and communicators, as well as the decision-makers, on the true effects of chemicals.

We must use every practical means available to minimise the environmental impact of chemicals used in agriculture. We must identify the particular practices or substances that can have an unacceptable impact and, where necessary, prohibit the practices in question. This has already been done in the case of many persistent organochlorines (e.g. DDT) and mercury-based fungicides.

Much research concerning potential environmental impact is completed prior to the marketing of new pesticide products, to meet statutory and other requirements. The position with regard to researching the actual impact of such products following use is less satisfactory. One wonders whether more progress could be made in this area at minimal cost if the scientific bodies concerned harnessed the resources and goodwill of organisations such as wildlife clubs, gun clubs, ornithology groups and so on in carrying out surveys, etc. While the impact of the now obsolete persistent organochlorine pesticides has been well researched, much work remains to be done on the environmental impact of other groups of pesticides, particularly on the more persistent ones.

The increasing use of fertilisers has shown that the environmental impact which follows their use varies considerably from situation to situation. Key factors include soil type, rainfall, height of water table, underlying geological formation, and climate. A particular practice may result in minimal environmental impact in one region, while in another that same practice may have unacceptable consequences. Solutions and proposals for solutions to existing problems must reflect this.

It is clear that nutrient enrichment of bodies of water has become a serious problem in many parts of Europe. Much research remains to be done to establish the extent to which nutrients added to soils in the form of fertilisers, end up in waters. Excessive use of fertilisers, apart from being wasteful and, in some cases, conducive to pathogenic infections, can have serious effects on fish life and other species. We need, therefore, to develop fertiliser products at realistic prices which maximise the availability of nutrients to plants over the necessary time-span, while at the same time minimising the possibility of nutrients being removed in drainage water.

This is neither the time nor the place to burden you with a detailed account of Ireland's position in relation to the use of chemicals in agriculture, but I should give a brief sketch of the situation. In Ireland, about 85% of good agricultural land is used for grass production. The usage of nitrogenous fertilisers has increased substantially in the last ten years while, in contrast, the usage of phosphatic fertilisers has decreased in the same period. The amounts of fertilisers used in Ireland are, on average, less than those used in other EEC countries, but it is expected that increases will occur regularly until there is a general achievement of the optimum application rates. The usage of pesticides and other agents such as growth regulators increased by 20% in the period 1975–1980 when approximately 1500 tonnes were used. While the level of usage of pesticides on a national basis is low, it is increasing at the rate of about 5% per annum. Recent surveys of river quality show that nitrate levels in surface and groundwaters are generally low and well within the limits set in the EEC Surface Water Directive.

In spite of the relatively low use of chemicals in agriculture and the fact that our environment is relatively clean, we must have concern for the future. We must ensure that, not only does the position not deteriorate, but that subsequent surveys will show an improvement.

It would be an oversimplification to attribute any deterioration that occurs in waters totally to agricultural chemicals. There have been cases of fish kills which were due to the careless use of pesticides. But there are other causes of pollution arising from modern agricultural practice as well as causes more directly associated with urbanisation and industry. I am confident that, with more effective implementation of our existing legislation and other systems, coupled with EEC Directives, we will bring these matters under control. The Member States of the Community already have an extensive range of legislative measures in place to ensure suitable quality of product, whether fertiliser or pesticide, and to promote their safe and effective use. Much of this legislation has been, or is being, harmonised. In relation to pesticides, my colleague the Minister for Agriculture is soon to present a Pesticide Bill to Parliament. When this is enacted, there will be provision for a comprehensive range of controls relating to quality approval in terms of safety and effectiveness, marketing and use of pesticides. These measures will complement the many existing control measures. Finally, we must not overlook the fact that an essential element in any campaign to prevent pollution from certain agricultural operations is the further development of advisory and educational services for farmers.

In conclusion, Mr Chairman, may I wish you and all the participants in the Symposium every success in your discussions over the next few days.

Your work will no doubt provide valuable guidance to all of those concerned at national and international level with the inter-relationship between chemicals, agricultural production and the preservation of natural resources and environmental quality. I congratulate the Commission for their initiative in convening the Symposium and express my appreciation of the work done by An Foras Forbartha and the National Board for Science and Technology in making the necessary preparations and arrangements.

SESSION I

Unintended and Undesirable Environmental Effects of Chemicals Used in Agriculture

Chairman: W. McCumiskey
(*National Institute for Physical Planning and Construction Research, Dublin, Ireland*)

1

Background Paper: Agrochemicals and Environmental Quality with Particular Reference to the EC*

F. P. W. WINTERINGHAM†

Formerly Head, Joint FAO/IAEA Chemical Residues Programme, Vienna, Austria

SUMMARY

Current trends in cultivated land, population, agrochemical usage and needs in the immediate decades ahead are briefly reviewed in the global as well as in the EC context. The needs and benefits of agrochemical usage are defined and the simultaneous problems of environmental quality protection highlighted. The unique position of the EC as a major agrochemical user and exporter, as well as a major overall importer, of many agricultural products is illustrated. It is concluded that this situation has important implications for the CEC and for the agrochemical industry.

1. INTRODUCTION

The present century has witnessed an explosion of science and derived technology which now affects all aspects of human life, the environment and its resources. The growing technological impact in beneficial terms such as in human health, life expectancy, services, communications, and in the range and availability of manufactured goods shows a remarkable similarity among many nations of otherwise different culture, government or economy. In non-beneficial terms the effects are also multinational: for example, the continuing decline of many non-renewable resources, rising

* Viewpoints possibly implied in this Background Paper do not necessarily represent those of either the CEC or the UN Agencies mentioned.

† Present address: Darbod, Harlech, Gwynedd LL46 2RA, UK.

world population coupled with rising human aspirations and demands *per capita* together with all their implications in terms of resource depletion and pollution, unwanted unemployment as a result of technological innovation, and the drift from rural to overcrowded urban habitats. All these trends are overshadowed by an escalating potential for multinational destruction and biological injury; here as a result of the technology impact upon 'defence' programmes. These facts suggest a declining relevance of unilateral national policies and cogently argue for a rapidly improved international communication and cooperation for optimising the impacts of modern technology on society, and despite many historical barriers of national, political or commercial rivalries [30, 53, 57].

International machinery already exists at regional levels (e.g. through the EC) and globally through the specialised Agencies of the UN. Moreover, there are signs of a growing political awareness of the essentially international nature of these problems and means for their control, for example, the need for systematic technology impact assessment [40] and a recognition of 'unemployment' as a function of advancing technology [36, 37].

2. BACKGROUND, NATURE AND SCOPE OF CEC SYMPOSIUM

The impact of modern technology upon agriculture has been no less dramatic than upon other industries. However, unlike other industries, agricultural production and contingent food supplies world-wide are literally vital for daily human survival. The evident abundance of food supplies within the more industrialised nations tends to obscure the fact that the present world population can now only be sustained by the technologically-based or 'intensified' agriculture with effectively diminishing land and water resources *per capita*. This intensification has involved the clearance of native forests and grassland, mechanisation, the introduction of selected or novel crop varieties, increasing use and dependence upon artificial irrigation and chemical aids to soil fertility, crop protection, and food harvest, storage and processing.

Moreover, and taking fully into account the difficulties of quantifying world population trends and local food needs [22, 41, 55], the fact remains that there are today hundreds of millions of people with insufficient or inappropriate food for the achievement of adequate standards of health and development. If present patterns of use were to continue then there is abundant evidence that, in order to sustain even present average levels of food supply *per capita* in the lower population scenarios of the future, agrochemical usage world-wide would considerably increase. However, the adoption of new agricultural practices and methods, which do not necessarily involve a growth in the use of agrochemicals, could offset this increase [5, 8, 9, 10, 17, 20, 47, 62].

On the other side of the coin, clearances for new agriculture and cultivation are themselves a major disturbance of the natural ecological cycles of nitrogen, carbon and water. These play a vital role in the maintenance and improvement of soil fertility, stability of soil against erosion and, in the longer term, in climatic stability [44, 45, 60]. Additionally, of the growing range and weights of agrochemicals (fertilisers, pesticides, etc.) only small fractions of pesticides reach the target pests, and in the case of fertiliser nitrogen especially, usually little more than half is usefully recovered in the crop. Inevitably, proportions of the agrochemical 'residues' find their way out of the target area and they can impair environmental quality or harm non-target wildlife populations. These side-effects, which include the selection of entire field populations of pesticide-resistant pests, can reach unacceptable levels in tropical and temperate agricultural areas [15, 27, 28, 33, 38, 42, 56].

It is against this background that the present CEC Symposium is important and timely. Unlike most international meetings concerned with agrochemicals, it reviews simultaneously both sides of the coin: needs and benefits as well as side-effects. Secondly, it considers economic aspects and ways of minimising undesirable side-effects without impairing essential agricultural productivity. Finally, it indicates optimal lines for agrochemical development and use taking into account not only EC interests but the EC's position in the global community. In short, it is an agrochemical technology impact assessment and attempts not to lose sight of the forest for technical trees.

3. AGROCHEMICAL STATUS AND TRENDS WITH PARTICULAR REFERENCE TO THE EC

3.1. Population and land

While the populations of Western Europe as a whole and of Community countries in particular have tended to stabilise in recent years the overall trends in 'arable' land *per capita* have followed the global one as indicated in Table 1 based on FAO statistics [21]. These data show that over the last

Table 1
Trends in population and 'arable' land[a]

	1969–70	*1979–80*
World—arable land (ha)	$1{\cdot}41 \times 10^9$	$1{\cdot}45 \times 10^9$
—total population	$3{\cdot}67 \times 10^9$	$4{\cdot}41 \times 10^9$
—agricultural workers + dependants	$1{\cdot}89 \times 10^9$	$2{\cdot}04 \times 10^9$
—arable land per head of total population (ha)	0·38	0·33
Western Europe—arable land (ha)	$9{\cdot}88 \times 10^7$	$9{\cdot}53 \times 10^7$
—total population	$3{\cdot}54 \times 10^8$	$3{\cdot}72 \times 10^8$
—agricultural workers + dependants	$0{\cdot}56 \times 10^8$	$0{\cdot}40 \times 10^8$
—arable land per head of total population (ha)	0·28	0·26
Belgium + Lux.—arable land per head (ha)	0·09	0·08
Denmark—arable land per head (ha)	0·54	0·52
Federal Republic of Germany—arable land per head (ha)	0·13	0·12
France—arable land per head (ha)	0·37	0·35
Greece—arable land per head (ha)	0·44	0·41
Ireland—arable land per head (ha)	0·39	0·30
Italy—arable land per head (ha)	0·26	0·22
Netherlands—arable land per head (ha)	0·07	0·06
Spain—arable land per head (ha)	0·62	0·55
United Kingdom—arable land per head (ha)	0·13	0·12
Total tractors in use—World agriculture	$15{\cdot}5 \times 10^6$	$20{\cdot}3 \times 10^6$
—European Community agriculture	$4{\cdot}5 \times 10^6$	$5{\cdot}5 \times 10^6$

[a] 'Arable land' includes land under temporary crops, temporary pasture or fallow, and land under permanent crops but not under permanent pasture [21].

full decade world population increased by 20% while that of Western Europe increased by 5%. However, and without exception, there was a decline in 'arable' or cultivated land per head of population globally, regionally and nationally. Also relevant is the growing interest in cultivated biomass as an alternative energy source and for providing botanical alternatives to certain fossil carbon products. This represents an additional future pressure on land for non-food crops with possible implications for the EC [52]; certainly for some countries [62]. The data of Table 1 also indicate the global and regional decline in employment on agricultural land when the numbers of agricultural workers (and their dependants) are expressed as a fraction of total relevant population or as the number per hectare of agricultural land in Western Europe. This trend and the tendency to bigger and fewer farms within the EC is well recognised [23]. These trends are expected consequences of the processes of agricultural intensification at farm level which involve mechanisation as well as artificial irrigation, agrochemical usage, etc. The number of tractors in use is an index of mechanisation within the EC as also shown in Table 1.

An overall index of agricultural intensification since the beginnings of agriculture some 10 000 years ago is the ratio: calorific value of harvested food/total energy input (calories) over and above manual labour (i.e. fuel, power, energy equivalents of agrochemicals, etc.). This ratio has declined from its obviously infinite value for the primitive hunter-gatherer to values of the order of 3 or less for cereals today [18, 62].

3.2. Agrochemical usage and trends

Adequate food, shelter and clothing are now surely an unquestionable and universal human right. Yet the necessary agriculture and forestry by which such needs are uniquely met are probably better appreciated in the Third World, whose people are closest to the consequences of agricultural or forestry shortcomings. Perhaps, understandably therefore, the vital roles of chemical aids to modern agriculture are even less widely appreciated. Agrochemicals (chemical fertilisers, insecticides, fungicides, herbicides, plant-growth regulators, soil- and post-harvest fumigants, etc.) play two essential roles in modern agriculture. Firstly, by increasing yield and quality as a result of soil fertility maintenance and improvement. Secondly, by the reduction or elimination of otherwise serious pre- and post-harvest losses due to pest and disease attack. The upward trends in global fertiliser usage of recent decades are not, therefore, surprising. Some comparative trends and projections are summarised in Table 2 based on FAO statistics and by assuming an effective rate of increase of 3 % per year for N, P and K in developed countries and 5 % per year for the developing countries until the year 2000 [16, 17, 43, 60].

Fertiliser nitrogen has been the single most important agricultural contributor in relation to the dramatic rise in agricultural productivity of recent decades [6, 8, 14, 26, 35, 38]. Between 1951 and 1970 the consumption of fertiliser-N 'more than doubled in Belgium and the Netherlands, tripled in Germany, Luxembourg and Italy, and showed more than a five-fold increase in France' [11]. In the UK fertiliser-N usage increased more than four-fold over the same period [38] and application rates in kg-N, -P or -K per hectare increased by some 20 % over the quinquennium 1977–82 [1, 29]. Current production and consumption of fertiliser-N, -P and -K within the EC are also indicated in Table 2.

Table 2
Trends and projections in the global use of fertiliser-N, -P and -K, and current (1982/83) production and consumption in the EC
(all data in Mt per yr)

	Global-N	*-P*	*-K*
1960	9·7	4·2	7·1
1970	28·7	8·2	12·9
1980	57·2	13·6	19·5
1990	83	19	27
2000	120	28	37
Current production and consumption in the EC[a]	*N*	*P*	*K*
Belgium + Luxembourg	0·76 (0·20)	0·46 (0·09)	— (0·14)
Denmark	0·16 (0·39)	0·13 (0·11)	— (0·15)
France	1·53 (2·20)	1·20 (1·63)	1·60 (1·74)
Federal Republic of Germany	0·99 (1·47)	0·56 (0·74)	2·22 (1·04)
Greece	0·38 (0·41)	0·19 (0·18)	— (0·05)
Ireland	0·24 (0·30)	0·01 (0·15)	— (0·19)
Italy	1·20 (0·97)	0·64 (0·66)	0·13 (0·38)
Netherlands	1·51 (0·46)	0·35 (0·08)	— (0·10)
United Kingdom	1·40 (1·56)	0·32 (0·46)	0·26 (0·52)

[a] Data kindly made available by Dr E. Meert, Secretary-General, CMC-Engrais, Brussels.

It is difficult to extract comparably meaningful data on pesticide usage because of the wide and changing spectra of the different chemicals needed and used in different countries, and because of the difficulties of quantifying the effects of pesticidal applications on crop yields and quality. Estimates on the basis of FAO and other statistics indicated an approximate global use of 1·5 Mt per year of active chemicals during the 1960s, usage by weight then increasing by some 10 % yr^{-1}, total usage being expected to level off at not less than 7 Mt yr^{-1} for various reasons, including the problems of pest resistance [61]. More recent data [31, 39] suggest a current growth rate of 4–5 % yr^{-1} in terms of trade value. These and other data [19] as combined in Table 3 clearly show the scope for increased usage in the less developed regions of the world. Thus, if the pesticide usage in 1978 per unit of agricultural land or *per capita* of North America or of Western Europe were matched in the other regions, this alone would result in a two- to four-fold increase in total world usage. In any event the North American and Western European regions represent the bigger users and exporters of agricultural pesticides.

Table 3
Agricultural land, population and pesticide usage in 1978

	Population (M)	*Arable + cult. land (M-ha)*	*Pest. use (G-US $)*
North America	242	231	3·0
Latin America	347	143	0·8
Western Europe	368	96	2·1
Eastern Europe + USSR	372	279	1·1
Africa + Middle East	549	262	0·4
Far East	1 183	266	1·3
		Total world usage	8·7

Note: 'G' = 10^9.

Despite the difficulties of quantifying crop losses and the effects of chemical pest control [12] some useful and timely appraisals have been made, classically by Cramer [13], and more recently Ahrens *et al.* [2]. Yield losses as a result of withholding crop protection chemicals for one and two consecutive years (in UK) have been reported as summarised in Table 4 [31].

A comparison of actual typical yields of various crops world-wide with those demonstrably obtainable under optimal conditions of soil fertility, pest control, irrigation, etc., suggests an immense potential for improved agricultural productivity by appropriate extension of agrochemical and related practices [7, 8, 43]. Such extension in the Third World would depend, of course, upon the necessary financial support [10], social adjustment, as well as upon education and training at farmer and extension service levels. Nevertheless, this potential has important implications for the EC-based agrochemical industry as a major agrochemical exporter in

Table 4
Reported yield reductions (%) in the absence of crop protection chemicals after one and two years in succession [31]

	One year	*Two years*
Cereals	24	45
Potatoes	27	42
Sugar beet	37	67

the global context. These implications can be seen in terms of responsibility as well as opportunity (see below).

3.3. Environmental quality

It is now increasingly recognised by scientists and technologists themselves that the accelerating scale and pace of unconstrained technological application have already become a threat to the human environment itself, a recognition first signalled globally by the United Nations Conference on the Human Environment [54]. One consequence has been the extension of classical toxicology to include the broader concepts of environmental toxicology or 'ecotoxicology'. This takes into account the effects and interactions between any or all environmental chemicals (and of physical agents such as unnatural levels of heat, ionising radiation, etc.) and the range of living micro- and macro-organisms. These organisms provide for maintaining local ecosystems and the entire biosphere within the relatively narrow limits favourable to human existence, especially in relation to the hydrological, carbon and nitrogen cycles which make agriculture, forestry and fisheries possible. Less fortunately, perhaps, this new awareness has also led to a virtual explosion of related scientific investigation, publication, stored information, programmes and lobbies of opinion. A balanced consideration of risk/benefit ratios for all environmental chemicals has, therefore, become difficult as well as important. Any 'risk' should be considered in the context of other associated risks if priorities are to be developed realistically [48, 64, 65].

3.3.1. Fertilisers

The major problem here is the increasing usage and dependence upon fertiliser-N applications coupled with (a) the associated losses to agriculture of added and native soil nitrogen, and (b) the appearance of undesirable levels of dissolved mineral nitrogen in derived surface and ground-water bodies. Within the EC the problem is a growing one and now so recognised [11, 51]. A decade of coordinated studies of these problems world-wide [24, 25, 63] has indicated the need for (a) improved soil-N and irrigation management (especially the tailoring of fertiliser-N applications to actual soil-plant needs) and, where feasible and locally acceptable, the introduction of more appropriate crop rotation practices; (b) systematic monitoring for longer-term changes in soil organic matter, for total and 'crop-available' nitrogen as well as for nitrates in derived water bodies; and (c) more attention to possible ways of reducing nitrate levels in public water supplies by supply mixing, controlled denitrification, etc.

3.3.2. Pesticides

Although rarely so identified a serious and persistent problem is the emergence of pesticide-resistant populations of agricultural pests (insects, fungi, weeds, rodents, etc.) in crop areas and under post-harvest storage conditions world-wide [15]. It is serious from the obvious point of view of crop-protection failure and, also, from those of environmental quality and of the agrochemical industry, the latter because of the ongoing need and costs of developing, evaluating and introducing alternative pesticides or control measures. These increasingly include the needs for new fungicides [50] and herbicides [32] as well as for insecticides. Many years of research into this problem have failed to indicate any simple solution but early experience demonstrated the advantages of early detection by simple and standardised field tests [15, 58, 66].

3.3.3. Research, development, monitoring and communication

A current compendium [3] lists a total of some 820 currently-used or recently superseded chemical pesticides. A decade ago the comparable total was about 550. At that time some 100 compounds had been reviewed under the FAO/WHO pesticide residues programme and maximal acceptable daily intake values (in food, etc.) established for about 70 of these compounds [59]. Since the total number of available chemical pesticides in 1950 was surely less than 100 these figures indicate the progress of agrochemical research and development. Indeed, despite the problems of pest resistance and despite the escalating costs and needs for developing new agrochemicals and formulations which are acceptable in the context of both human health and environmental quality their availability continues, thanks to effective research and cooperation between independent institutes and industry [4, 47]. These figures also indicate the scale of the problems of adequate laboratory and field evaluation before release on the one hand and the need for constantly updated, balanced and responsible publication of benefits and possible environmental risks on the other hand.

The value of laboratory tests for predicting the behaviour of new agrochemicals under field conditions, and the need for systematic monitoring for environmental residues and their possible effects using comparable methodology, are now well recognised [46]. The often slow or delayed appearance of obvious ecotoxicological effects may be too late for the introduction of effective or economic countermeasures. Such problems especially call for international cooperation and the harmonisation of testing guidelines and criteria of acceptability. In particular, they call for clear and agreed guidelines for the development and use of laboratory tests for predicting the environmental behaviour of new agrochemicals and formulations. This clearly has important implications for the CEC in relation to Europe. But there are also implications for the CEC in the global context as illustrated below.

4. CEC PROGRAMMES IN THE GLOBAL CONTEXT

The unique position of the EC as a major exporter of agrochemicals and significant net importer of many agricultural products has important implications for any 'integrated EC research and development programme'. The relatively high level of EC-based agrochemical exports is shown by the data of Table 5 [39].

The agricultural and forestry *net* import status of the EC is implied by the comparative regional estimates summarised in Table 6 based on FAO statistics [20]. These data relate to 'Western Europe' and, therefore, include some non-EC countries but the overall pattern for the EC would be similar. The agro-*chemical* export/agro-*product* import status of the EC might be seen as significant in the so-called 'circle of poison' concept. This relates to the export of 'unregistered' or 'dangerous' pesticides by 'Western' nations to Third World countries leading to unacceptable residues in food later imported by the country which originally exported the pesticide [49]. However 'justified' or 'unjustified', the publication of this concept is a further incentive for the CEC to seek improved and internationally extended harmonisation of the criteria of acceptability and safe usage of

Table 5
Comparative agrochemical exports by major exporting countries in 1983[a]

	Export by volume (kt)	
Federal Republic of Germany	164	>68% (EC countries listed)
France	126	
United Kingdom	111	
Netherlands	66	
Belgium	?	
Non-EC countries	219	<32%
	686 + ?	100

[a] Data kindly made available by Mr Alan Woodburn of Wood, MacKenzie & Co., London [39].

Table 6

Inter-regional dependence as indicated by *net* export (+) or *net* import (−) of key agricultural, forestry and fisheries products in 1979 (estimates based on FAO statistics [20])
(units in k-metric tons unless otherwise indicated)

	W. Europe	*E. Europe + USSR*	*N. America dev'ed*	*Oceania dev'ed*	*Africa dev'ing*	*Latin America*	*Near East dev'ing*	*Far East dev'ing*	*Asian cent. plan. econ.*
Bananas + citrus	−3 700	−1 300	−2 300	−25	+1 000	+4 500	−75	+750	+210
Botanical oils[a]	−800	−280	+400	−75	−230	+340	−520	+1 250	−160
Animals[b]	−2 100	−1 600	−200	−2	+1 900	+400	−8 300	−3 600	+3 800
Cocoa + coffee + tea	−2 600	−470	−1 470	−80	+1 600	+2 370	−230	+710	+125
Cotton (lint)	−1 100	+80	+1 400	+20	+250	+680	+640	−690	−770
Eggs (in shell)	+40	+60	+25	+1	−45	−15	−50	−70	+60
Fish + products	−500	−60	−360	−6	−310	+1 030	−160	+500	+50
Forest products (in units of km^3)	−33 000	+18 000	+37 000	+6 100	+5 650	−260	−4 600	+31 600	−6 000
Meat	−600	+60	−140	+1 800	−140	+470	−620	−220	+180
Milk (dried)	+400	−40	+5	+125	−15	−95	0	−75	0
Rice (milled)	−450	−920	+2 200	+230	−1 700	−560	−1 900	+700	+1 570
Rubber (natural)	−900	−430	−840	−50	+125	−185	−36	+2 950	−275
Soybeans	−15 000	−2 300	+20 600	0	−11	+2 900	−170	−700	−1 370
Wheat + flour[c]	+3 100	−10 800	+47 200	+6 900	−7 300	−6 500	−10 300	−8 000	−10 950
Wool (greasy)	−380	−190	−10	+700	0	+75	−8	−39	−21

[a] Soybean + groundnut + palm + coconut.
[b] Bovines + sheep + goats + pigs (in units of k-heads).
[c] A comparison of the above data with those of a similar analysis for 1977 [62] suggests an apparently growing dependence by the Eastern European and other developing regions upon wheat imports—mainly from North America.

agrochemicals world-wide. Such harmonisation already has the support of the agrochemical industry [34].

5. CONCLUSIONS

1. Because of global and regional trends in cultivated land *per capita* agrochemical usage has become an indispensable part of modern agricultural productivity. It seems likely to increase in the immediate decades ahead if present levels of productivity and contingent food production *per capita* are to be sustained. This takes fully into account research, likely improvements in agricultural practices, pest management, etc. But it also takes into account the often under-estimated time needed to bring research to the stage of sufficiently tested and large-scale application under field conditions for an effective impact [62].

2. The need for evaluating, monitoring and limiting pesticide residues in food has been recognised for many years, internationally, and certainly by countries of the EC. Recognition of a similar need for agrochemical residues generally in the context of environmental quality is more recent and has particular implications for the EC because of relatively high population density, effectively limited land and water resources, and high dependence upon agrochemical usage.

3. Some problems of agrochemical residue behaviour and effects on environmental quality already exist within EC countries. On the other hand, there is good evidence that these can be contained without impairing essential agricultural practices. However, balanced and updated publication of risk/benefit ratios is important.

4. The European Community is in a unique position as a major user and exporter of agrochemicals and as a net importer of many important agricultural products. It places a special burden upon the CEC and upon the EC-based agrochemical industry to seek improved harmonisation of the criteria for acceptably safe distribution and usage of agrochemicals world-wide. In short, the global situation represents both considerable opportunity and responsibility for the future.

REFERENCES

1. ADAS (1982). Fertilizer use on farm crops in England and Wales. Agricultural Development and Advisory Service (ADAS), Rothamsted Experimental Station and Fertilizer Manufacturers' Association, UK. Report SS/CH/11, pp. 1–13, Tables A1–A15.

2. Ahrens, C., Cramer, H. H., Mogk, M. and Peschel, H. (1984). Economic impact of crop losses. *Proc. 10th Int. Congr. Plant Prot., UK*, 1983, Lavenham Press, UK, pp. 65–73.
3. BCPC (1983). *The Pesticide Manual—A World Compendium*, British Crop Protection Council (BCPC), Lavenham Press, UK, pp. i–xvii and 1–695.
4. BCPC (1984). *Proc. 10th Int. Congr. Plant Prot.*, UK, 1983, Lavenham Press, UK, Vols I–III.
5. Bixler, G. and Shemilt, L. W. (Eds) (1983). *Chemistry and world food supplies: The new frontiers. Perspectives and recommendations.* Proc. Conf., Manila, 1982, Int. Rice Research Inst., Manila, pp. i–xvii and 1–169.
6. Boyd, D. A. and Needham, P. (1976). Factors governing the effective use of nitrogen. *Span*, **19**(2), 68–70.
7. Boyer, J. S. (1982). Plant productivity and environment. *Science*, **218**(4571), 443–8.
8. Brady, N. C. (1982). Chemistry and world food supplies. *Science*, **218**(4575), 847–53.
9. Brandt, W. (Chairman) (1980). A programme for survival. Report of an independent Commission on International Development Issues, Pan Books, London and Sydney, pp. 1–304.
10. Brandt, W. (Chairman) (1983). Common crisis North–South: Cooperation for world recovery. Second report of an independent Commission on International Development Issues, Pan Books, London and Sydney, pp. 1–174.
11. CEC (1975). Ecological consequences of applying modern production techniques to agriculture. Report of the CEC, Brussels.
12. Chiarappa, L., Gonzalez, R. H., Moore, F. J., Strickland, A. H. and Chang, H. C. (1975). The status and requirements of the FAO International Collaborative Programme on Crop Loss Appraisal. *FAO Plant Prot. Bull.*, **23**(3/4), 118.
13. Cramer, H. H. (1967). Plant protection and world crop production. Bayer-Pflanzenschutz, Leverkusen, FRG, pp. 1–524.
14. Dibb, D. W. and Walker, W. M. (1979). 200 Bushel corn: How some did it. *Crops & Soils Mag.*, **32**(2), 6–8.
15. FAO (1970). Pest resistance to pesticides in agriculture: Importance, recognition and countermeasures. FAO, Rome, pp. i–vi and 1–32.
16. FAO (1978) (1977). Annual fertilizer review. FAO, Rome, Statistics Series No. 19, pp. i–xix and 1–115.
17. FAO (1979). Agriculture: Toward 2000. FAO, Rome, Conf. Doc. C79/24, pp. i–xxxvii and 1–257.
18. FAO (1979). Energy for world agriculture. FAO, Rome, Agriculture Series No. 7, pp. i–xxvi and 1–286.
19. FAO (1979) (1978). *Production Yearbook*, FAO, Rome, Statistics Series No. 22, pp. i–v and 1–287.
20. FAO (1981). The state of food and agriculture—1980. FAO, Rome, Agriculture Series No. 12, pp. i–v and 1–181.
21. FAO (1981) (1980). *Production Yearbook*, FAO, Rome, Statistics Series No. 34, pp. i–v and 1–296.
22. FAO (1984). World food report—1984, FAO, Rome.

23. FAO/EEC (1980). Monthly report from the FAO-EEC Liaison Officer, No. 18, based on the European Economic Community 1979 Annual Report.
24. FAO/IAEA (1974). Effects of agricultural production on nitrates in food and water with particular reference to isotope studies. IAEA, Vienna, Panel Proc. Series, pp. 1–158.
25. FAO/IAEA/GSF (1980). Soil nitrogen as fertilizer or pollutant. IAEA, Vienna, Panel Proc. Series, pp. 1–398.
26. FAO/SIDA (1972). Effects of intensive fertilizer use on the human environment. FAO, Rome, Soils Bull. No. 16, pp. i–viii and 1–360.
27. FAO/UNEP (1976). Impact monitoring of agricultural pesticides. FAO, Rome, pp. i–iv and 1–108.
28. FAO/UNESCO (1974). Ecological assessment of pest management and fertilizer use on terrestrial and aquatic ecosystems. Final reports (2). MAB Report Series, Nos 15 and 24.
29. FMA (1982). Fertilizer statistics—1982. Fertilizer Manufacturers' Association (FMA), UK, Table 7.
30. FOUNDATION FOR ENVIRONMENTAL CONSERVATION (1977). The Reykjavik imperative on the environment and future of mankind. *Environ. Conservation*, **4**(3), 161.
31. GIFAP (1980). News from GIFAP—Industry facts. GIFAP, Brussels, *Bull.*, **2**(6), 1–2.
32. GRESSEL, J. (1984). Spread and action of herbicide tolerances and uses in crop breeding. *Proc. 10th Int. Congr. Plant Prot.*, UK, 1983. Lavenham Press, UK, pp. 608 *et seq.*
33. GTZ (1979). Pesticide residue problems in the Third World. Report by the Deutsche Gesellschaft für Technische Zusammenarbeit GmbH (GTZ), Eschborn, pp. 1–60.
34. HAYES, A. (1983). International challenges to the agrochemical industry. *Proc. 10th Int. Congr. Plant Prot.*, UK, 1983. Lavenham Press, UK, pp. 29–33.
35. HINISH, W. W. (1980). Soil fertility. *Crops and Soils Mag.*, **32**(4), 7–11.
36. ILO (1984). New technologies: Their impact on employment and the working environment. ILO, Geneva, Report published by Tycooly International Publishing Ltd, Dublin.
37. JALLADE, J.-P. (1981). Employment and unemployment in Europe. *Proc. Hague Conf. sponsored by the European Cultural Foundation*, Trentham Books, UK, pp. i–xiii and 1–234.
38. KORNBERG, H. (Chairman) (1979). Agriculture and pollution. Royal Commission on Environmental Pollution—7th Report. HMSO, London, pp. i–x and 1–280.
39. LEWIS, M. and WOODBURN, A. (1984). *Updated Agrochemical Overview Section of Agrochemical Service*, Wood, MacKenzie & Co., May, pp. 1–16.
40. OECD (1983). Assessing the impacts of technology on society. OECD, Paris, pp. 1–80.
41. PEARCE, F. (1984). Africa's drought revisited. *New Scientist*, **103**(1419), 10–11.
42. PIMENTEL, D. (1971). Ecological effects of pesticides on non-target species. Presidential Office of Science and Tech., USA, pp. 1–220.
43. RUSSELL, R. S. and COOKE, G. W. (1983). Contributions of chemistry to

removing soil constraints to crop production. *Proc. Int. Conf. on Chem. & World Food Supplies*, Manila, 1982, Pergamon Press, Oxford, pp. 1–20.

44. Salati, E. and Vose, P. (1983). Depletion of tropical rain forests. *Ambio*, **12**(2), 67–71.
45. SCOPE (1976). Nitrogen, phosphorus and sulphur—global cycles. SCOPE Ecological Bull., No. 22, pp. 1–192.
46. SCOPE (1984). Tests to predict the behaviour of environmental chemicals. SCOPE Monograph. In press.
47. Shemilt, L. W. (Ed.) (1983). Chemistry and world food supplies: The new frontiers. Proc. Conf., Manila, 1982, Pergamon Press, Oxford, pp. i–xvi and 1–664.
48. Smeets, J. (1978). Ecological aspects of the European Environmental Action Programmes. *Ecotox. & Environ. Safety*, **2**(2), 143–50.
49. Smith, R. F. and Bottrell, D. G. (1984). Transnational crop protection projects. *Proc. 10th Int. Congr. Plant Prot.*, Lavenham Press, pp. 35–42.
50. Staub, T. and Sozzi, D. (1984). Recent practical experience with fungicide resistance (as in ref. 49), pp. 591 *et seq.*
51. Stewart, W. D. P. and Rosswall, T. (Organizers) (1982). *The Nitrogen Cycle*, Royal Society, London, pp. i–ix and 1–274.
52. Strub, A., Chartier, P. and Schleser, G. (Eds) (1983). Energy from biomass. 2nd EC Conf., FRG, 1982. Applied Science Publishers, London, pp. 1–1148.
53. Thomas, L. (1984). Scientific frontiers and national frontiers—A look ahead. *Foreign Affairs*, **62**(4), 966–94.
54. UN (1972). Report of the UN Conf. on the Human Environment, Stockholm, 1972. Conf. Doc. A/CONF/48/14, pp. i–vii and 1–122 + Annexes.
55. UN (1974). The world food problem. UN Conf. Report, Rome, pp. 1–237.
56. USDA/EPA (1976). Control of water pollution from cropland. Washington, DC, Vol. I (Guidelines), pp. 1–111; Vol. II (Overview), pp. 1–187.
57. Wilson, J. T. (1977). Overdue: Another scientific revolution. *Nature, Lond.*, **265**, 196.
58. Winteringham, F. P. W. (1969). International collaborative programme for the development of standardized tests for resistance of agricultural pests to pesticides. *FAO Plant Prot. Bull.*, **17**(4), 73–5.
59. Winteringham, F. P. W. (1975). Fate and significance of chemical pesticides: An appraisal in the context of integrated control. *EPPO Bull.*, **5**(2), 65–71.
60. Winteringham, F. P. W. (1979). Agroecosystem—chemical interactions and trends. *Ecotox. & Environ. Safety*, **3**(3), 219–35.
61. Winteringham, F. P. W. (1980). Food and agriculture in relation to energy, environment and resources. *Atomic Energy Review*, **18**(1), 223–45.
62. Winteringham, F. P. W. (1983). Biomass cultivation and harvest: Global trends and prospects. *Outlook on Agriculture*, **12**(1), 21–7.
63. Winteringham, F. P. W. (1984). Soil and fertilizer nitrogen: review of studies world-wide with particular reference to plant nutrition and environmental quality protection. FAO/IAEA/GSF, Vienna. IAEA Technical Report, Series No. 244, pp. i–x and 1–107.
64. Winteringham, F. P. W. (1972). Foreign chemicals and radioactive substances in food and environment: A comparative and integrated approach to the problems. *Kemian Teollisuus* (*Finland*), **29**(9), 561–74.

65. WINTERINGHAM, F. P. W. (1984). Environmental toxicology in relation to biomass cultivation and harvest: An editorial. *Reviews in Environmental Toxicology* (*Elsevier*), **1**, 1–4.
66. WINTERINGHAM, F. P. W. and HEWLETT, P. S. (1964). Insect cross-resistance phenomena: Their practical and fundamental implications. *Chem. & Industry*, 1512–18.

2

Review of Environmental Problems Caused by the Use of Fertilisers and Pesticides

J. H. KOEMAN

Department of Toxicology, Agricultural University, Wageningen, The Netherlands

SUMMARY

A general overview is given concerning the possible impact pesticides and fertilisers may have on the natural environment. The various aspects are discussed in connection with basic functions of the environment. Special attention is paid to the need for a regional assessment of risks posed by these chemicals. Ecosystems vary largely from one region to another, as a consequence of climate, geology and physiography. Therefore the vulnerability of the systems may vary largely as well. Considering the already existing impact especially by pesticides and the many uncertainties concerning possible undesirable effects, it seems essential to strengthen the efforts with regard to the development of more sophisticated agricultural management strategies. In connection with pest management such strategies are already available but unfortunately they have not yet been adopted widely.

1. INTRODUCTION

The use of both pesticides and fertilisers shows an increasing trend on a world-wide basis. This development relates to the need to increase the world food production for the ever-growing world population. There are in general two ways to achieve an increase in production in agriculture, first by increasing the acreage of arable land and secondly by intensification of production per unit area. The expansion of arable land will provide an average by one of the proposed 4% annual production growth rate. Increased yields per hectare are required to account for the remainder [1].

Strategies for yield increase may vary from rather simple methods to prevent pre- and post-harvest losses to very sophisticated techniques, requiring external inputs and large cash expenditures. Two basic means to increase yields are the use of selected high-yielding varieties of crops and the reduction of environmental constraints that hamper production through the application of fertilisers, irrigation schemes, pest management (including the use of pesticides), improved storage, etc. [2].

A serious drawback of these developments is that the measures taken to increase productivity in agriculture may cause an array of undesirable effects on the environment. It is the object of this paper to give an overview of the problem. As numerous publications and reviews have been published on the subject in recent years the main aspects of the problem will be discussed in a broad sense.

2. ENVIRONMENTAL EFFECTS OF PESTICIDES

Environmental effects of chemicals can be classified according to the values one may recognise in the functions of the environment. In the present paper a system for classification of functions was adopted which is based on a system which was adopted by Van der Maarel and Dauvellier [3] (see also ref. 2). According to this system the following main classes of functions of the natural environment for the human society can be distinguished: (i) functions related to production and productivity, (ii) regulation functions, and (iii) carrier and information functions. The production function refers to the supply of material and energy from the natural environment to society, either for direct survival (e.g. food, oxygen) or for construction and textiles (e.g. wood, fibre).

The regulation functions comprise those ecological processes which annihilate disturbances brought about by human activities (waste processing, assimilation of carbon dioxide, and the restoration of over-exploited populations of certain economic species of animals and plants).

Carrier functions refer to the environment as substrate for human activities (e.g. recreation).

The information functions indicate both the use of information derived from the natural environment for research and education as well as the potential information with resource value such as genetic sources of useful plants and animals.

The moral rights for survival of plant and animal species are not taken into consideration in the present paper. This is largely a matter of religious and cultural concern. A global consensus on this ethical problem does not exist. It seems prudent, therefore, only to consider those environmental values which are relevant to the survival of the human species in an objective sense.

2.1. Interference with productive functions

In connection with the possible impact of pesticides on the productive functions of our environment one should, firstly, consider the direct impact of pesticides on economic species such as fish and edible crustaceans. Cases of large-scale mortality have been reported from many places. It is noteworthy that many of these are caused by an injudicious use or deliberate misuse of pesticides. Especially in certain tropical countries it has, for instance, become common practice to abuse pesticides to catch fish in lakes and rivers.

Other effects of productive functions are damage to pollinating insects and effects on organisms, which serve as food for economic species. The pollinators comprise an array of species much larger than bees alone. Many economic tree species (e.g. fruit trees) but also forest trees fully depend on insects for reproduction [4].

The presence of pesticides in water systems may depress the populations of aquatic invertebrates that form an important source of food for fish and aquatic birds. The possible occurrence of long-term suppressive effects of pesticides on the total biomass has not yet been studied in sufficient detail but certainly deserves further attention, especially in areas where the frequency of pesticide application is comparably high.

Pesticides may have profound effects on communities of animals and plants in soils. The composition of these communities may change considerably, as has been demonstrated in various studies. Certain changes in the microflora may pose problems of major concern, for instance, when the application of pesticides against one pest or disease leads to a shift to non-susceptible pests and diseases. This has, for example, been demonstrated for carbendazim precursors like benomyl (e.g. [5]). Domsch *et al.* [6] have also pointed to the impact which fumigants, fungicides and herbicides may have on equilibria of fungi and other soil micro-organisms. Changes in these equilibria may lead to an increased uptake of nutrients by crops but also to a reduced uptake. Decreased nitrification and cellulose breakdown have also been reported [7].

2.2. Interference with regulating functions

Pest resurgence as a result of predator or parasite extermination appears to be a common phenomenon. The importance of the natural enemies of pests and diseases in agricultural crops is recognised more and more. Parasites and predators may exert considerable natural regulation and they may increasingly form the basis of important new control measures in the context of Integrated Pest Management (IPM) programmes (e.g. [8], [10], [15]). In certain parts of the world, for instance with cotton in the southern parts of the USSR, biological control is already practised on a large scale (e.g. by release of Trichogramma species).

Pesticides may also affect various regulatory functions of natural systems. The purification function of the aquatic environment largely depends on the presence of metabolic activity exerted by a variety of plant and animal species, including primary and secondary producers as well as decomposers. Micro-climatic changes may occur in cases where the plant cover is altered by the use of herbicides. Subsequent changes in soil condition may cause soil degradation, erosion and a decreased water retention capacity of soils. Again possible effects of this nature have not yet been studied in detail. However, full attention should be given to the subject, for instance, in areas with potentially vulnerable soil conditions.

Perhaps the problem of development of pesticide resistance in invertebrate crop pests, pathogenic fungi and certain weeds should also be reported as an interference with regulation functions, as the result often is that the pests become much more predominant ('the environment hits back'). The number of resistant arthropod species recorded in 1965 was 182, and by 1981 the number had increased to at least 428 [9]. The conventional remedy in case of resistance was to use higher dosages and more frequent applications, causing a still higher selection as, for instance, in Nicaragua, where the average number of insecticide applications in cotton rose to an average of 25–30 per season. Nevertheless, cotton production decreased with plant protection costs varying from 26–40% of total production expenditures [10].

2.3. Interference with carrier and information functions

In many parts of the world the diversity of natural ecosystems within or in the vicinity of agricultural habitats has become impoverished markedly. This implies that valuable species either have become rare or have virtually disappeared. As a consequence many habitats lose their natural attractiveness and recreational potential for people who like to enjoy the natural amenities. The introduction of certain applications of insecticides has, for instance, given rise to a serious impact on populations of birds. Especially the treatment of seeds with insecticides and fungicides (e.g.

organomercurials) has caused large-scale mortality and population decline in birds of prey, waders, ducks and geese [11]. In a number of countries certain applications have been banned and replaced by others for this reason. As a consequence the abundance of some of the species concerned has increased steadily after the measures were taken. Populations of migratory birds are of special concern, as they are dependent on a wide range of habitats and especially vulnerable at the time of migratory movements. Pesticides may, for instance, diminish the amount of food available for the restoration of their condition in the winter habitats before spring migration.

Another aspect connected with the information function is the possible loss of plant and animal species, which have resource value for human generations to come. There are numerous species of plants in the world which have a potential as a food or medicinal resource and which are not yet exploited for these purposes at the present time. Many wild species also have resource value for plant breeding. This applies especially to the wild ancestors of the crop species, which presently form the main source of food for the human populations. It will be obvious that the use of herbicides requires special attention in this context. Specific cases have not yet been reported, but it is a fact that, due to a variety of factors, the diversity of plant life is decreasing in many countries. A contribution by pesticides therein can certainly not be excluded.

3. ENVIRONMENTAL EFFECTS OF FERTILISERS

Fertilisers such as the N-, P_2O_5- and K_2O-fertilisers are compounds of a different nature as far as their impact on the environment is concerned. Their primary effect is that they increase or maintain soil fertility. The main impact could be that they also exert a fertilising effect on the non-agricultural parts of the environment, thus giving rise to eutrophication of lakes, rivers, etc. The enrichment of soils with fertilisers may also influence the trophic status of soils in borderline or island habitats in agricultural territories.

N-fertilisers also have a certain toxic potential as nitrates may be converted to nitrite, which may interfere with respiratory functions in both warm-blooded and cold-blooded animals through induction of met-haemoglobinaemia. Another toxicological aspect concerns the possible occurrence of toxic impurities in fertiliser preparations such as cadmium.

It is at present extremely difficult to assess the exclusive role of fertilisers in connection with changes in environmental quality as the same nutrients as well as organic compounds enter into the same environment through domestic effluents and the use or disposal of animal manures.

Especially in areas where intensive animal husbandry is practised, as is the case in various countries of the European Community, the use and disposal of animal manure seems to pose a greater problem than the use of fertilisers. In certain places in The Netherlands the nitrate content of ground water is rising steadily. This concerns especially arable lands with sandy soils with a high input of animal manure. In one place in the province of Gelderland the nitrate levels in ground water have increased from approximately 10 mg per litre in 1970 to 30 mg per litre at the present time [12]. Although still well below the WHO standards of 50 mg nitrate per litre, it can be anticipated, considering the present rate of increase, that marginal situations will develop within a few decades. It is also anticipated that P-levels will start to increase considerably within five to ten years [13].

There does not seem to be an adequate solution to the manure problem at the present time. Replacing fertiliser with manure over larger areas than is the case at present is being considered, in order to reduce the problem.

Some remarks should be made about cadmium. Over larger areas within the European Community cadmium levels in soil show a rising trend. This toxic metal enters into the soil through various pathways, including wet and dry aerial deposition, sewage sludge and P-fertilisers. Selective measures against fertilisers only will certainly not solve the problem. A general policy comprising all uses and occurrences of cadmium is required. Otherwise within 100–200 years situations may arise where arable lands are no longer suitable for the production of crops for human consumption. For the fertiliser industry this may imply that methods need to be developed for the removal of cadmium.

4. REGIONAL ASPECTS OF ENVIRONMENTAL EFFECTS BY PESTICIDES AND OTHER CHEMICALS

An important fact that is still frequently overlooked is that the ecosystems in the world show a tremendous variability. As a consequence of differences in climate, geology and physiography, both the ecosystem structure and performance vary. This implies that in one location a pesticide is broken down fairly quickly (e.g. at high ambient temperatures and rich microbial life in soil and sediment) while this does not occur or at a much lower rate in other places. Other variables are the composition of ecosystems (e.g. the presence or absence of vulnerable species), differences in life cycle phenomena (e.g. spawning periods and breeding seasons), the geographical distribution of the biota in relation to the agricultural locations (e.g. the availability of undisturbed habitats from which affected species may recover) and other land use activities that may put additional stress on the ecosystems concerned. Finally, there is a large variation in the way pesticides as well as fertilisers are applied (application method, formulation, dose rate and frequency of application). It is therefore virtually impossible to extrapolate experience gained with regard to environmental effects of these chemicals from one locality to another. A proper assessment of risks will always require that local circumstances are considered in detail. There is an urgent need in our society to pay more attention to this aspect, as the present policy with regard to the appraisal of environmental hazards of chemicals is too general and therefore often leads to wrong decisions.

5. THE NEED FOR IPM STRATEGIES

It is increasingly being recognised that agricultural management should become much more sophisticated than it is at present, generally speaking. With regard to pest management one should, for instance, benefit as much as possible from the presence of naturally occurring or introduced biotic factors, which help to limit the abundance of pest organisms and the incidence of diseases. This concept is an element of the strategy known as Integrated Pest Management (IPM), which is defined as 'a pest management system that, in the context of the associated environment and the population dynamics of the pest species, utilises all suitable techniques and methods in as compatible a manner as possible and maintains the pest population at levels below those causing economic injury' [14]. The progress made in this field is still slow, in spite of the fact that remarkable results have been obtained in many places. An important obstacle is that IPM technology has not been adequately researched and developed in many places. For instance, in the US only 5% of the total federal budget for pest control supported progress in this field in recent years [15]. Another restraint is that it is difficult to sell this strategy to farmers, who are accustomed to the simpler chemical control strategy. It must also be said that the phytopharmaceutical industry does not seem to be much interested in this development, maybe because they fear that it will markedly reduce the sales of their products. The latter will certainly be the case when IPM is implemented on a wide scale. Pesticides will then be used much more carefully and only in cases where it is likely that economic threshold values will be surpassed. Considering the environmental impact pesticides already have and the many uncertainties involved, the IPM approach deserves full attention throughout the world and much more progress should be made in its development than has been the case so far. In the same way comparable developments are needed with regard to a judicious use of fertilisers in the future.

REFERENCES

1. FAO (1979). Agriculture towards 2000. Twentieth Session. Rome, 10–29 November 1979.
2. Balk, F. and Koeman, J. H. (1984). Future hazards from pesticide use. *The Environmentalist*, **4** (supplement) No. 6.
3. Van der Maarel, E. and Dauvellier, P. L. (1978). Naar een globaal ecologisch model voor de ruimtelijke ontwikkeling van Nederland. Deel 1,2. Studierapporten Rijks Planologische Dienst. Ministerie van Volkshuisvesting en Ruimtelijke Ordening.
4. Ware, G. W. (1980). Effects of pesticides on non-target organisms. *Residue Reviews*, **76**, 179–201.
5. Van der Hoeven, E. P. and Bollen, G. J. (1980). Effect of benomyl on soil fungi associated with rye. 1. Effect on the incidence of sharp eye spot caused by Rhizoctonia cerealis. *Neth. J. Plant Pathology*, **86**, 163–80.
6. Domsch, K. H., Jagnow, G. and Traute-Heidi Anderson (1983). An ecological concept for the assessment of side-effects of agrochemicals on soil microorganisms. *Residue Reviews*, **86**, 65–105.
7. Pimentel, D. and Goodman, N. (1974). Environmental impact of pesticides. In: *Survival in Toxic Environments* (M. A. Q. Khan and J. P. Bederka, Eds), Academic Press, London.
8. Clements, R. O. (1984). Control of insect pests in grassland. *Span*, **27**, 77–9.
9. Georghiou, J. P. (1981). The occurrence of resistance to pesticides in arthropods. An index of cases reported through 1980. FAO, Rome.
10. Brader, L. (1979). Integrated pest control in the developing world. *Annual Reviews of Entomology*, **24**, 225–54.
11. Koeman, J. H. (1979). Chemicals in the environment and their effects on wildlife. In: *Advances in Pesticide Science. Part 1. World Food Production—Environment—Pesticides* (H. Geissbuhler, Ed.), Pergamon Press, Oxford, New York.
12. De Haan, F. A. M. (1984). Personal communication, Agricultural University, Wageningen.
13. Lexmond, Th. M., Riemsdijk, W. H. van and Haan, F. A. M. de (1982). Fosfaat en koper in gebieden met intensieve veehouderij. Serie Bodembescherming. No. 9 Staatsuitgeverij, Den Haag.
14. FAO (1967). Panel of Experts on Integrated Pest Control.
15. Bottrell, D. G. and Smith, R. F. (1982). Integrated pest management. *Environmental Science and Technology*, **16**, 282–8A.

3

Pathways of Nitrate and Phosphate to Ground and Surface Waters

A. DAM KOFOED

State Experimental Station, Askov, Vejen, Denmark

1. INTRODUCTION

Agriculture and environment are inextricably connected. The natural environment is the workplace of agriculture, and through the years it has been formed by agricultural activities. That is the way it has been and that is the way it is all over the world, wherever there is agricultural activity.

The development of agriculture over the last 20–30 years has been characterised by very large changes, and direct employment in agriculture has been halved.

Many of the changes in the agricultural structure were made possible through technological advances and by using better varieties, earlier crop planting, by the use of fertilisers and chemical control of plant diseases and pests. All this has been and still is a basis for the very large harvest yields/ha which now can be obtained, and this is the foundation for the food supply of the world population of 4·4 billion (Table 1) [1].

However, the development that has taken place has had its price. There is probably no doubt that the pollution potential has been increased.

In this connection one should realise that the goal of agricultural policy has been higher production and increased efficiency, but continued increase in production has only been possible through intensification of the farming system with a simultaneous increase of pollution risks.

Table 1
The world agricultural area and population (1980)

	Arable area ($\times 10^6$ *ha*)	*Permanent grass* ($\times 10^6$ *ha*)	*Total population* ($\times 10^6$)	*Inhabitants per ha arable area*
EEC	55	46	271	4·9
The rest of Europe	86	40	213	2·5
USSR	232	374	266	1·1
USA	191	238	228	1·2
The rest of North and Central America	81	115	147	1·8
South America	126	450	241	1·9
Africa	181	784	470	2·6
China	99	220	995	10·1
India	169	12	684	4·0
The rest of Asia	187	372	899	4·8
Oceania	46	466	23	0·5
The world total	1 453	3 117	4 437	3·1

2. AGROCHEMICAL–ENVIRONMENTAL INTERACTIONS

2.1. Which nutrients should be considered when the talk is about pollution and leaching?

A number of different elements are necessary for plant growth. Those that should be considered are C, H, O, N, P, K, Mg, Ca, S, Cl, Fe, Cu, Mn, Zn, Mo, B, and perhaps Na and Co. When the talk is about pollution with fertilisers, mainly N and P and a few heavy metals are involved. Investigations have demonstrated, however, that plants contain many more elements as they take up whatever is dissolved in the soil solution to the degree that the conditions for this apply. Plants cannot 'discern' whether the elements they take up are useful or not.

Using neutron activation analysis of barley and wheat grains from our now 90-year-old experiments at Askov, 29 different elements have been found besides those 15–17 which are essential to the plants. For instance, not only heavy metals such as Cd and Pb but also Ag and Au have been detected.

When we talk about essential plant nutrients, it is about macro nutrients, which are taken up in considerable amounts (50–100–200 kg ha^{-1} or more), and micro nutrients, which are only taken up in very small amounts.

Table 2
Estimates of the amount of nutrients in the soil

Element	*Content in the plough layer* (‰)	*Content in the plough layer* ($kg\,ha^{-1}$)	*Extract-able* ($kg\,ha^{-1}$)	*Consumed by crops* ($kg\,ha^{-1}\,yr^{-1}$)	*Reserve, especially*
N	2	6 000	50	50–300	Humus
P	0·4	1 200	500	15–30	Humus (Ca phosphates)
K	12	36 000	300	50–250	Clay minerals
S	0·3	900	15	5–25	Humus
Mg	1	3 000	100	10	Clay minerals
Ca	3	9 000	3 000	20	Feldspar
Cu	0·02	60	10	0·03	Clay minerals
Mn	0·5	1 250	10	0·3	Feldspar
Mo	0·001	2·5	1	1	Clay minerals
B	0·003	7·5	2	0·1	Tourmaline

The soil's own supplies of readily available plant nutrients are not large, and without fertilisers the reserves are used up and the soil is exhausted.

Høj [2] indicates the content of plant nutrients in the soil under Danish conditions as shown in Table 2.

In our long-term experiments, which include treatments which for more than 90 years have received no FYM or commercial fertiliser, the soil has only been able to supply the plants in a four-course rotation with 42 kg N, 5 kg P and 25 kg K ha^{-1}, whereas, for instance, a grain crop removes about 100–125 kg N ha^{-1}, 20–30 kg P ha^{-1} and 60–80 kg K ha^{-1}. Therefore plant nutrients have to be supplied in the form of animal manure or other organic manure or mineral fertiliser.

2.2. The basis of fertilisation

The real basis for modern fertilisation planning is the compensation or maintenance fertilisation that is expressed in the principle which was formulated by Justus V. Liebig in 1842. This states that in order to maintain plant production, one has to supply as much plant nutrient as the plant removes, and this plant nutrition is taken up in an inorganic form as mineral matter (as ions).

Thus, it is possible to increase the yield by supplying the mineral which is at a minimum until another element or another growth factor becomes limiting. However, this holds true only to a certain limit.

Modern fertilisation and planning rest on compensation fertilisation and the law of the minimum and on the law of diminishing returns. In the longer term what the crops remove must be supplied.

This has led to the following evolution in the use of commercial fertilisers [3]:

From 1880 it is recognised that P deficiency in the soil can be met by the use of industrially produced P-fertilisers.
From 1910 it is recognised that K deficiency can be met by increased use of K-fertilisers.
From 1920 with the invention and development of technical N-synthesis increased use of N-fertiliser is possible.
From 1940 the use of micro elements is increased after their importance has been recognised.
From 1960 the use of multi-compound fertilisers increases.
From 1970 there is increased attention towards the environmental consequences of fertilisation.
From 1975 interest in the use of more refined fertilisation practice increases by the availability of crops with higher yield potentials.
From 1980 both before and after 1980, the increasing cost of energy has resulted in increasing fertiliser prices, and in the 1980s we have computer programmed fertilising.

2.3. Pathways and transformations

In the context of the environment attention is particularly directed towards nitrogen (N) and phosphorus (P) besides certain heavy metals such as cadmium (Cd) and lead (Pb), nitrogen being most important.

Nitrogen gas (N_2), which constitutes 78·1% of the air, is used only by certain micro-organisms (N_2-fixing micro-organisms) and by the chemical fertiliser industry, which convert it to ammonia by the process of N_2-fixation. Small additional amounts of nitrogen are fixed as oxides by lightning and during high-temperature combustion. The nitrogen thus fixed eventually reaches the soil and is taken up by plants, which are eaten by man, and by the animals which produce a portion of his food. Such inputs of nitrogen from the air are counterbalanced by nitrogen losses, mainly from the nitrates that are formed as the end-product of the various biological transformations undergone by fixed nitrogen. Nitrates are decomposed by a process known as denitrification, and the nitrogen which they contain is returned to the atmosphere. Together, these gains, transformations and losses of nitrogen constitute the nitrogen cycle.

There is now an almost countless number of publications about N and the N-cycle. A general view of the N-cycle is shown in Fig. 1.

From an agricultural point of view, it is vital that there is an appropriate and reasonable balance between input and output of a given element, so that the economical and ecological optimum can be as close as possible under a given set of circumstances.

The biological transformations of nitrogen include the following important processes [5]:

(a) Assimilation of N in inorganic form, mainly as ammonium and nitrate by plants and micro-organisms, whereby organic nitrogen compounds are formed.
(b) Heterotrophic transformation of organic nitrogen from one organism to another (organic N in plants to animal protein).
(c) Ammonification, whereby organically bound nitrogen is decomposed to NH_4^+ or NH_3.
(d) Nitrification by oxidation of ammonium to nitrite and nitrate.
(e) Denitrification by reduction of nitrate to N_2O or N_2.
(f) Nitrogen fixation, whereby N_2 is reduced to ammonium and organic compounds by a number of micro-organisms.

The most important micro-organisms, contributing to the assimilation in water and soil, are autotrophic algae and higher plants, but in the soil bacteria also play an important role in the assimilation of inorganic N, which can be mineralised as well as immobilised.

Mineralisation takes place through ammonification and nitrification. Bacteria and fungi take part in this process. Nitrification happens mainly through the action of aerobic bacteria, which obtain their energy through the oxidation of ammonia to nitrite and nitrate.

In the long-term experiments at Askov, including a treatment where no FYM or mineral fertiliser has been used, the crop rotation (winter wheat, beet roots, spring barley and clover grass) has annually removed 40–45 kg N ha^{-1} of mineralised N.

Under other conditions maybe more than 100 kg N ha^{-1} is mineralised from the soil organic N pool, but if the soil humus content does not decrease a similar amount must be immobilised. By intensive afforestation very intensive mineralisation takes place.

In experiments using ^{15}N-tagged nitrogen fertiliser, 20 % of the applied N has been found incorporated (immobilised) in the soil organic N pool.

Denitrification takes place through the action of bacteria, which reduce

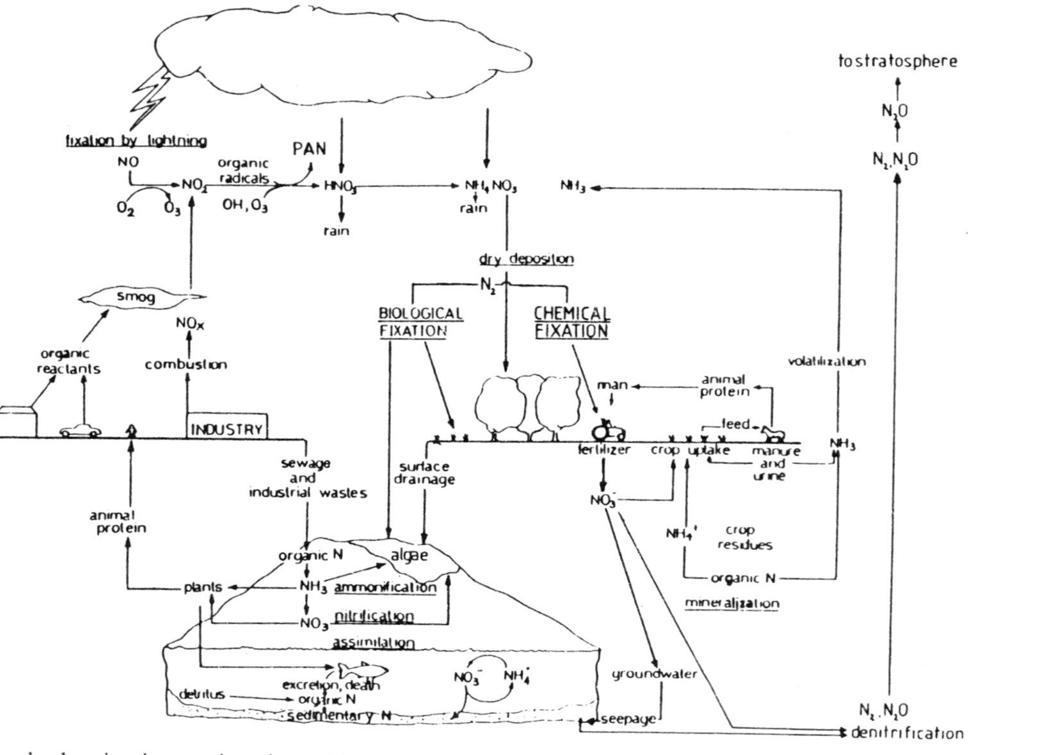

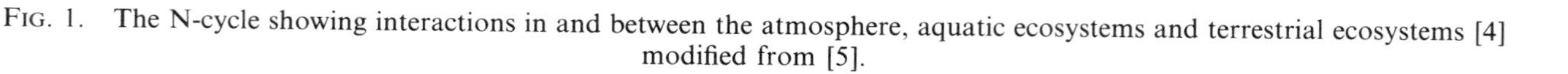
FIG. 1. The N-cycle showing interactions in and between the atmosphere, aquatic ecosystems and terrestrial ecosystems [4] modified from [5].

nitrate under anaerobic conditions, or by chemical reduction in the deeper soil horizons.

Nitrogen fixation by certain bacteria and algae is an important source for the plant's nitrogen supply.

Jansson [6] has described in detail the pathways of the nitrogen cycle in the soil, and has shown, as described in Fig. 2, that the soil contains a large passive organic N pool and a smaller active organic N pool besides the inorganic N in the form of NO_3^- and NH_4^+.

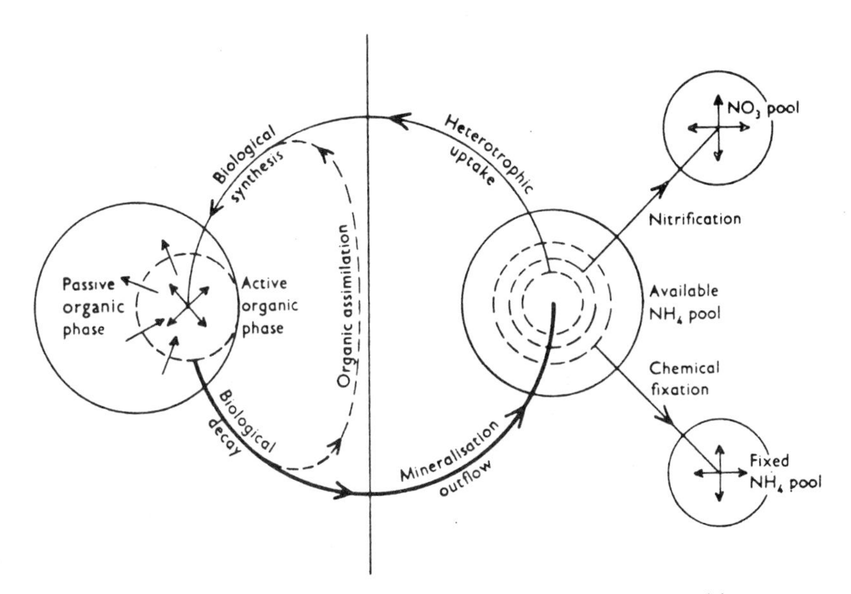

FIG. 2. Pathway of the internal nitrogen cycle in a closed decomposition system. The inorganic phase represents an incompletely nitrifying system.

The NO_3^- present in the system is exposed to leaching as the nitrate ion shows practically no absorption in the soil colloids.

Volatilisation of NH_3 can occur in the upper soil horizon. The possibilities exist when large amounts of nitrate-rich plant residues are decomposed.

Denitrification that occurs under anaerobic conditions results in a loss of nitrogen.

As a result of the risk of loss from the soil NO_3^- reservoir by leaching and denitrification, it is not good practice to sustain a considerable nitrate pool.

2.4. Fertiliser consumption

N in fertiliser occurs as NH_4^+, NO_3^- and amide nitrogen, while P is present either as water-, citrate- or mineral-acid soluble P. Water-soluble P-fertilisers are the most frequently used.

Ammonium nitrogen constitutes 30–35% of the total-N in FYM (solid) and 60–65% of the total-N in slurry. The rest is found in organic compounds.

As mentioned above, fertilisation is the basis for a sufficiently large and adequate plant production.

Fertilisation aims at

(a) supplementing the natural supply of nutrients;
(b) replacing nutrients that have been taken up by plants or otherwise lost;
(c) increasing the production potential of the soil.

2.4.1. Mineral fertiliser

The total world consumption of mineral fertilisers is shown in Table 3.

The world consumption of N has increased, while P and K consumption has been relatively constant during recent years. There has been a not inconsiderable increase in the consumption of mineral fertilisers in the developing countries over the past ten years (Fig. 3).

There is no doubt that the use of fertiliser will become an even more necessary and important factor in world agriculture than it is today.

Within the EEC, the consumption of mineral fertiliser has developed as depicted in Table 4.

2.4.2. Farmyard manure

For many years animal manure was the principal means of supplementing plant nutrients of the soil. It still has great significance.

Table 3
The world consumption of mineral fertilisers
(million tonnes)

	1978–79	*1979–80*	*1980–81*	*1981–82*
Nitrogen, as N	53·8	57·3	60·6	60·4
Phosphorus, as P	13·1	13·6	13·7	13·5
Potash, as K	20·3	19·9	20·1	19·9

From *FAO Fertilizer Yearbook* (1983).

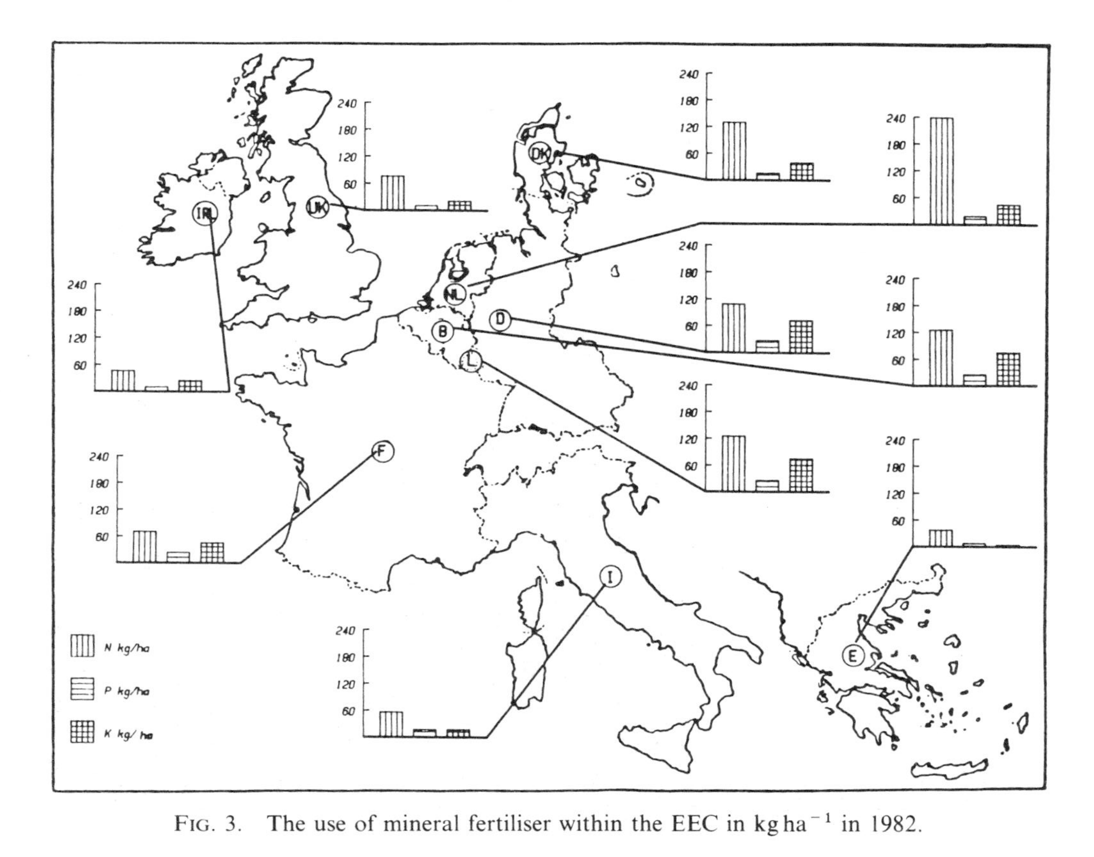

FIG. 3. The use of mineral fertiliser within the EEC in kg ha^{-1} in 1982.

Table 4
The nitrogen consumption within the EEC countries
(1000 t N; 1982)

Locality	*1971–72*	*72–73*	*73–74*	*74–75*	*75–76*	*76–77*	*77–78*	*78–79*	*79–80*	*80–81*	*81–82*
Belgium + Luxembourg	167	167	165	185	182	194	190	205	196	194	197
England	930	789	874	927	1 045	1 110	1 177	1 222	1 314	1 240	1 386
France	1 525	1 588	1 833	1 555	1 708	1 815	1 832	1 978	2 135	2 147	2 193
The Netherlands	374	376	412	435	453	430	447	442	486	483	477
Ireland	98	132	130	133	153	168	230	264	248	275	275
Italy	625	692	672	672	724	700	801	1 043	1 107	1 012	981
Luxembourg	12	13	14	—	—	—	—	—	—	—	—
Federal Republic of Germany	1 131	1 189	1 101	1 201	1 228	1 323	1 325	1 354	1 477	1 551	1 323
Denmark	308	329	365	300	339	349	374	379	394	374	376
Greece	—	—	—	—	—	—	—	—	—	333	335

Figure 4(a) shows the calculated t ha^{-1} of FYM (solid + liquid), and N, P and K in FYM within the EEC based on livestock populations (1 livestock unit = 1 cow 550 kg = 16 t FYM/year) and Danish analyses. The distribution of holding sizes is also shown in Fig. 4(b).

The conditions of animal production naturally vary within the EEC. In Table 5 is shown the density of livestock in various European districts.

The highest concentration of animals is found in Belgium and The Netherlands. If the animals could be distributed evenly, most places would have no problems regarding the soil loading, but this is not the case. There is a clear tendency in Western Europe towards larger but fewer animal producers both as regards cattle and pig production. The distribution is seen in Figs 5(a) and 5(b) for cattle and pigs, respectively. In a number of cases, the expansion of the animal livestock has taken place on small farms and here problems can arise in relation to the appropriate use of the animal manure.

If the price of animal manure is relatively high (its value depends on the price of mineral fertiliser), relatively less is used per ha. But if the situation is such that money has to be spent in order to get rid of the manure, larger amounts are used per ha. However, this has a number of consequences that include surplus manuring, risk of unreasonable accumulation of plant nutrients and, if the storage capacity is small, manure has to be applied at an unfavourable date with the risk of surface run-off and contingent pollution.

On farms with a more harmonious relation between animal production and acreage, the problems are not the same, but naturally there remains a

Table 5
Districts of the European Community with the highest density of livestock[a]

District		*Density of livestock*	*District*		*Density of livestock*
Name	*Country*	*Total LU (100 ha AA)*	*Name*	*Country*	*Total LU (100 ha AA)*
Antwerpen	B	365·87	Vejle	DK	158·00
West-Vlaanderen	B	354·96	Lombardia	I	157·42
Noord-Brabant	NL	325·92	Bornholm	DK	155·97
Gelderland	NL	325·71	Ribe	DK	151·05
Oost-Vlaanderen	B	300·43	Brabant	B	150·96
Limburg	B	289·96	Luxembourg	B	150·18
Limburg	NL	287·07	Fyn	DK	149·12
Utrecht	NL	260·69	Århus	DK	148·59
Overijssel	NL	227·35	Ringkøbing	DK	147·37
Oldenburg	D	200·83	Detmold	D	147·19
Finistère	F	194·36	Cheshire	UK	146·19
Viborg	DK	188·95	Radnorshire	UK	145·90
Liège	B	176·57	Lancashire	UK	144·84
Munster	D	176·37	Sønderjylland	DK	144·55
Côtes du Nord	F	172·15	Mayènne	F	144·16
Osnabruck	D	168·79	Northern Ireland	UK	141·59
Friesland	NL	167·89	Denbighshire	UK	140·77
Flintshire	UK	165·09	Zuid-Holland	NL	140·54
Ille-et-Vileine	F	163·41	Berlin (West)	D	140·42
Nordjylland	DK	160·32	Hainaut	B	140·08

[a] EEC Study No. 51 (September 1978).

need for an appropriate economical use of the manure and consideration of environmental aspects.

As regards fertiliser and pollution, organic manure is a very important factor, and the content of organic matter also is of considerable importance. Thus, the BOD_5 can be 25–50 000 (limits for water to be discharged into freshwater recipients are 20–30).

It is beyond doubt that better handling and use of animal manure would contribute to decreased pollution of ground water as well as other water areas.

Sewage sludge as an N–P fertiliser does not play any large role in agricultural fertilisation (Table 6) [7]. The content of heavy metals is important in this connection.

The nutrient content of sewage sludge is compared to other organic fertilisers in Table 7 [7].

The total content of nitrogen in sludge as a percentage of wet matter is comparable to cattle and pig slurry. The phosphorus content is higher, but varies according to the treatment used in the sewage plant. The potassium content is low compared to the other organic manures.

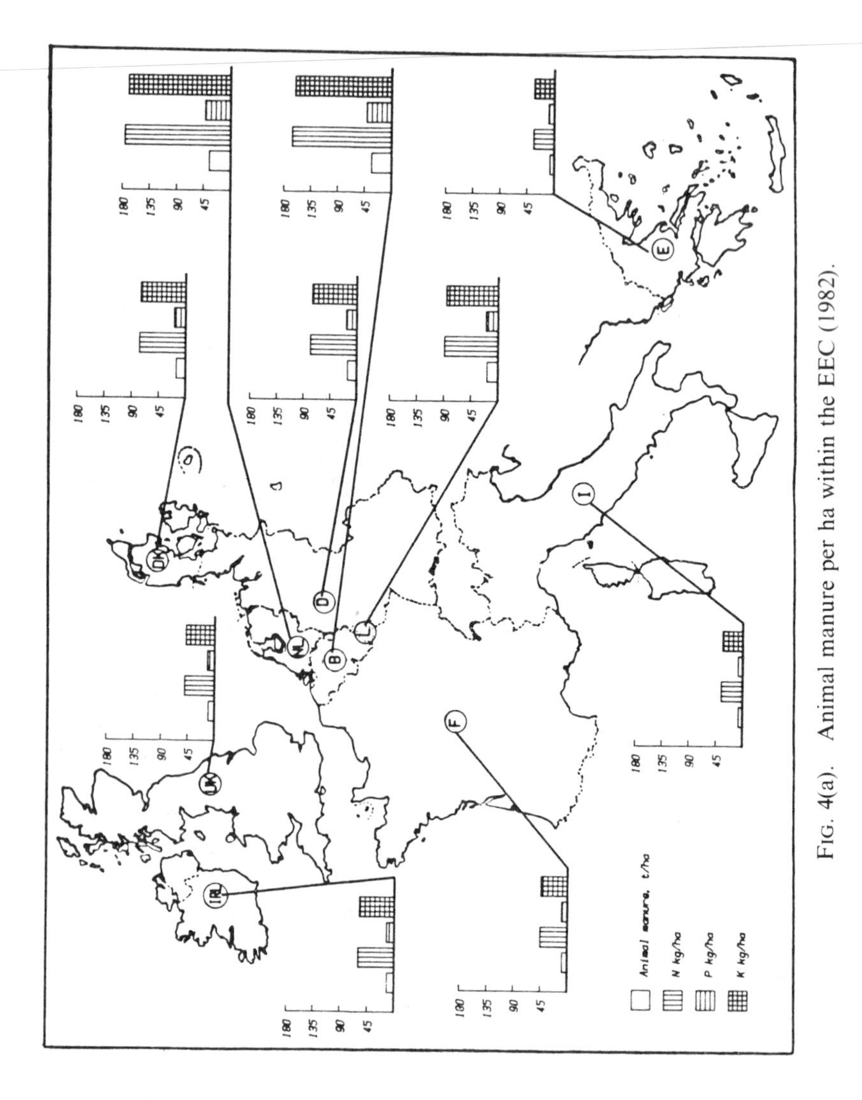

FIG. 4(a). Animal manure per ha within the EEC (1982).

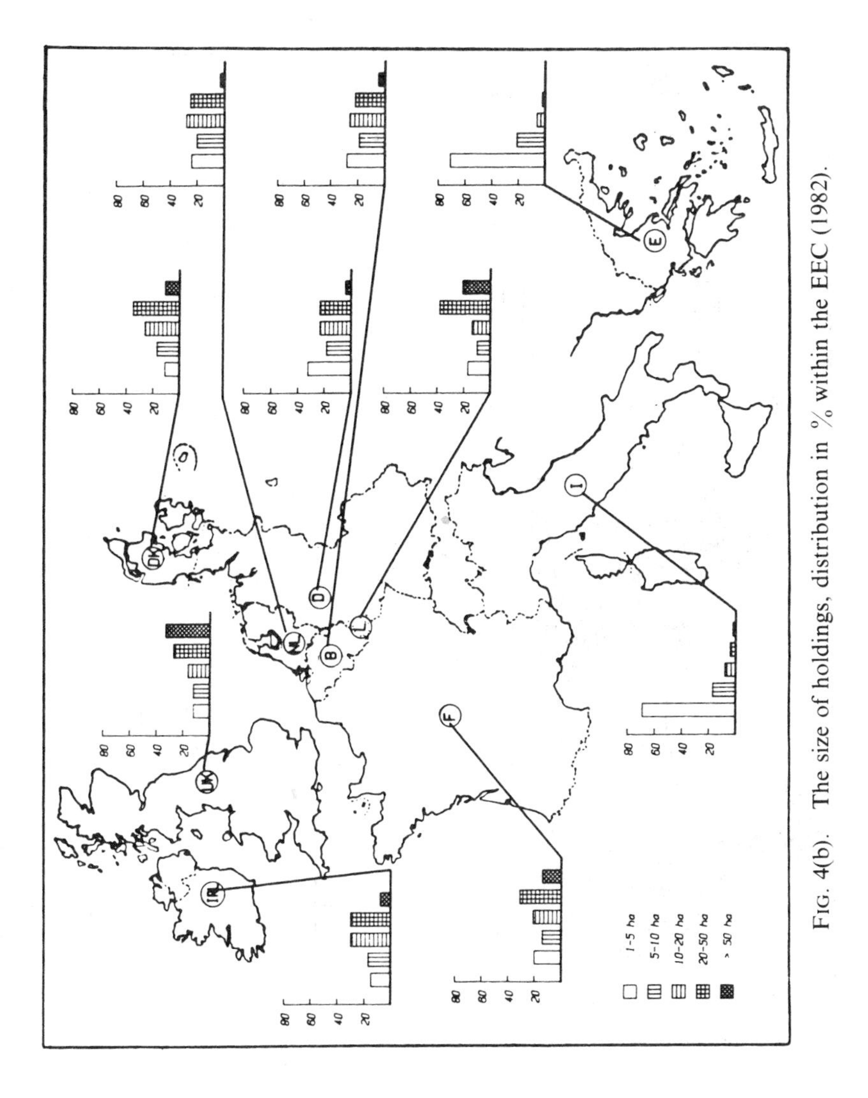

FIG. 4(b). The size of holdings, distribution in % within the EEC (1982).

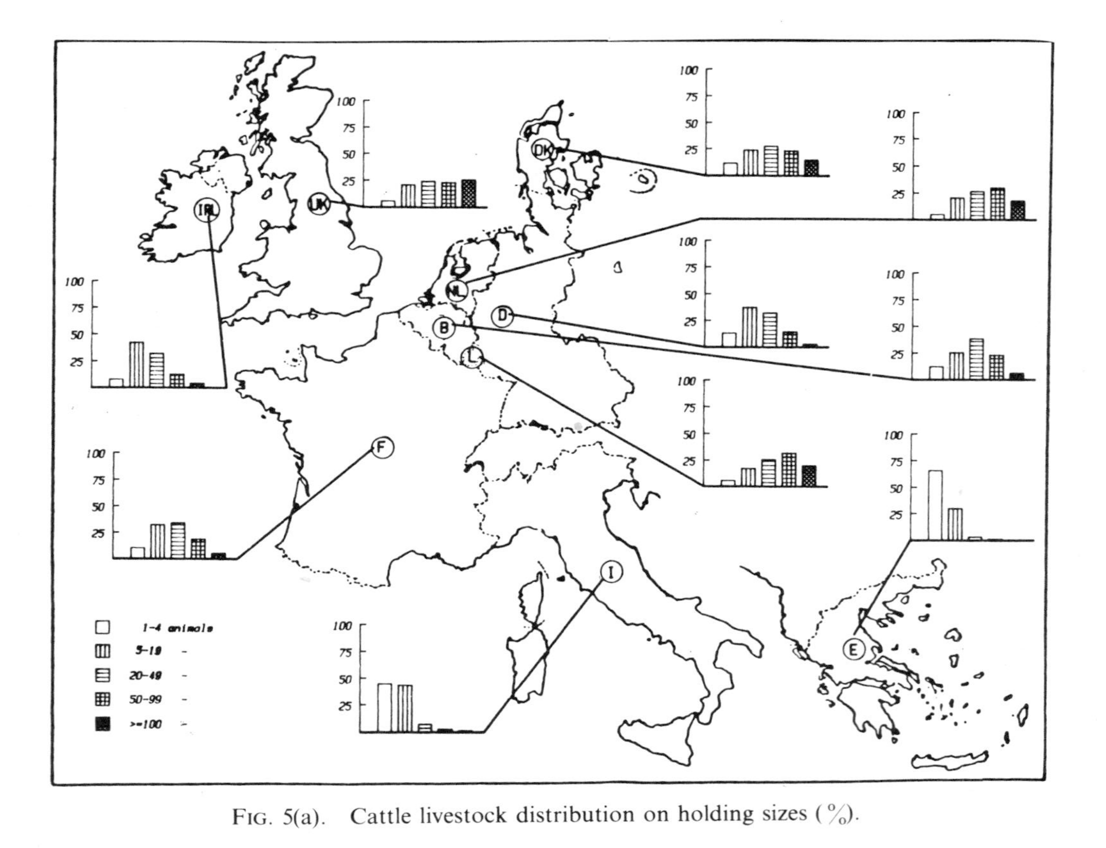

FIG. 5(a). Cattle livestock distribution on holding sizes (%).

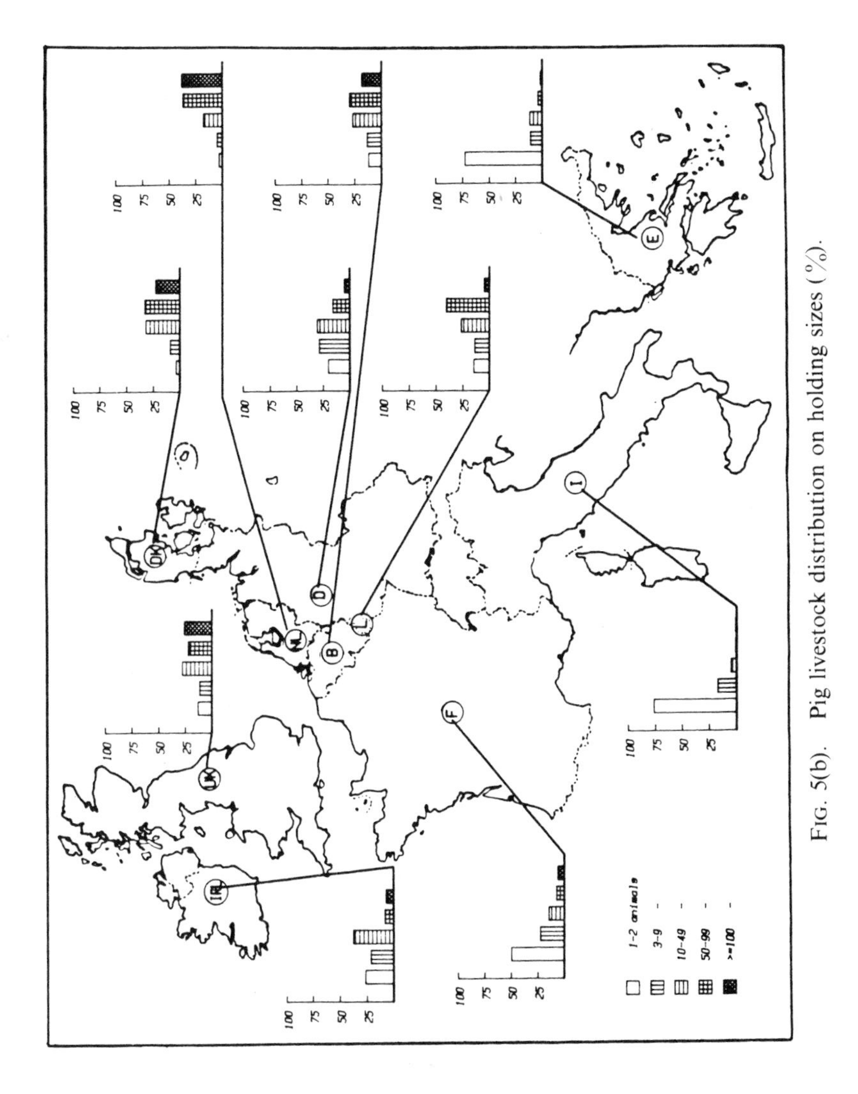

FIG. 5(b). Pig livestock distribution on holding sizes (%).

Table 6
Production of sludge and disposal in agriculture (1978–80) [7]

	Production of non-stabilised sludge in 1000 t dry matter	*Disposal in agriculture (%)*
Sweden	240	41
Norway	55	18
Finland	117	21
Denmark	127	48
Federal Republic of Germany	2 200	39
France	840	33
Belgium	70	15
The Netherlands	201	34
United Kingdom	1 500	40
Ireland	20	4
Switzerland	147	61

2.5. Environmental effects of fertiliser

The most important plant nutrients related to pollution are, as said before, N and P. With N the focus is on pollution of drinking water and on eutrophication.

An EEC directive of 1980 recommended 25 mg NO_3 litre^{-1} (5 mg N litre^{-1}) as 'permissible' for drinking water but up to 50 mg NO_3 litre^{-1} as 'acceptable'. The problems concerning the pollution of drinking water and the health aspects in relation to NO_3^- will not be discussed here.

Table 7
The nutrient content in sludge, compost and animal manure

Source	*Dry matter*	*% in wet matter*			*Organic matter*
		Total N	*Total P*	*Total K*	
Sludge	25	0·5	0·4	0·02	15
Compost (sludge + household waste)	60	0·5	0·3	0·1	20
Farmyard manure	27	0·6	0·2	0·4	16
Cattle slurry	8	0·4	0·1	0·4	6
Pig slurry	5	0·5	0·1	0·2	6
Chicken 'slurry'	15	1·1	0·5	0·4	15

Considering P, the environmental concern is mainly related to eutrophication.

Pollutants from agriculture enter natural waters in three ways:

in drainage water;
in eroded soil;
from animal excreta.

The potential pollutants in this respect are [8]:

(a) Plant nutrients in solution which, when added to natural waters, encourage the undesirable growth of aquatic micro-flora (eutrophication).

(b) Organic materials, which are residues of plant or animal life, including fresh excreta dropped in streams by livestock; organic manures escaping to streams or washing off the surface of land or down fissures; silage effluent and the effluent from farm or factory washing and processing of crops. Annual leaf fall and plant debris carried by wind are important sources of nutrients to bodies of water with large surface area and small flow. All these materials supply (i) combined carbon which may assist organic growth in water, (ii) perhaps other growth promoting substances, and (iii) both 'major' and 'minor' plant nutrients. The forms in which organic wastes supply nutrients may be important. For example, silage effluent and other plant wastes are rich in ammonium-N, which stimulates certain micro-organisms.

Erosion can be an important source of pollution by agriculture as it enables direct transfer of nutrients from fertilisers to water supplies. Even in areas where erosion seems unimportant, dust is continually removed from areas used for agriculture and some is deposited in water. The constituents of soil which become sediment may become more active because they remain wet (for example, phosphate in anaerobic muds formed from soil may become more soluble than it was on 'dry' land). The contribution of erosion to water pollution is difficult to quantify and few data exist.

Some definitions and terminologies relevant to this review are summarised below:

Watershed (drainage basin; catchment area). Part of the surface of the earth occupied by a drainage system, consisting of a surface stream, or a body of standing water, together with all tributary surface streams and bodies of standing surface water.

Stream. A body of flowing water, usually in a natural channel.

Run-off. Precipitation that falls on land and ultimately appears in surface streams and lakes; run-off is classified according to its source.

Surface run-off (overland flow). The part of rainwater or snow-melt which flows over the land surface to streams.

Sub-surface run-off (storm seepage). The part of precipitation which infiltrates the surface soil and moves toward streams as ephemeral, shallow, perched groundwater above the main groundwater level. In many agricultural areas sub-surface run-off may be intercepted by artificial drainage systems which accelerate its movement to streams.

Groundwater run-off (base run-off). The part of precipitation that has passed into the ground, has become groundwater, and is subsequently discharged into a stream channel or lake as spring or seepage water.

Point sources. These enter at discrete and identifiable places and can be quantified and their impact on the receiving water measured. Point sources include effluents from industrial and sewage treatment plants.

Diffuse sources. These can be only partially estimated quantitatively; their effect may be diminished but usually not eliminated. Diffuse sources can be divided into:

(a) Natural sources such as eolian loading, and eroded material from virgin lands, mountains and forests.

(b) Artificial or semi-artificial sources which are directly related to human activities, such as fertilisers, eroded soil materials from agricultural and urban areas, and wastes from intensive animal rearing operations.

The N and P reaching streams from natural diffuse sources is a baseline against which the loads from man-made sources may be compared.

2.6. Leaching of nitrogen

Due to the structure of the soil, the absorption capacity and the ability to form relatively insoluble compounds, the soil acts as a kind of filter. Only water-soluble compounds can leach to greater depths, and mechanical filtering prevents the leaching to any greater depths of dissolved organic matter.

The absorption capacity of the soil is important in withholding cations. NH_4^+ is held by this process but not NO_3^-, which is easily soluble and is transported like Cl^- with water down the soil profile.

Very recently many investigations on the leaching of nitrogen have been carried out in a number of countries.

The amount of NO_3^- leached depends on the following:

(a) The amount of water percolating through the soil, through cracks and crevices (quickly), and through pores (slowly). This amount depends on the balance between precipitation, soil water-holding capacity and evapotranspiration.
(b) The amount of nitrate in the soil whether it originates from biological processes during mineralisation of nitrogen from the soil pool or from applied fertiliser.
(c) The soil type, as it is usually expected that greater losses occur from sandy soils than from clay soils.
(d) The cropping system, as smaller losses occur under grass than under crops with a shorter growing period.

Regardless of the origin of the NO_3^-, it can be leached whether it comes from mineral fertilisers, plant residues or other types of organic matter (humus), or animal manure.

2.6.1. The leaching of N following fertiliser application

An example of the relation between leaching of N and the use of fertiliser is shown in Table 8 [9]. During the period 1973/74–1975/76, when spring barley was grown on the clayey soil, the discharge through drains, which constitutes 40–50 % of the total precipitation, has varied between 6 and 249 mm, and the leaching from the 1 N treatment (110 kg N ha^{-1}) has varied between 0·4 and 20·9 kg N ha^{-1}.

The amount and distribution of the precipitation in the growing period thus has a decisive influence on plant growth. Too little precipitation or too poor a distribution retards plant growth and thereby nitrogen uptake, and the leaching of nitrogen increases. Irrigation can either increase or decrease nitrogen loss. Crops covering the soil during the whole year can take up nitrogen during the whole period in which nitrogen is mineralised.

Spring-sown crops that are harvested during the summer and autumn period cannot take up N mineralised early or relatively late in the year. Therefore, there will be a greater leaching from, for instance, a grain crop than from a grass crop.

The significance of the soil type for the leaching of nitrate is primarily connected with the pore geometry and the hydraulic conductivity. Sandy soil has large pores, the highest hydraulic conductivity and the smallest water-holding capacity, and therefore leaching from this soil is usually greatest. The denitrification capacity is greatest on clayey soils, and this also contributes to reduced leaching.

Table 8
The supply of nitrogen, run-off, nitrogen leaching and nitrogen concentration

Year	*Crop*	*Supply* (*kg N ha*$^{-1}$) *at 1 N*	*Run-off* (*mm*)	*Nitrogen leaching* NO_3^--*N* (*kg ha*$^{-1}$)				*Nitrogen concentration* NO_3^--*N* (*mg litre*$^{-1}$)			
				0 N	*½ N*	*1 N*	*1½ N*	*0 N*	*½ N*	*1 N*	*1½ N*
Clayey soil (JB7) Sdr. Stenderup											
1973/74	Barley	110	148	12·8	14·3	16·1	17·6	8·6	9·7	10·9	11·9
1974/75	Barley	110	249	15·1	17·2	20·9	31·1	6·1	6·9	8·4	12·5
1975/76	Barley	110	6	2·0	0·9	0·4	0·1	—	—	—	—
1976/77	Rape	150	98	3·4	4·9	5·6	15·4	3·4	5·0	5·7	15·7
1977/78	Wheat	150	207	16·2	18·6	21·3	37·7	7·8	9·0	10·3	18·2
1978/79	Barley with under-sown grass	135	162	4·8	7·0	14·2	20·4	2·9	4·3	8·8	12·6
1979/80	Grass	120	225	0·4	1·5	2·4	9·5	0·2	0·7	1·1	4·2
1980/81	Grass	55	406	6·6	8·6	16·6	24·0	1·6	2·1	4·1	5·9
1981/82	Wheat	160	242	8·4	10·0	23·8	30·9	3·5	4·1	9·8	12·8
Mean of nine years		122	194	7·7	9·2	13·5	20·7	4·0	4·7	7·0	10·7
Sandy soil (JB4) Agervig											
1978/79	Barley with under-sown grass	103	131	11·7	9·1	9·8	14·3	8·9	6·9	7·4	10·9
1979/80	Clover grass	240	145	10·9	7·9	9·5	19·8	7·5	5·4	6·5	13·7
1980/81	Clover grass	300	466	18·8	13·2	24·4	77·0	4·0	2·8	5·2	16·5
1981/82	Barley	70	260	41·8	43·9	44·8	48·5	16·1	16·9	17·2	18·7
Mean of four years		178	251	20·8	18·5	22·1	39·9	8·3	7·4	8·8	15·9

Lysimeter experiments at Askov [10] using labelled and unlabelled N in fertiliser yielded the information about uptake and leaching summarised in Table 9.

In the first run-off period, 1974/75, about 3% of the ^{15}N-labelled fertiliser was leached; the rest of the leached nitrogen must have originated from the soil N pool. A total nitrogen balance after three years gave the following result for the ^{15}N-labelled fertiliser: 58% of the added ^{15}N-labelled ammonium nitrate fertiliser is taken up by the crops, 21% is immobilised in the soil and 5% is leached, making a total of 84% of the added fertiliser nitrogen, while the rest is assumed to have been denitrified.

Similar experiments have been carried out at the Letcombe Laboratory [11], where ^{15}N-labelled fertiliser was also used in lysimeters. Here it was found that leaching may remove, on average, from 2 to 5% of the fertiliser applied to grass swards.

By far the largest amount of leached nitrogen originates from mineralisation of assimilated nitrogen present in plant roots and micro-organisms and from the soil humus.

The average leaching through drains at Sdr. Stenderup was 7·7 kg after a run-off of 194 mm when no N was applied, increasing as increasing amounts of N were applied, but it increased markedly only when fertilisation was above the economical optimum ($1\frac{1}{2}$ N).

On the sandy soil the same is true. Here leaching is greater but the rotation included a clover grass mixture ploughed down in winter 1980/81 to be followed by spring barley, and in the following winter, 1981/82, there was very heavy leaching.

Table 9

Nitrogen supply, total leaching and leaching of ^{15}N-labelled fertiliser: lysimeter experiments

Applied ^{15}N-labelled fertiliser in 1974 ($g\,m^{-2}$)[a]	*Barley 1974–75 total leaching ($g\,m^{-2}$)[a]*	*Leaching, % of applied*		
		Barley 1974–75	*Barley 1975–76 (^{15}N-labelled)*	*Spring rape 1976–77*
0	6·8	—	—	—
5·4	6·4	2·9	1·5	0·3
10·8	6·1	2·8	1·5	0·3
16·2	6·6	2·8	1·5	0·4

[a] $g\,m^{-2} \times 10 = kg\,ha^{-1}$.

Table 10

The loading of the soil with increasing amounts of slurry, yield in c.u. ha^{-1}, average of 1–8 experimental year, field experiments on clayey soil 1973–80, Askov Experimental Station

Crop	*Experimental year*	*NPK in mineral fertiliser*			*Tonnes slurry ha^{-1}*		
		$\frac{1}{2}$	*1*	*2*	*25*	*50*	*100*
		(*yearly application*)			(*yearly application*)		
Beet	1 and 5	104	122	139	94	106	124
Barley	2, 4, 6 and 8	42	53	52	49	59	57
Grass	3 and 7	37	55	69	37	49	66
Mean	—	56	71	78	57	68	76

1 NPK in mineral fertiliser = 80 kg N ha^{-1} to barley, 160 kg N ha^{-1} to roots and grass.
P and K as applied in 25, 50 and 100 t of slurry.
One crop unit (c.u.) is equivalent to the fodder value of 100 kg barley.

2.6.2. Leaching of N after slurry

The effect of leaching of nitrogen from cattle slurry has been examined in lysimeter as well as field experiments on sandy soil and clayey soil in a four-field crop rotation with fodder beet, barley, Italian ryegrass and barley [12].

Table 10 shows the average yield in crop units (1 c.u. = fodder value of 100 kg barley). Fodder beet and grass were able to utilise relatively large amounts of fertiliser, while the yield of barley was not increased by applying more than 50 t ha^{-1} slurry.

The nitrogen balance data summarised in Table 11 [12] show that after application of 100 t cattle slurry ha^{-1} yr^{-1} there is a 'difference' of 277 kg N ha^{-1} which has not been accounted for. Of this amount, a little will have been volatilised—incorporation by ploughing took place within 1 h, denitrification has taken place, and a certain amount has been leached. The amount removed in this way has not been determined in this experiment, but the percolation down the soil profile has been determined (Fig. 6). The experiments also include FYM and mineral fertiliser.

The extent of nitrogen leaching after using increasing amounts of cattle slurry has been examined in lysimeter experiments with the same experimental layout as in the above-mentioned field experiments with the results shown in Table 12 [12].

Table 11
The loading of the soil with increasing amounts of slurry, N balance $kg\,ha^{-1}$, 1973–80, Askov Experimental Station

Crop	*Experimental year*	*NPK in mineral fertiliser*			*Tonnes slurry ha^{-1}*		
		½	*1*	*2*	*25*	*50*	*100*
		(yearly application)			*(yearly application)*		
Beet	1 and 5	152	199	266	142	179	234
Barley	2, 4, 6 and 8	56	79	111	69	105	133
Grass	3 and 7	95	170	275	95	143	200
Uptake	Average/year	90	132	191	94	133	175
Applied	—	60	120	240	113	226	452
Difference	Average/year	−30	−12	49	19	93	277

Application of 40–120 kg N ha^{-1} as CAN to spring barley resulted in leaching of 30–50 kg N ha^{-1} on the sandy soil and 40–60 on the clayey soil. Using 25 t of cattle slurry ha^{-1} yr^{-1} in the crop rotation, the leaching after barley was 70 kg N ha^{-1}, after 50 t it was 80 kg and after 100 t cattle slurry it was the same on sandy and clayey soil, namely 120 kg N ha^{-1} yr^{-1}. After beet leaching was less and it was least after grass. Leaching increases

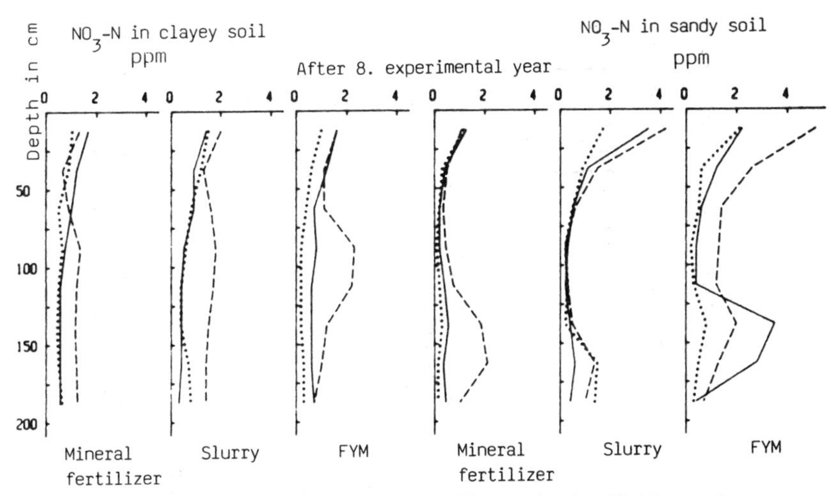

FIG. 6. The nitrate content in the soil at different depths. Field experiments on clayey soil 1973–80 and sandy soil 1974–81. ·····, ½ NPK or 25 t of animal manure; ——, 1 NPK or 50 t of animal manure; – – –, 2 NPK or 100 t of animal manure.

Table 12
N-uptake, -leaching and -balance: lysimeter experiments, Askov 1974–81 ($g\,m^{-2}$)

	Experimental year	Sandy soil						Clayey soil					
		NPK in mineral fertiliser			*Tonnes slurry ha⁻¹*			*NPK in mineral fertiliser*			*Tonnes slurry ha⁻¹*		
		½	*1*	*2*	*25*	*50*	*100*	*½*	*1*	*2*	*25*	*50*	*100*
Uptake in													
Beet	1 and 5	10	17	30	11	13	21	20	27	36	19	19	29
Barley	2, 4, 6 and 8	5	9	13	6	9	13	8	10	16	9	13	17
Grass	3 and 7	6	11	22	7	11	17	9	13	23	8	13	21
Leaching													
Beet		1	2	2	2	3	4	2	2	3	3	3	4
Barley		3	4	5	7	8	12	4	5	6	7	8	12
Grass		<1	<1	<1	3	3	6	<1	1	1	1	1	4
Balance													
Supply		6	12	24	11	22	44	6	12	24	11	22	44
Uptake		7	12	20	7	10	16	11	15	23	11	14	21
Leaching		2	3	3	4	6	8	3	3	4	4	5	8
Difference		−3	−3	1	0	6	20	−8	−6	−3	−4	3	15

1 N in mineral fertiliser: 8 $g\,m^{-2}$ to barley and 16 $g\,m^{-2}$ to beet.
1 N in animal manure: 5 kg cattle slurry m^{-2}.

slightly for all crops after 50 t slurry ha^{-1} yr^{-1}, but is very large after 100 t ha^{-1} yr^{-1} to barley.

It is important to note that after yearly application of 50 t cattle slurry ha^{-1} there is only a limited residual amount left for accumulation in the soil and for possible denitrification, but with the high amount of slurry there is a considerable residual amount. Summarising all these data indicate that in normal farming practice with appropriate fertilisation some 50–60 kg N ha^{-1} are leached from the root zone. Fertilisation above the economical optimum results in greatly increased N-leaching.

FYM (solid) and slurry result in leaching of N if used at rates which do not accord with good fertilisation practice. Leaching is highest after crops having a short growth period.

Based upon Danish investigations [13] it can be concluded that the annual amount of nitrogen leaving the root zone will be

50–65 kg N ha^{-1} from sandy soils;
40–50 kg N ha^{-1} from clayey soils; and
5 kg N ha^{-1} from woodlands.

By adequate and appropriate use, mineral fertiliser, FYM and slurry can be utilised without any long-term unfavourable effects.

Loss of nitrogen in cultivation is a loss to the farmer, and can also cause pollution. In the interest of good farming and to minimise pollution, it should be as small as possible.

2.7. Denitrification

Denitrification is a process whereby NO_3^- is converted to gaseous compounds N_2O and N_2 (NO_x). However, there is a considerable loss as emissions of NO_2 and NO (NO_x) due to human activities.

Table 13 shows the percentage distribution of NO_x for the United Kingdom (calculated at 700 000 t N yr^{-1} [4]) and Denmark (77 000 N ha^{-1} [14]).

Table 13

Percentage distribution of NO_x from various sources

	UK	*DK*
Power plants and district heating	46	51
Industry	24	14
Traffic	25	30
Other sources	5	5

Table 14

Annual nitrogen losses by denitrification: percentage of applied nitrogen in fertilisers

Soil type	*Nitrogen type*	*Loss (%)*	*Crop*	*Period (years)*	*Country*	*Reference*
Lysimeter investigations						
6	NH_4	13	grass	4 (58–61)	DK	Lindhard (1980)
6	NO_3	20	grass	4 (58–61)	DK	Lindhard (1980)
6	NH_4[a]/NO_3[a]	16	spring	3 (74–76)	DK	Kjellerup and Kofoed (1983)
Field investigations						
All	All	20	arable	general	NL	Kolenbrander (1980)
All	All	15	grass	general	NL	Kolenbrander (1980)
	NH_4[a]	17–18	maize	2 (76–77)	USA	Olson (1980)
	NH_4[a]	10–18	w-wheat	1 (79–80)	USA	Olson (1982)
6	NH_4/NO_3	17	s-barley	1 (78–79)	DK	Aslyng and Hansen (1982)[d]
6	NH_4/NO_3	16[b]	s-barley	1 (81)	DK	Vinther (1983)
6	NH_4/NO_3	23[b]	s-barley	1 (81)	DK	Vinther (1983)
6	NH_4/NO_3	36[c]	s-barley	2 (81–82)	DK	Andersen *et al.* (1983)
1	NH_4/NO_3	6	s-barley	2 (81–82)	DK	Andersen *et al.* (1983)

[a] ^{15}N.
[b] 120 and 30 kg N applied, respectively, 0–8 cm soil depth denitrification.
[c] One of the years had a large summer denitrification.
[d] Based on Hvelplund and Østergaard (1980).

About 100 000 t should be added to the Danish figure for NH_3 volatilisation in handling and use of FYM. Some 15–20 kg N ha^{-1} is returned annually in precipitation.

Microbial denitrification in the soil takes place under anaerobic conditions and is increased by the presence of abundant organic matter which stimulates biological activity, causing an increased use of oxygen.

Loss of nitrogen by denitrification can be quite considerable and varies much. Andersen *et al.* [15], in laboratory investigations of intact, freshly collected soil samples from sandy soils, demonstrated losses of 5–10 kg N ha^{-1} yr^{-1} while samples from clayey soils showed a loss of 30–70 kg N ha^{-1} yr^{-1}.

Very large variations in gaseous N loss have been found in other areas in the spring months of 1982 and 1983. Losses in April–May from a soil fallowed without fertiliser for several years were 3–5 kg N ha^{-1}. In 1983 the nitrogen loss within the same two months on a soil with a normal crop rotation was 40 kg N ha^{-1}. It is noted that spring 1983 was unusually wet.

In field experiments Lind and Christensen [16] have shown, using measurements of concentrations from surface emissions, that organic matter stimulates the biological activity with a consequent increased use of oxygen. In field experiments with organic and inorganic fertilisers, more N_2O was found from a treatment supplied with 100 t cattle slurry or 100 t FYM than from a treatment supplied with 25 t cattle slurry or from a treatment supplied with fertiliser. This is especially true during springtime, when there is more NO_3^- in the soil and the process is favoured by high water content. Later in the growing period the difference becomes less.

Aslyng and Hansen [17] give the following most important factors for microbial denitrification in the soil:

Nitrate content.
Availability of carbon to the micro-organisms.
The degree of water saturation.
Temperature.
pH.
Soil type.
Cropping system (fallow/cropped soil).

The same authors have collected results from long-term balance studies in lysimeters and from field studies, and give the N losses due to denitrification (Table 14).

Nitrate which is not taken up by the plants or immobilised or denitrified can, when it leaves the root zone, be chemically reduced, as demonstrated by experiments with intact soil samples from soil profiles sampled at different depths as shown in Table 15 [18].

Samples from several depths in the profiles confirm that nitrate reduction has taken place. Chemical analyses of the drill samples showed that when there was a high content of ferrous iron (Fe^{2+}) there was no nitrate or only very little.

The capability of a soil to convert nitrate to gaseous nitrogen compounds depends on the denitrification capacity of the profile and its reducing potential and the transport time of water in the soil.

Table 15

Nitrate reduction in subsoils: results from laboratory experiments with added nitrate under oxygen-free conditions

Locality	*Depth (m)*	*Soluble Fe^{2+} (ppm)*	*% NO_3^- converted to $N_2O + N_2$ in 45 days*		*ppm NO_3-N (measured in the profile under the root zone)*
			10°C	*25°C*	
Herlufmagle	8	48	4	36	0
Clay soil profile	12	41	22	43	0
Homogeneous	18	66	30	51	0
Bramming Sand soil profile Layered with clay	Mean all depth	3–8	2	51	0·5–2
Skælskør	10	40	6	8	0·3
Inhomogeneous Sandy	16	125	5	8	0·3

If most of the groundwater in Denmark has a low nitrate content it can only be explained by the reduction of NO_3^- during the percolation of the water through the soil layers from the root zone to the groundwater and possibly also partly in the groundwater itself.

Most often denitrification is described as a negative process, but in rivers, lakes and marine areas the process can be of importance in removing surplus nitrate from the water and, as just mentioned, this can have a positive effect on the quality of the groundwater.

In [4] Hahn and Crutzen (1982) are cited for the following amounts and sources of atmospheric nitrogen oxides listed in Table 16, where denitrification in soil only constitutes a part of the sources.

Table 16

Atmospheric species of oxidised N: sources, removal processes, lifetimes and diffusion distances in east–west, south–north and vertical directions over which concentrations are reduced to 30% by chemical reactions

Species	*Source*		*Annual emission rate (Mt)*	*Removal by*	*Mean atmospheric lifetime*	*Transfer distances* Δx; Δy; Δz *(km)*
	Primary	*Secondary*				
$NO + NO_2$	combustion		10–20	OH	1·5 days	1 500; 400; 1
	jet aircraft		0·25	dry deposition		
	biomass burning		10–40			
	lightning		3–4			
	soils		0–15			
		oxidation of N_2O	0·3–0·8			
HNO_3		$NO_2 + OH$	22–80	rain	~3 days	3 000; 600; 1·5
N_2O	combustion		1·8	photolysis	~100 yr	global
	biomass burning		1–2	in		
	oceans		1–10	stratosphere		
	soils, natural		0·7–7			
	soils, manipulated		1·5–3	other sinks?		
	N fertiliser application		0·4–4			

2.8. Nitrification inhibitors

Chemical compounds that prevent the conversion of ammonium to nitrate have been developed during recent years. Their effect is to inhibit the biological conversion of ammonium to nitrate, and as NH_4^+ is not exposed to leaching to the same extent as NO_3^- they have significance in connection with nitrogen fertilisation, especially for early application, as leaching of nitrogen may be reduced. A considerable amount of experimental work is presently taking place in Europe as well as in the USA with the increasing numbers of different inhibitors that are available. The results differ considerably depending on the chemical compound, the crop and the time of application.

Provisional Danish results with N-Serve, Dwell and Didin to cattle slurry have shown that slurry mixed with inhibitors and applied in early autumn (1982) had no effect on the following crop of spring barley [19], probably because the soil temperature was high enough to cause the compounds to be decomposed relatively rapidly, and the ammonium nitrogen present in the slurry was nitrified in the usual way and leached out. Winter application of slurry (December) with added inhibitors gave, on the contrary, a better effect than slurry without addition. With spring application (March) the addition of inhibitors was without significance.

Continued experimental investigations will form the basis for a closer evaluation regarding practical use.

2.9. Prognosis for nitrogen use

In recent years there has been, in many countries, increased interest in methods of predicting the crop's nitrogen requirement and correct timing of application, but nitrogen demand and nitrogen uptake vary considerably and are difficult to predict. There is variation between fields and between years depending on soil type, preceding crop, rooting depth and the amount of inorganic N present in the root zone.

The methods of the prognosis rest upon the foundation that the optimum nitrogen application depends on the N pool in the soil (plant available N just before the growth period), N-mineralisation and uptake of applied nitrogen fertiliser. The accuracy of the N prognosis thus depends on the certainty with which the three factors are determined. The nitrogen pool includes the amount of plant available N in the soil to 1 m depth just before the growing season. The nitrogen mineralisation in the growing period must be predictable before it can contribute to an evaluation of the requirement, but it can vary considerably. Further, one has to know the uptake percentage of the nitrogen fertiliser, of the pool in the soil and mineralisation. Work currently aims at improving methods for these investigations and evaluations, and as yet they are not adequate.

The Agricultural Advisory Service in Denmark has in recent years developed a comprehensive EDP model for planning fertiliser use. This has been well received by the most advanced farmers.

2.10. Phosphorus

In phosphoric fertiliser, P is mainly present as water-soluble primary calcium phosphate [$Ca(H_2PO_4)_2$]. In the dry matter of pig slurry there is 1–2 % P, of which 75–85 % is present as inorganic phosphate—especially slightly soluble calcium phosphates. The remaining P is organically bound; 20–30 % of the organic P is present as phytate phosphorus originating from the fodder. Grain, especially oats, contains phytate-P. Two to three per cent of the total P in pig slurry is present as micro-organisms [20].

P in the soil can be divided into three main groups:

(a) P in the soil solution.
(b) Adsorbed (labile or available).
(c) Slowly available P, mainly slightly soluble P-salts and organic P.

The slightly soluble soil phosphate reaches equilibrium with the soluble fraction only very slowly but there is no sharp line between these fractions because the majority of the soluble phosphate ions are adsorbed. The adsorbed ions are exchanged with phosphate ions in the soil solution, when it is diluted as a result of plant uptake.

The amount of easily soluble phosphate which is adsorbed, and P in the soil solution, is often denoted exchangeable phosphorus, plant available phosphorus or the labile phosphorus pool. The labile pool is supplied by desorption of phosphate by other ions (HCO_3^-, silicate, OH^-), by dissolution of slightly soluble phosphate, and by mineralisation of organic phosphorus from the non-labile pool.

A very small part of the phosphate is present in a soluble state in the soil. Only about 10–15 % P from applied phosphorus fertiliser, containing water-soluble P, is taken up by the crop during the year of application. The rest contributes to a pool that will be partly available during the following years; less available to the plants but more available than the native soil phosphate. About 50 % of the phosphorus reserves in a well-managed soil originate from fertiliser phosphorus. Soil analyses form the basis for an evaluation of the soil P-status.

The leaching of phosphorus through drainage from cultivated soils amounts to only 50–300 g ha^{-1} as phosphate ions have a very low mobility in soil.

In Denmark we figure that 50–100 g ha^{-1} is leached on clay soils. On very light sandy soils with almost no binding capacity it may well be higher, and the same goes for bog soils.

2.11. Other elements

The leaching of other elements has also been examined in the experiments at Sdr. Stenderup mentioned in Table 8. Results are presented in Table 17.

There are very large differences between the amount of the examined elements leached but for the heavy metals it is quite low. Leaching is also very dependent on the run-off.

Table 17

	P	*K*	*Na*	*Ca*	*Mg*	*Cl*	*SO_4-S*
Leaching of nutrients (kg ha^{-1})							
1973/74	0·04	0·5	12·6	149	6·1	50	32
1974/75	0·18	1·1	24·4	276	11·7	90	59
1975/76	0·00	0·0	0·6	5	0·2	2	1
1976/77	0·03	0·3	9·8	97	4·3	34	26
1977/78	0·03	0·9	22·5	221	9·5	75	59

Leaching of heavy metals (g ha^{-1})	*Cu*	*Cd*	*Pb*	*Ni*	*Cr*	*Mn*	*Co*	*Fe*
1974/75	17	2	17	30	17	165	3	—
1976/77	8	0·2	1	<10	<10	5	<5	167

2.12. Surface run-off

Surface run-off and leaching of N and P are quite difficult to determine as many factors influence these processes.

In field lysimeters with 4·5% slope, Uhlen [21] found the losses of plant nutrients by surface run-off and in drainwater as shown in Table 18.

When 60 t of FYM was mixed into the soil the surface run-off was 0·3 kg P ha^{-1}, but it was 10 times higher when it was applied to the soil surface. In drainage water, 0·1 kg P ha^{-1} was found to be leached.

In connection with surface run-off it is important that the manure is mixed into the soil, and that the correct amount is used for the crops in question.

Table 18
Nutrient losses following surface applied or mixed in farm manure in autumn 1973: field lysimeter on 4·5% slope (G. Uhlen)

	Run-off (mm)	*Total-N*	NH_4*-N*	NO_3*-N*	*Total-P*	PO_4*-P*	*K*	*Cl*
		(kg ha^{-1})						
Added in 60 tons of FYM		147	24	—	50	—	50	(21)
			Surface run-off (30/12/73–31/12/74)					
No manure	227	9	1·0	5	2·6[a]	0·1	9	7
60 t Mi	192	9	1·3	5	1·5	0·3	6	12
60 t Sa	237	12	3·5	6	4·9	3·4	13	12
			Drainage water (30/12/73–31/12/74)					
No manure	340	—	0·2	77	0·7	0·1	7	23
60 t Mi	353	—	0·1	88	0·6	0·1	7	36
60 t Sa	370	—	0·1	86	0·5	0·1	7	47

[a] Farm manure, either mixed into the soil (Mi) or surface applied (Sa), reduced soil erosion in winter and summer run-offs. Eroded soil particles contained 0·12% total-P.

In German experiments [22] on surface run-off at soil slopes of 9–12%, less P-leaching was found than stated above. The mixing of the slurry with the soil in these experiments also caused reduced leaching. It is interesting that these experiments also considered the sediment run-off as seen in Table 19.

The smallest soil erosion (sediment kg ha^{-1}) was found when the slurry was mixed in with a cultivator.

Summarising the problems about surface run-off and fertiliser use, the

Table 19
Surface run-off, sediment [22]

		Sediment (kg ha^{-1})	*N* (kg ha^{-1})	P_2O_5 *(DL)* (kg ha^{-1})	K_2O *(DL)* (kg ha^{-1})
Without slurry		3·9	0·14	0·07	0·05
Surface application	30 m^3 slurry	796·0	1·87	0·22	0·48
	60 m^3 slurry	411·8	1·16	0·49	0·37
Cultivator	30 m^3 slurry	94·1	0·37	n.b.	0·13
	60 m^3 slurry	15·7	0·05	0·005	0·01
Rotary cultivator	30 m^3 slurry	792·2	2·56	0·48	0·75
	60 m^3 slurry	415·7	0·87	0·08	0·22

following conclusions were reached at an EEC seminar in Wexford, Ireland, 20–22 May 1980:

1. Pollution caused by surface run-off due to heavy precipitation can not be avoided and in this connection there is no great difference between animal manure and mineral fertiliser, but in the winter period there is a difference.
2. Animal manure spread in the winter period is exposed to leaching and surface run-off. In addition, the soil is often saturated with water and the evaporation is small. This involves a greater risk of surface run-off, particularly if the soil is frozen. Under such conditions the spreading of manures should be avoided.
3. Manures can be spread with the least risk of surface run-off when the soil is passable for slurry tankers or manure spreaders.
4. When animal manure is mixed with the soil, the risk of surface run-off is limited. The spreading of manures near streams or lakes can cause a risk of direct discharge to these waters.
5. The use of manure in excess of plant needs causes accumulation and an increased risk of leaching.

3. FERTILISER AND POLLUTION

Pollution of water areas by fertilisers is due to the fact that N and P are essential nutrients for the organisms in the aquatic environment.

The plant and animal population in a river or lake varies both with regard to numbers and to composition in accordance with the amounts of available nutrients. Lakes or rivers can be classified into three types, namely:

Those that are rich in nutrients (eutrophic).
Those with a medium level of nutrients (mesotrophic).
Those with low levels (oligotrophic).

Lakes and in the highlands usually have clean and clear water with a small population of fauna and flora, while lakes in the lowlands are nutrient-rich with more turbid water and with a larger growth of animals and plants.

In lakes, etc., fed by rivers there is a natural tendency towards eutrophication as a result of accumulating N and P supplied by the rivers. Eutrophication therefore is a process that takes place even without the presence of man, but human activity can accelerate the process. Cultivation and fertilisation of the soil increase the nutrient content in drainage water as well as in run-off, and to this is added waste from sewage plants. The effect of such enrichment is mainly seen in water areas where there is no current or only a small current. Alga bloom in the surface and prevent light reaching aquatic plants at greater depths, where they die. The decomposing vegetation removes oxygen from the water and this, together with the removal of food sources, causes a rapid reduction in the populations of fish and other living organisms. The water area becomes lifeless and structureless, it is polluted and eutrophied.

It is usually assumed that the main cause of very large algal growth is abundant N and P, but other elements can be involved as well. These factors that induce algal growth are complex and not fully explained. Some water areas 'bloom' with vigorous algal growth. This has been known for many decades, but their intensity and frequency seem to have increased in recent years.

In connection with water pollution, the discussions about N and P involve unwanted growth of algae and other plants in:

(a) Stagnant waters, lakes, inlets and connected streams.
(b) Streams leading to the sea.
(c) The sea and open bays, straits and belts.

Pollution with N and P can originate from:

(a) Controllable sources such as emissions from sewage plants and from fish farms and industry.
(b) Non-controllable sources such as
 run-off from areas under agriculture and forestry;
 illegal discharges of dunghill waters, liquid manure, silage effluent;
 run-off from road systems, atmospheric fall-out, etc.

From an environmental point of view, it is these aspects that are of interest, and during recent years there have accordingly been efforts from a number of European countries to minimise the unwanted effects.

3.1. Global and regional pathways for N and P

The large amounts of N and P that are brought into the cycle via agricultural production enter as a part of the global or regional cycles.

3.1.1. Global cycles

Rosswall [23] has constructed and quantified the global cycles of N and P as shown in Figs 7 and 8.

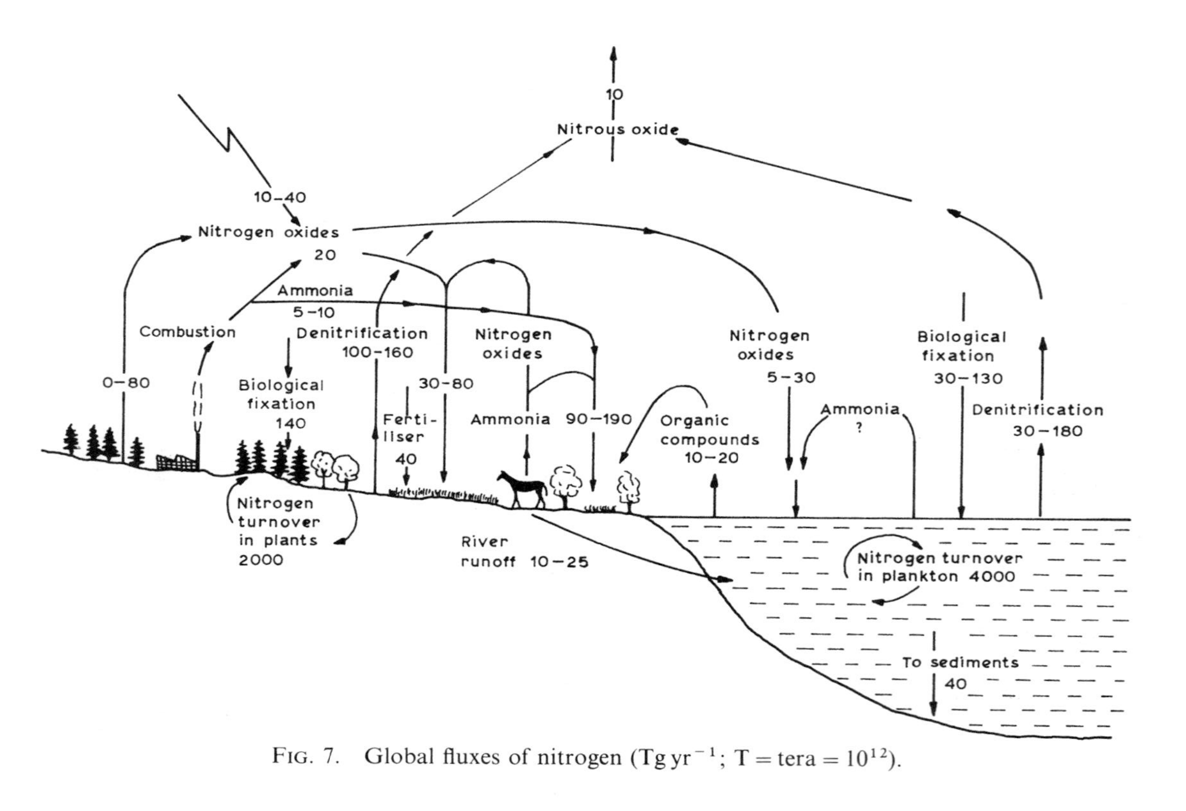

Fig. 7. Global fluxes of nitrogen ($Tg\,yr^{-1}$; $T = tera = 10^{12}$).

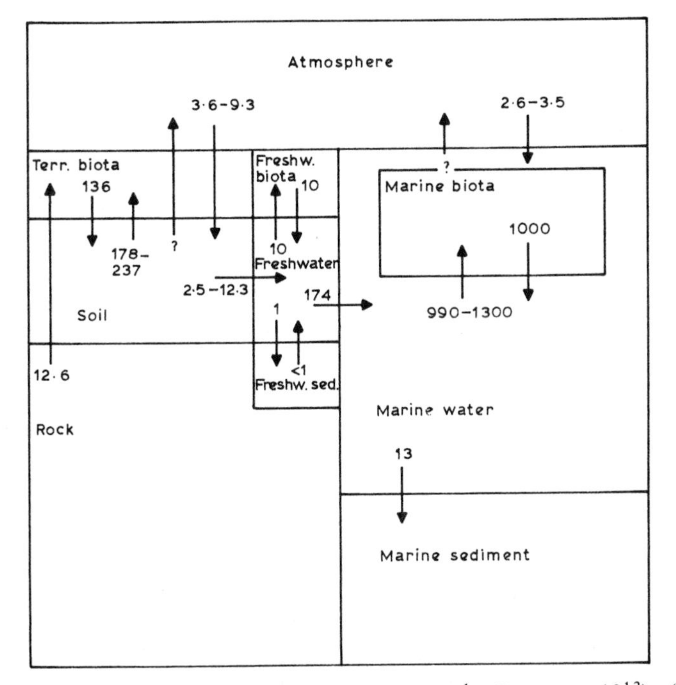

FIG. 8. The global fluxes of phosphorus (Tg yr^{-1}; T = tera = 10^{12}). (From Pierrou, U., in *Ecological Bulletin* No. 22, SCOPE Report No. 7, on N-, P- and S-global cycles. Swedish National Science Research Council, 1976.)

The transport of N and P to rivers and seas is given by Rosswall as shown in Table 20.

The amount of nitrogen supplied to rivers and seas in inorganic form will vary considerably from one area to another and depends on a number of factors, as mentioned earlier.

Nitrogen and phosphorus in river water in organic form may indicate additions through surface run-off and/or erosion.

Exceedingly large quantities of N and P circulate, but the quantities that are transported with water currents to streams, rivers and coastal areas and open water are small compared to the other transport fluxes.

The quantities of nutrients that are added to the nutrient cycle through agricultural activities induce a number of consequences which did not apply in the extensive agriculture that was found years ago along with the much lower population densities.

Nutrients that are essential to cultivated plants are also essential to

Table 20
Global transports of nitrogen and phosphorus with rivers to the sea ($Tg\,yr^{-1}$)

Inorganic	NO_3^-	5–11	P	14%	
	NH_4^+	<1			
					3–12
Organic	N	8–13	P(1)	10%	
			P(5)	76%	

Nitrogen (Söderlund and Svensson, 1976).
Phosphorus (Pierrou, 1976; Porter, 1975).
$T = tera = 10^{12}$.

marine organisms but as sea water is rich in, for example, Na, K, Mg and Ca it is mainly the changes in the sea water content of N and P that are important, rarely Br, I or B or other trace metals.

In recent years concern about sea pollution has increased and focused on N and P due to the consequences of eutrophication. These nutrients normally limit plant production in marine waters since they occur only in very small amounts, organic N often being less than 0·02 mg $litre^{-1}$ and inorganic P less than 0·002 mg $litre^{-1}$.

The concentrations of inorganic nutrients in the marine waters depend on a number of processes of addition and removal. These vary throughout

Table 21
Load distribution related to source (averages the nearest 5% over the period studied)

	Total-N	*Total-P*
Sewage plants	30	75
Direct discharges from industries	5	10
Fish farms	(no)	(no)
Contribution from rural areas	40	5
Urban run-off	(no)	(no)
Scattered residences plus discharges from farm areas (liquid manure, etc.)	25	10
	100	100
Point sources	35	85
Non-point sources	65	15
	100	100

the year according to climatic conditions such as light, temperature, precipitation, wind and currents.

The removal of nutrients from the waters takes place through uptake in plants, by transport or by chemical transformation, so that they can no longer be used.

3.1.2. Regional addition

As an example of estimates of regionally added N and P to marine areas, Fig. 9 shows the additions to Danish and Swedish waters during the period 1975–81 [24].

The total N load of the water areas in question has been estimated at 150 000 t yr^{-1}, of which the contribution from atmospheric fall-out is about 20 %. The load arising from the Danish land area is about 45 % and that from the Swedish land area about 35 %.

The total P load has been estimated at about 14 000 t yr^{-1}, of which the atmospheric contribution is about 5 %. The contribution from the Danish land area is about 60 % and from the Swedish land area about 35 %.

In both 1980 and 1981 there was a considerably higher nitrogen leaching from both Danish and Swedish land areas than in the preceding five years.

Leaching from the Danish land area is distributed among a number of different sources as shown in Table 21.

In this analysis the following definitions have been used:

Point sources
- Discharges from sewage plants.
- Direct discharges from industry (i.e. not via sewage plants).
- Direct discharges from fish farms (i.e. not via sewage plants).

Non-point sources
- Surface run-off from urban areas.
- Contributions from the open land.
- Area contributions
  - from cultivated areas;
  - from woods and moorlands.
- Scattered residential areas.
- Discharges from animal production (run-off from dunghills, etc.).

In connection with pollution control, there are increased efforts to reduce the load of N and P.

In Denmark, an estimation [25] of the nitrogen supply and nitrogen leaching from cultivated soils has been carried out. This was based on

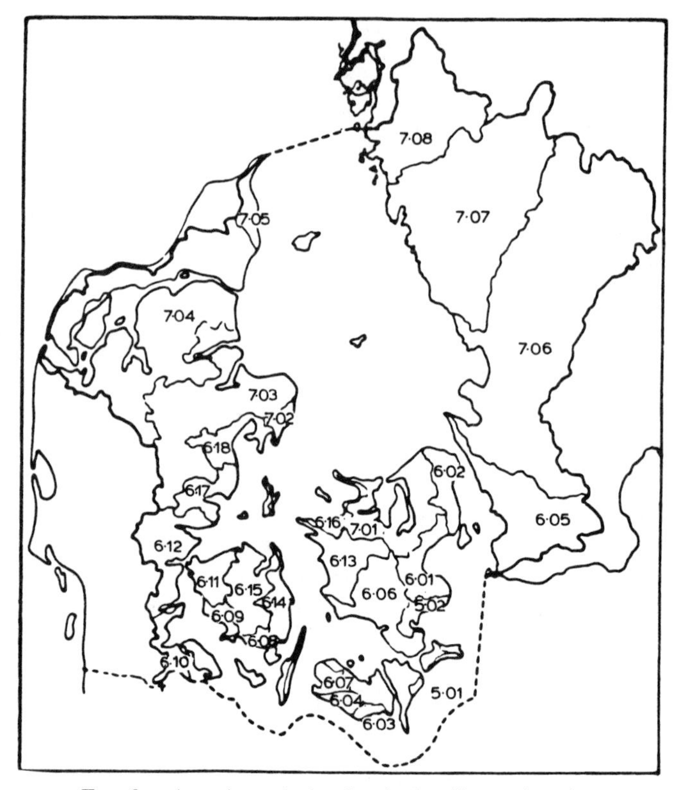

FIG. 9. Area boundaries for the loading estimation.

optimum fertilisation of the various crops, where animal manure was distributed first and then mineral fertiliser as summarised in Table 22.

Values are mean values for each county and naturally include considerable variations and show, as an average situation, that fertilisation practice is reasonable considering the choice of crops and production structure, but within individual areas with large concentrations of animals or on individual holdings there are problems with too large amounts of animal manure.

4. CONCLUSION

The intensive farming which is seen within the EEC will continue to be the foundation of agriculture, but it should take place with due consideration

Table 22
Nitrogen requirement, nitrogen use and nitrogen leaching: mean values of 1978–82

County (Denmark)	*Area (ha)*	*Optimum fertiliser*	*Minimum fertiliser*	*Animal manure*		*Surplus fertiliser*	*Leaching*
				×0·4	*Total*		
		(kg N ha^{-1})					
Kbh. Frbg. Rosk.	131 348	142	130	17	42	5	47
Vestsjælland	210 307	136	112	21	53	−3	45
Storstrøms	249 828	130	119	16	39	4	40
Bornholm	37 038	140	102	28	70	−11	42
Fyn	250 872	149	121	29	73	0	48
Sønderjylland	291 113	177	134	40	99	−4	54
Ribe	211 302	195	138	46	115	−11	55
Vejle	205 706	163	140	37	93	13	57
Ringkjøbing	322 799	171	141	39	98	9	65
Århus	296 850	154	135	31	77	11	57
Viborg	276 125	171	126	44	110	−3	57
Nordjylland	422 839	167	136	39	98	7	63

Table 23
The average yield of grain during 50–116 years in experiments conducted in some European countries (hkg grain ha^{-1})

Experimental locality and year span	*Number of years*	*Grain species*	*Unfertilised*	*Animal manure*	*Mineral fertiliser*
England, Rothamsted, 1852–1961	110	barley M	8·8	29·6	24·6
England, Rothamsted, 1852–1967	116	wheat M	9·6	24·4	23·6
Germany, Halle, 1879–1958	80	rye M	14·1	25·3	24·8
Denmark, Askov clay field, 1894–1968	75	grain S	12·8	24·2	29·2
Denmark, Askov sand field, 1894–1968	75	grain S	8·1	19·0	24·9
Austria, Grossenzersdorf, 1906–1975[a]	70	barley M	15·0	33·2	33·0
Austria, Grossenzersdorf, 1906–1975[a]	70	rye M	13·9	23·1	27·7
Austria, Grossenzersdorf, 1906–1975[a]	70	rye S	24·3	29·6	31·7
Poland, Skierniewice, 1925–1973	49	rye M	12·8	21·0	22·0
Poland, Skierniewice, 1924–1973	50	rye S	15·7	—	30·6
England, Woburn, 1877–1926	50	wheat M	7·7	15·5	14·8
England, Woburn, 1877–1926	50	barley M	8·5	20·7	11·2
Average 50–116 years			12·5	24·2	24·8

M = monoculture.
S = crop rotation.
[a] The experiment at Grossenzersdorf has been conducted since 1906 according to the plan, but harvest yields are only from the last 16 years; the other experimental data were lost during the Second World War.

of environmental quality. Without the use of fertilisers yields would decrease substantially. An example is shown in Table 23 [26].

In relation to fertiliser use and environmental quality protection, the main question is: what is the pollution limit? Nobody can be interested in a limit which is so low that profitable farming cannot be carried out. The main point is a reasonable consideration of benefits and drawbacks where, for example, the concept of good agricultural practice must play a central part and a realistic recognition of the impact of the individual factors on the economical and the environmental aspects in the short as well as the long term.

REFERENCES

1. SKOVBÆK, J. (1984). *De danske Landboforeninger*, Bilag til foredrag, Viborg.
2. HØJ, K. (1984). Gødskningens indflydelse på udbytte, plantekvalitet og miljø. Under trykning. Statens Forsøgsstation St. Jyndevad, 6360 Tinglev, DK.
3. KUNTZE, H. und WOLFGANG, V. (1980). *Statusbericht Düngung*, Landwirtschaftsverlag GMCH, 4400 Münster-Heltrup.
4. The Nitrogen Cycle of the United Kingdom, A study group report. The Royal Society, 1983.
5. Nitrates: An Environmental Assessment, National Academy of Sciences, Washington, DC, 1978.
6. JANSSON, SV. L. (1958). Tracer studies on nitrogen transformations in soil with special attention to mineralisation–immobilisation relationships. *Annals of the Royal Agricultural College of Sweden*, **24**, 101–361.
7. KOFOED, A. DAM (1983). Optimum use of sludge in agriculture, Proceedings of EEC Seminar, Uppsala (Ed. S. Berglund, R. D. Davis and P. L'Hermite).
8. COOKE, G. W. (1975). Technical Bulletin 32. Agriculture and Water Quality, Ministry of Agriculture, Fisheries and Food, London.
9. KJELLERUP, V. and KOFOED, A. DAM (1983). Kvaelstof og planteproduktion. *Tidsskrift for Planteavl, Specialserie*, 58–62.
10. KJELLERUP, V. and KOFOED, A. DAM (1983). Kvælstofgødningens indflydelse på udvaskning af plantenæringsstoffer fra jorden (lysimeterforsøg med anvendelse af ^{15}N (Nitrogen fertilization in relation to leaching of plant nutrients from soil, Lysimeter experiments with ^{15}N). *Tidsskrift for Planteavl*, **87**, 1–22.
11. Royal Commission on Environmental Pollution (1979). Her Majesty's Stationery Office, London, 1979.
12. KOFOED, A. DAM (1983). Husdyrgødning og kunstgødning i planteproduktion. *Bilag til Statens Planteavlsmøde*, 10–12, Statens Planteavlskontor.
13. HANSEN, L. (1983). Kvælstoftab til dræn- og grundvand. *Bilag Statens Planteavlsmøde*, 18–22, Statens Planteavlskontor.
14. Miljøstyrelsens NP-O rapport (1984). Luftforureningslaboratoriet, København, Denmark.

15. ANDERSEN, C., EILAND, F. and VINTHER, F. (1983). Økologiske studier af jordbundens mikroflora og fauna i dyrkningssystemer med reduceret jordbehandling, vårbyg og efterafgrøde (Ecological investigations of the soil microflora and fauna in agricultural systems with reduced cultivation, spring barley and catch crops). *Tidsskrift for Planteavl*, **87**, 257–96.
16. LIND, A.-M. and CHRISTENSEN, S. (1983). Kvælstof og planteproduktion, *Tidsskrift for Planteavl Specialserie*, 18–26.
17. ASLYNG, H. C. and HANSEN, S. (1984). *Nitrogen Balance in Crop Production, Simulation Model NITCROS*, Hydrotechnical Laboratory, The Royal Veterinary and Agricultural University, Copenhagen.
18. LIND, A.-M. and PEDERSEN, M. B. (1976). Nitrate reduction in the subsoil, I and II. *Tidsskrift for Planteavl*, **80**, 64–76.
19. KJELLERUP, V. (1984). Stoffer der hæmmer kvælstoftab fra gødninger, *Ugeskrift for Jordbrug*, nr. 11.
20. GERRITSE, R. G. (1977). Phosphorus compounds in pig slurry and their retention in the soil. EEC Seminar: *Utilization of manure by land spreading*, Modena, Italy, 1976, EUR-5672e, 257–66.
21. UHLEN, G. (1981). Surface run-off and the use of farm manure. Proceedings of an EEC Workshop, Wexford.
22. VETTER, H. and STEFFENS, G. (1981). Leaching of nitrogen after spreading of slurry. Proceedings of an EEC Workshop, Wexford.
23. ROSSWALL, T. (1983). Globala balanser av kväve och fosfor. *Kungl. Skogs- och Lantbruksakademiens Tidsskrift*, Stockholm.
24. Iltsvind og fiskedød (1984). *Miljøstyrelsen Strandgade*, 29, 1401 København K.
25. Kvælstoftilførsel og kvælstofudvaskning i dansk planteproduktion 1984. Landbrugsministeriet, 1. afdeling, 6. kontor, Arealdatakontoret, Vejle, DK.
26. STAPEL, C. (1979). Biologisk-dynamisk landbrug og nogle alternative landbrugsmetoder som led iverdens fødevareproduktion. *Tidsskrift for Landøkonomi*, **166**, 151–94.

4

The Impact of Agricultural Loads on Eutrophication in EEC Surface Waters

M. VIGHI
Faculty of Agriculture, University of Milan, Italy

and

G. CHIAUDANI
Department of Biology, University of Milan, Italy

SUMMARY

A review is given on eutrophication problems in the surface waters of the EEC countries. Phosphorus is generally considered the key element which controls or limits the productivity of waters. Phosphorus loads deriving from different sources are calculated by means of a theoretical approach, based on statistical data, for the whole European Community and for the different countries. In general agricultural loads are much lower than domestic loads.

Costs and benefits of different phosphorus reduction strategies are presented. The possibility of the reduction of agricultural non-point sources has been compared with the reduction of urban point sources. Finally, as an example, the case history of Northern Adriatic coastal waters is described. The application of a simple predictive model allows us to compare the water quality improvement resulting from different phosphorus reduction strategies.

1. INTRODUCTION

In the early 1960s it became obvious that eutrophication was emerging as one of the most relevant causes of water quality deterioration. The resultant increase in fertility in affected lakes, reservoirs, slow-flowing rivers and certain marine coastal waters provoked symptoms that often adversely affect the vital uses of water such as for a water supply, fisheries and recreation. Aesthetic qualities may also become impaired.

The concern about eutrophication led the OECD to implement research on this subject. Since 1973 18 Member Countries have been cooperating in the OECD Cooperative Programme on Eutrophication.

The aim of this paper is to define the role of agriculture in eutrophication in the countries of the European Economic Community and also to evaluate the need and the realistic possibility of control measures on agricultural loads.

2. THE IMPACT OF EUTROPHICATION IN THE EEC COUNTRIES

In the EEC countries eutrophication has been observed and it is present in different degrees of importance. Table 1 shows, for the various countries, the type of water bodies in which the problem has been identified, its extension and gravity.

From the table it appears that Italy is the most affected country with serious problems on a national scale in natural lakes, reservoirs and marine coastal waters. A survey of 65 of the most important Italian lakes demonstrated that more than two thirds are involved in more or less serious

Table 1
The eutrophication problem in the EEC countries

	Natural lakes	*Reservoirs, rivers and irrigation systems*	*Estuaries, lagoons*	*Marine coastal waters*
Belgium		+		
Denmark	+			+
France	+ +			+
Germany (FRG)	+	+ +		+
Greece		+	+	+
Ireland	+	+		
Italy	+ +	+ +	+	+ +
Luxembourg				
The Netherlands	+	+ +	+	
United Kingdom	+ +	+ +		

+ = identified problem, + + = serious problems on a national scale.
(Modified from Vollenweider [18].)

eutrophication processes [6,11]. Regarding marine coastal waters, eutrophication in the Northern Adriatic Sea can be considered the most relevant water quality problem in the whole Mediterranean area [3, 5].

In France eutrophication greatly affects natural lakes. Lake Geneva, located between France and Switzerland, was a naturally oligotrophic lake until the 1950s. At present it is highly eutrophied and this condition is of particular concern if one considers that more than 700 000 people utilise its water for drinking [9].

In the most important lake of Germany, Lake Constance, near the border of Switzerland and Austria, phosphorus concentration was, in the 1930s, about 10 μg litre^{-1}, which is a level typical of oligotrophy. In 1978 phosphorus concentration had risen to about 90 μg litre^{-1} indicating a very high level of eutrophication [9]. As for Lake Geneva, Lake Constance is an important source of drinking water. Many reservoirs in Germany are also heavily eutrophied.

In Denmark intensive studies on eutrophication were done from 1973 to 1975. More than 61 lakes have been analysed in an attempt to evaluate their trophic status and many of them are in eutrophic or hypereutrophic conditions [16].

In the United Kingdom several problems have been identified in natural lakes and reservoirs. Lough Neagh, the largest body of water in the British Isles, situated in Northern Ireland, was identified already ten years ago as being one of the most eutrophic of the world's major lakes [19].

The problem is also serious for reservoirs in The Netherlands and a certain degree of eutrophication has been identified in the marine coastal waters of Denmark, France, Germany and Greece, particularly in the gulf of Salonika.

3. LOAD OF NUTRIENTS DERIVING FROM DIFFERENT SOURCES

Although nitrogen and other plant nutrients are involved in eutrophication processes, phosphorus is generally considered the key element which controls or limits the productivity of waters.

Furthermore, phosphorus inputs are generally more susceptible to control measures than nitrogen inputs. Also water bodies which are originally nitrogen limited can possibly be made phosphorus limited if the phosphorus inputs are reduced sufficiently [12].

As a consequence the control of eutrophication can be considered at present as synonymous with the control of phosphorus loadings.

Man-made sources of phosphorus may be divided into two categories: point sources and diffuse sources. A point source may be industrial waste, sewage, waste from animal husbandry or storm sewer outlets. Any soluble substance containing phosphorus and deposited on the drainage basin such as fertilisers, animal droppings, plant remains, waste products or eroded soil from land uses, can contribute to the diffuse sources. In order to plan interventions directed at controlling eutrophication it is of primary importance to know the respective weight of all different sources of phosphorus load.

In Italy the Water Research Institute evaluated the total national phosphorus load which flows into surface waters and the relative weight of the different sources by means of a theoretical approach [2]. The contribution of different phosphorus point and non-point sources to the total load was evaluated on the basis of statistical data.

The domestic point loadings were obtained by evaluating both the metabolic contribution and the consumption of polyphosphates which are added as builders in synthetic detergents. Agricultural loadings were evaluated by considering the national consumption of synthetic fertilisers. Moreover animal production was considered and the amount of animal manure, deposited on cultivated land, as a natural fertiliser, or directly discharged in sewages was calculated. It was evaluated that about 10% of animal manure is directly discharged in surface waters resulting in a point source, while 60% is deposited on cultivated land and the remainder is treated or discharged in septic tanks [7].

For all types of cultivated land a mean loss coefficient of 3% of the total amount of phosphorus applied to the soil was adopted.

Industrial loadings, in relation to eutrophication, are not as relevant as for other kinds of water pollution and are restricted to some particular activities (i.e. food industry, phosphoric acid production, etc.). A precise evaluation of industrial phosphorus loading is not easy but, according to

Table 2
Comparison between the theoretical and the experimental evaluation of phosphorus load in the drainage basins of the two major Italian rivers

	Theoretical load ($t\ P\ yr^{-1}$)	*Experimental load* ($t\ P\ yr^{-1}$)
River Po	17 582	15 600
River Adige	1 192	1 300

an estimate by the OECD [17], could be considered as high as about 10% of domestic point loading.

Finally, the contribution of non-cultivated lands was evaluated by assuming a mean phosphorus loss coefficient obtained as a function of the different geological characteristics of the soil and of the different vegetation.

Atmospheric load (wet and dry precipitation) is quite difficult to quantify but in general it could be considered negligible with respect to the principal point sources. Some authors [1, 13] suggest a mean value of 0·05 kg P yr^{-1} ha^{-1} of the water body's surface. Therefore, in this evaluation, atmospheric load has not been included.

According to all these considerations, the following coefficients were adopted to evaluate the theoretical phosphorus load for the Italian territory [2]:

Non-cultivated land:	0·1 kg P ha^{-1} yr^{-1}
Cultivated land:	3% of total phosphorus applied (synthetic and natural fertilisers)
Domestic loads:	0·64 kg P per capita, per year (metabolic and detergent phosphorus)
Industrial loads:	10% of domestic loads
Animal farms (directly discharged):	
Horses:	0·590 kg P per capita, per year
Cattle:	0·455 kg P per capita, per year
Pigs:	0·280 kg P per capita, per year
Sheep:	0·075 kg P per capita, per year
Poultry:	0·003 kg P per capita, per year

The reliability of this theoretical approach in calculating phosphorus loadings was verified by applying it to some river drainage basins for which experimental data on phosphorus loading were available. In particular the method was applied to the drainage basins of the River Po and the River Adige, the most important Italian rivers. The comparison, shown in Table 2, between the theoretical evaluation and the experimental data, was very satisfactory [2].

The same theoretical approach, revised for the different national situations, was applied to the various countries of the European Community.

Data on the population, the animal production, the consumption of synthetic fertilisers refer to 1982 [8].

Table 3
Phosphorus loadings from different sources in the EEC countries

	Agriculture		*Animal farms*		*Domestic point sources*		*Industry*		*Non-cultivated land*		*Total*	
	($t\ P\ yr^{-1}$)	%	($t\ P\ yr^{-1}$)	%	($t\ P\ yr^{-1}$)	%	($t\ P\ yr^{-1}$)	%	($t\ P\ yr^{-1}$)	%	($t\ P\ yr^{-1}$)	%
Belgium	2 392	17	2 863	21	7 490	55	750	5·5	161	1·2	13 656	100
Denmark	2 802	25	3 966	35	3 890	35	389	3·5	141	1·3	11 188	100
France	27 372	32	15 044	17	38 770	44	3 880	4·4	2 310	2·6	87 376	100
Germany (FRG)	14 360	17	13 255	16	48 690	59	4 870	5·9	1 270	1·5	82 445	100
Greece	2 721	23	1 775	15	6 270	53	627	5·3	396	3·3	11 789	100
Ireland	2 987	34	3 134	36	2 230	26	223	2·6	130	1·5	8 704	100
Italy	10 064	17	7 481	13	36 250	62	3 625	6·2	1 260	2·1	58 680	100
Luxembourg	112	21	120	22	262	49	26	4·9	13	2·4	533	100
The Netherlands	3 003	15	5 397	26	11 020	53	1 100	5·3	209	1·0	20 729	100
United Kingdom	9 566	17	10 014	17	33 610	59	3 360	5·9	562	1·0	57 112	100
EEC	75 379	21	63 049	18	188 482	54	18 850	5·4	6 452	1·8	352 212	100

Table 3 shows phosphorus loadings deriving from different sources and the percentages with respect to the total load for each country and for the European Community as a whole. Agricultural loads include diffuse phosphorus loadings deriving from the leaching of both synthetic fertilisers and animal manure distributed on cultivated land. Animal wastes deriving from animal farms and directly discharged in surface waters are considered separately as a point source.

It should be noted that the data of Table 3 do not take into account eventual national legislative measures, introduced after 1982, which are able to reduce phosphorus loads from different sources. This is the case of the law that has been in force in Italy since 1984 and that imposes the reduction (up to 5% in weight) of the maximum phosphorus content in synthetic detergents.

Table 3 indicates that agricultural loads are in general quite a deal lower than domestic loads. Their percentage values in all of the European Community are about 21% and about 54% respectively. Nevertheless a more precise observation of the table demonstrates that there are very different situations in the various countries. In particular agricultural diffuse loads have a greater impact in Ireland, France and Denmark with percentage values respectively of 34%, 32% and 25%.

The contribution of animal farm point sources is also very high in Denmark and Ireland. In these two countries the total load deriving from all agricultural practices (fertilisers and animal discharges) has been estimated as high as about 60% and 70% respectively.

Area phosphorus loadings expressed as kg P km^{-2} were also calculated and are shown in Table 4, indicating a wide range of values for the different countries. High values are characteristic of The Netherlands and Belgium, the lowest values are present in Ireland and Greece. In general the area loading of phosphorus reflects the population densities (Fig. 1).

Table 4

Area phosphorus loadings, expressed as kg P per km^2 of national territory, in the different EEC countries

	kg P km^{-2}		*kg P km^{-2}*
Belgium	447	Ireland	124
Denmark	259	Italy	196
France	159	Luxembourg	206
Germany (FRG)	331	The Netherlands	503
Greece	89	United Kingdom	234

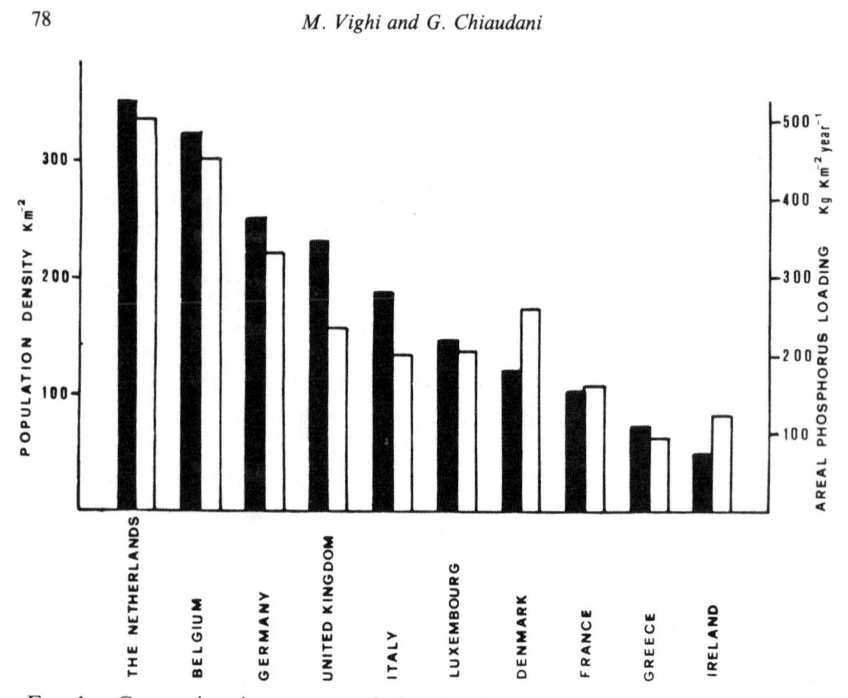

FIG. 1. Comparison between population density (solid bars) and area phosphorus loadings (open bars) in the EEC countries.

Nevertheless the evaluation of the area loading on a national basis might not be sufficient to identify critical areas and a more detailed evaluation would be useful. The Italian situation could be taken as an example. Figure 2 shows the total phosphorus loadings which reach different marine hydrographic basins. About 50 % of the total national loading originates in the Po Valley and reaches the Northern Adriatic Sea. Moreover most of the Italian lakes, more or less affected by eutrophication, are located in the Po Valley.

4. COSTS AND BENEFITS OF POSSIBLE CONTROL MEASURES

A complete and appropriate measure for the control of water pollution in order to resolve eutrophication problems, should in theory be obtained only considering all the principal sources of phosphorus load.

However, the development and the implementation of phosphorus load

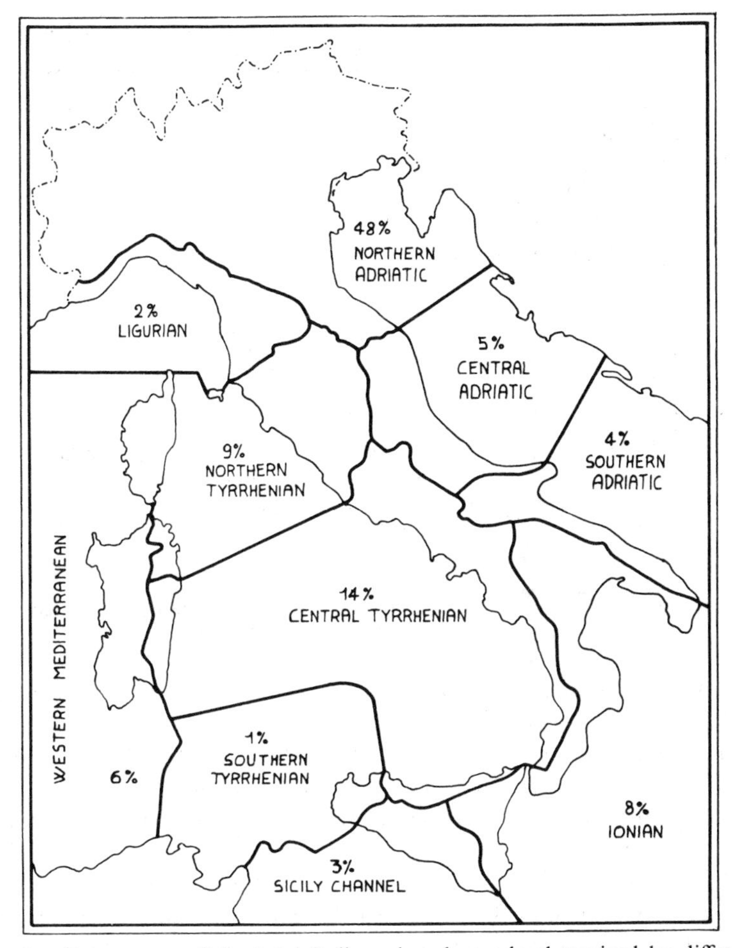

FIG. 2. Percentages of the total Italian phosphorus load received by different marine hydrographic basins (from Chiaudani *et al.* [5]).

control measures is not purely a scientific and technical problem. Economic considerations play a dominant role in these matters. The measure of success of eutrophication control is the most economical achievement of the trophic response in a water body.

It is obvious that the reduction of phosphorus load deriving from point sources is relatively easy to obtain; on the contrary the reduction of diffuse loads, and in particular of agricultural loads, is much more difficult and expensive (Fig. 3). Therefore it is essential to estimate realistically the cost of

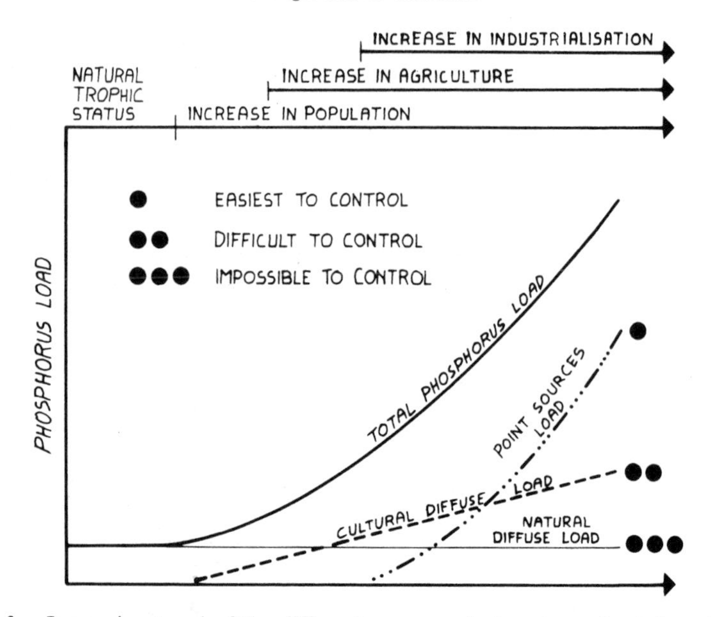

FIG. 3. Increasing trend of the different sources of phosphorus load (based on [14]).

the control measures for the different phosphorus sources, so that control priorities may be established on a sound economic basis.

In a comparison between the different strategies of phosphorus control it is also important to consider that, after an initial effort, additional point source control could become more expensive than certain diffuse source control measures.

As an example Table 5 indicates the estimated costs of different phosphorus reduction alternatives [14]. The data shown refer to one particular drainage basin but it could give an order of magnitude of the necessary costs and therefore could be applied more generally. Costs were evaluated in 1978 and must be analysed only in comparative terms between different strategies.

The reduction of municipal point sources considers two different levels of chemical treatment of sewages.

Regarding the reduction of rural non-point sources, level 1 of Table 5 corresponds to a sound management which avoids excess fertilisation and reduces soil erosion of all agricultural lands. This kind of intervention could reduce the phosphorus load by 10% at no extra cost.

Level 2 of Table 5 corresponds to the implementation of a more

Table 5
Estimated costs of phosphorus reduction alternatives from OECD (1982)

Remedial measure options	*\$ kg^{-1} phosphorus reduction*
Urban point sources:	
Reduction of municipal sewage treatment plant effluent concentration:	
(a) 1·0 mg litre^{-1} to 0·5 mg litre^{-1} (about 90% reduction)	8·0
(b) 0·5 mg litre^{-1} to 0·3 mg litre^{-1} (about 95% reduction)	95·5
Rural non-point sources:	
Level 1 Sound management on all agricultural lands, avoiding excess fertilisation, reducing soil erosion (10% phosphorus reduction)	—
Level 2 Level 1 measures, plus buffer trips, strip cropping, improved municipal drainage practices, etc., depending on region (25% reduction in phosphorus losses on soils requiring treatment)	64·3
Level 3 Level 2 measures at greater intensity of effort (to achieve 40% reduction in phosphorus losses on soils needing treatment)	174·0

advanced rural non-point source control which includes buffer trips, strip cropping and improved drainage practices.

This type of treatment is able to achieve a reduction of 25% of the phosphorus losses, but would greatly increase the unit cost of phosphorus reduction to \$64·3 kg^{-1}. The costs which are necessary in order to reduce the concentration of municipal treatment plant effluents up to 1 mg litre^{-1} (this reduction corresponds to about 90% of the initial concentration) are lower. Nonetheless the measures shown in level 2 are still less expensive than the \$95·5 kg^{-1} for the reduction of waste water treatment effluent concentration below 0·5 mg P litre^{-1}.

Level 3 of Table 5 includes all the measures of level 2 but at a greater intensity. This corresponds to a reduction of 40% in phosphorus losses and could be considered the maximum level realistically achievable in the control of rural non-point sources. The unit cost of phosphorus reduction would increase to \$174·0 kg^{-1}. Under such conditions the reduction of the sewage treatment effluent from 0·5 to 0·3 mg litre^{-1} would be an attractive alternative. In addition to the high costs of such advanced agricultural practices in terms of phosphorus control, it should be kept in mind that even the implementation of the lower level to diffuse phosphorus source control measures may be difficult (if not impossible) in certain cases because it depends on the full cooperation of individuals and institutions.

It follows that, also in those countries where agricultural phosphorus loads are particularly relevant, it is unlikely that such difficult and expensive interventions could be applied in a short time on the whole national territory.

On the other hand all the countries could have areas in which eutrophication is a particularly serious water quality problem. It is necessary to identify the 'hot points' and to have a detailed evaluation of all the phosphorus loads from point and non-point sources in these areas. This case by case evaluation could allow an estimate of costs and benefits deriving from the reduction of agricultural loads.

5. A CASE HISTORY: THE NORTHERN ADRIATIC COASTAL WATERS

The case history of the coastal waters of Emilia-Romagna (Northern Adriatic Sea) could be taken as an example for this case by case evaluation.

This highly eutrophied coastal area is affected mainly by the phosphorus input of the River Po. The River Po's drainage basin includes almost all of Northern Italy, a territory with a high degree of urbanisation, industrialisation and agricultural activities.

Phosphorus loads from different sources were quantified exactly both through theoretical evaluations [2] and experimental data [10]. Domestic load represents about 56% of the total phosphorus load and agricultural diffuse load is about 19%. The remainder is represented by industrial load, animal farms and leaching from non-cultivated lands. These values do not differ much from those obtained for the Italian territory as a whole.

The practical conclusion of a long term research project, which began in 1976, and was designed to study the eutrophication of the Northern Adriatic Sea, was the development and the application of a simple model able to describe the fate (diffusion and utilisation) of nutrients in sea water [4]. The model allows us to predict changes in water quality which derive from different phosphorus reduction scenarios in the River Po drainage basin.

In Fig. 4 the predicted trophic conditions deriving from different phosphorus reduction measures (B, C, D) are shown in comparison with the present trophic status (A).

Figure 4B corresponds to a reduction of 40% of agricultural diffuse

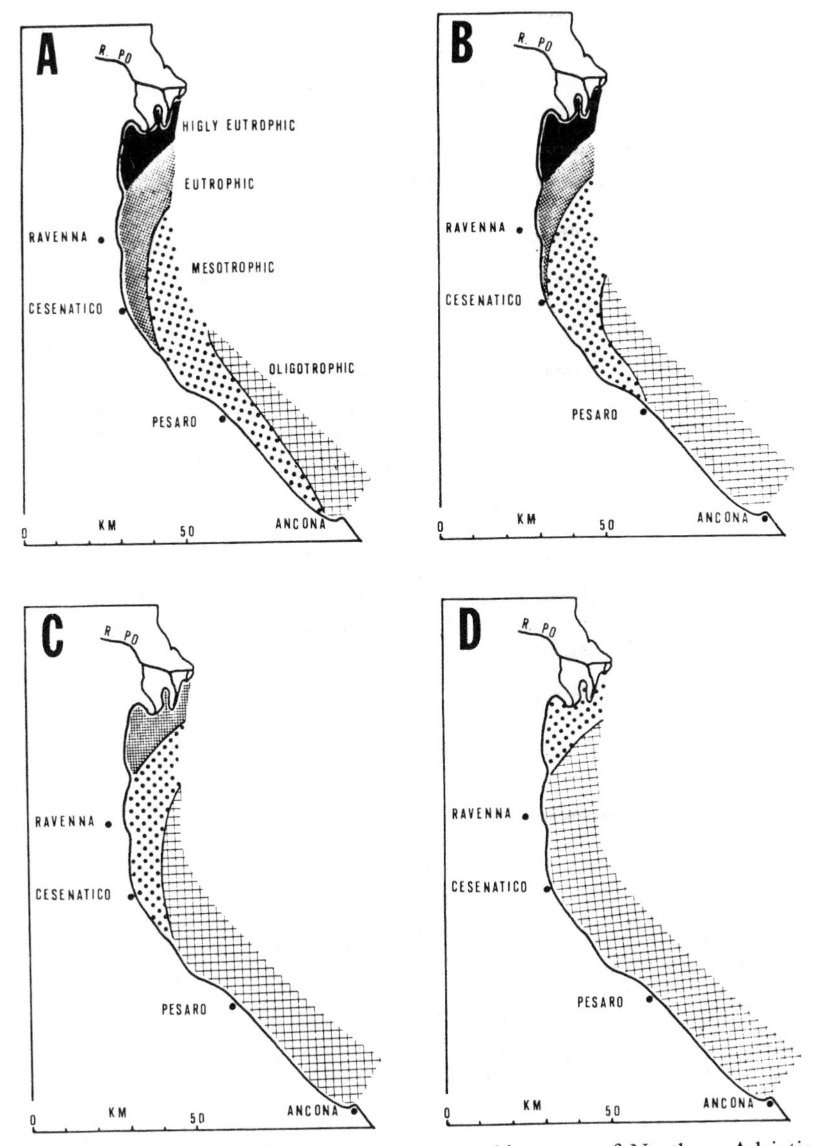

FIG. 4. Comparison between the present trophic status of Northern Adriatic coastal waters (A) and the predicted trophic levels deriving from different phosphorus reduction strategies: reduction of 40% of agricultural diffuse sources (B); reduction of 95% of urban point sources (C); reduction of 95% of all point sources and of 40% of diffuse sources (D).

sources. This is level 3 of the measures on agricultural sources shown in Table 5, and considered as the maximum objective which is realistically achievable. Only a slight improvement of the water quality with respect to the present condition results.

Figure 4C corresponds to a reduction of 95 % of domestic point sources. This kind of intervention could restore a large part of Italian Northern Adriatic coastal waters to an acceptable trophic condition. Therefore this latter intervention should be the first priority in a program of control measures which should be undertaken in the Po Valley in order to reduce eutrophication in the sea waters.

Nevertheless, in a long term strategy, the reduction of all man made phosphorus sources should be considered keeping in mind the main objective of restoring this economically important coastal area to a trophic status nearer to its natural conditions. This is shown in Fig. 4D, which corresponds to a complete intervention on all phosphorus loads, with the reduction of 95 % of all point sources and of 40 % of agricultural diffuse sources.

In conclusion, intervention measures on agricultural non-point sources, due to their high cost and to the relatively low water quality improvement achievable, should be considered a second priority with respect to other phosphorus control in this area.

REFERENCES

1. Bartsch, A. F. (1972). Role of phosphorus in eutrophication. National Environmental Research Center, U.S.E.P.A., Corvallis.
2. Chiaudani, G., Gerletti, M., Marchetti, R., Provini, A. and Vighi, M. (1978). Il problema dell'eutrofizzazione in Italia. Quaderni dell'Istituto di Ricerca sulle Acque N. 42. Milano.
3. Chiaudani, G., Gaggino, G. F., Marchetti, R. and Vighi, M. (1982). Caratteristiche trofiche delle acque costiere adriatiche: campagne di rilevamento 1978–80. CNR-Progetto Finalizzato Promozione della Qualità dell' Ambiente, AQ/2/14. Roma.
4. Chiaudani, G., Gaggino, G. F. and Vighi, M. (1982). Previsione dello stato trofico delle acque costiere dell'Adriatico settentrionale in funzione di variazioni del carico eutrofizzante. Atti del V Congresso dell'Associazione Italiana di Oceanografia e Limnologia, 323–339.
5. Chiaudani, G., Marchetti, R. and Vighi, M. (1980). Eutrophication in Emilia-Romagna coastal waters (North Adriatic Sea, Italy): a case history. *Prog. Wat. Tech.*, **12**, 185–192.
6. Chiaudani, G. and Vighi, M. (1982). L'eutrofizzazione dei bacini lacustri italiani: situazione attuale e possibilità di risanamento. *Acqua e Aria*, **4**, 361–378.

7. ERVET (Ente Regionale per la Valorizzazione Economica del Territorio) (1977). La utilizzazione e la depurazione dei liquami degli allevamenti suinicoli, Bologna.
8. EUROSTAT (1983). Review 1973–82.
9. FRICKER, H. (1979). OECD Cooperative Programme for Monitoring of Inland Waters. Regional Project: Alpine Lakes, OECD.
10. IRSA (Istituto di Ricerca Sulle Acque) (1977). Indagine sulla qualità delle acque del fiume Po. Quaderni dell'Istituto di Ricerca Sulle Acque N. 32. Roma.
11. IRSA (Istituto di Ricerca Sulle Acque) (1980). Indagine sulla qualità delle acque lacustri italiane. Quaderni dell'Istituto di Ricerca Sulle Acque N. 43, Roma.
12. LEE, G. F. (1973). Role of phosphorus in eutrophication and diffuse source control. *Water Research*, **7**, 111–128.
13. LOHER, R. C. (1974). Characteristics and comparative magnitude of non-point sources. *J. Water Poll. Control Fed.*, **46**, 1849–1972.
14. OECD (1982). Eutrophication of waters; Monitoring assessment and control, OECD, Paris.
15. OGLESBY, R. T., HAMILTON, L. S., MILLS, E. L. and WILLING, P. (1973). Owasco lake and its watershed. Technical Report Cornell University Water Resources and Marine Sciences Center, Ithaca, N.Y.
16. RIEMANN, B. and MATHIESEN, H. (1977). Danish research into phytoplankton primary production. *Folia Limnologica Scandinavica*, **17**, 49–55.
17. VOLLENWEIDER, R. A. (1968). Scientific fundamentals of the eutrophication of lakes and flowing waters, with particular reference to nitrogen and phosphorus as factors in eutrophication. OECD, Paris.
18. VOLLENWEIDER, R. A. (1979). L'eutrofizzazione: il problema in scala mondiale. Atti del Convegno sulla Eutrofizzazione in Italia. CNR AC/2/45-70, Roma.
19. WOOD, R. B. and GIBSON, C. E. (1973). Eutrophication and Lough Neagh. *Water Research*, **7**, 173–187.

5

Nitrat im Trinkwasser: Grundlagen für einen Grenzwert

F. SELENKA

Institut für Hygiene der Ruhr-Universität Bochum, Bundesrepublik Deutschland

ZUSAMMENFASSUNG

Nitrat wird unter bestimmten Bedingungen durch Bakterien in Wasser, Boden und Lebensmitteln sowie im menschlichen Organismus zu Nitrit umgewandelt. Nitrit kann beim jungen Säugling Blausucht (Methämoglobinämie) hervorrufen, wenn die Nahrung nitratreiches Gemüse enthält oder mit nitrathaltigem Trinkwasser zubereitet wird. Bei Konzentrationen unter 50 mg NO_3^-/Liter trat die Krankheit bisher extrem selten auf. Meist lagen die Konzentrationen über 100 mg NO_3^-/Liter.

Ein anderer unerwünschter Effekt des Nitrits besteht in seiner Fähigkeit, kanzerogene N-Nitrosoverbindungen—z.B. Nitrosamine—zu bilden. Die Bildung erfolgt in nitrathaltigen Lebensmitteln, vor allem beim Erhitzen, aber auch im menschlichen Organismus. Eine wichtige Rolle spielt dabei das in der Mundhöhle ständig entstehende Nitrit. Wie groß die Gefährdung des Einzelnen durch Nitrosamine ist, ist noch ungenügend bekannt. Die durchschnittliche alimentäre Aufnahme von Dimethylnitrosamin (NDMA) z.B. beträgt in der Bundesrepublik Deutschland 0·3 bis 0·5 µg/Tag. Dies liegt nur um den Faktor 200 bis 300 niedriger, als die im Tierversuch minimal wirksame Konzentration. Wenn man berücksichtigt, daß beim Menschen unterschiedliche Empfindlichkeit, unterschiedliche Verzehrsgewohnheiten und möglicherweise auch eine zusätzliche endogene Bildung bestehen, erscheint dieser Sicherheitsabstand relativ klein.

Da die Nitrosaminbildung von den Nitritkonzentrationen abhängt und letztere wiederum in direkter Korrelation zu der aufgenommenen Nitratmenge steht, erscheint es sinnvoll, die über Trinkwasser und Lebensmittel erfolgende Nitrataufnahme zu begrenzen. Die Herabsetzung des Grenzwertes für Nitrat in Trinkwasser auf 50 mg/Liter ist eine sinnvolle prophylaktische Maßnahme.

1. EINLEITUNG

Wenn die Frage gestellt wird, welche Bedeutung das Nitrat für die menschliche Gesundheit besitzt, dann muß zuerst darauf hingewiesen werden, daß das Nitrat eine wichtige Rolle bei der Sicherung unserer Ernährungsgrundlage spielt. Ein unterernährter Mensch ist eines großen Teiles seiner Lebensqualität beraubt, wenn er nicht überhaupt an den direkten Folgen der Unterernährung—vor allem im Kindesalter—stirbt. In zahlreichen Regionen der Welt ist dieser Aspekt von elementarer Bedeutung. Die Ernährung muß aber nicht nur quantitativ ausreichen. Sie muß auch qualitativ so beschaffen sein, daß sie nicht zu Gesundheitsschäden führt. Das für die pflanzliche Produktion essentielle Nitrat bereitet aber gerade hierbei Probleme.

Nitrat selbst ist wenig toxisch. Nach Moeschlin [12] führen erst 500 bis 1000 mg beim Erwachsenen zu geringen Reizerscheinungen; deutliche Vergiftungssymptome treten bei 1000 bis 2000 mg auf. Dies gilt für die Gabe von Reinsubstanz. Wird das Nitrat aber als Bestandteil von Lebensmitteln aufgenommen, so werden auch höhere Nitratmengen toleriert. Dies hängt mit der relativ langsamen Verdauung der Lebensmittel und der dadurch zeitlich verzögerten Resorption von Nitrat zusammen. Während sich ein Teil des Nitrats noch im Speisebrei des Magens und Dünndarms befindet, ist ein anderer Teil bereits durch die Nieren wieder ausgeschieden.

Der eigentliche Schadstoff ist das Nitrit. Nitrit kann überall, wo bakterielles Leben möglich ist, aus Nitrat entstehen. Die reduktive Umwandlung findet in Wasser und Boden, in Lebensmitteln und auch im tierischen und menschlichen Organismus statt. Abiotische Nitratreduktionen, die gelegentlich im Grundwasser oder in verzinkten Wasserleitungsrohren beobachtet werden, spielen demgegenüber nur eine geringe Rolle. Nitrit kann beim jungen Säugling Blausucht auslösen oder cancerogene N-Nitrosoverbindungen entstehen lassen, die alle Altersgruppen betreffen. Je höher die Nitratkonzentrationen, um so größer ist die Möglichkeit einer durch Nitrit ausgelösten Gesundheitsbeeinträchtigung. Aus diesem Grund ist eine Begrenzung des Nitratgehaltes in Wasser und in bestimmten Lebensmitteln erforderlich.

2. VORKOMMEN VON NITRAT IN TRINKWASSER

Nitrat ist in naturbelassenen Gebieten nur in Spuren bis hin zu wenigen Milligramm pro Liter im Grundwasser nachweisbar. Durch Mobilisierung des bodeneigenen Stickstoffs infolge bestimmter landwirtschaftlicher Maßnahmen, vor allem aber bei hoch dosierter Zufuhr von Stickstoff in Form von organischem Dünger oder Mineraldünger, kommt es zur Auswaschung von Nitrat in den Untergrund. Zahlreiche Faktoren, wie Bodenbeschaffenheit, Niederschlagsmenge, Zeitpunkt der Stickstoffanwendung, Art der landwirtschaftlichen Nutzung u.a., beeinflussen dieses Geschehen. Das absinkende, nitratbeladene Sickerwasser lagert sich auf die Oberfläche des Grundwassers auf und bildet eine relativ stabile Schicht, die mit dem Grundwasserstrom abfließt.

Neu hinzutretendes Sickerwasser wird wiederum aufgelagert, so daß das ursprüngliche Wasser immer tiefer abgedrängt wird. In feinporigem Bodenmaterial bleiben die Schichten über lange Strecken relativ ungemischt. Wird in einen derartigen Grundwasserleiter ein Brunnen niedergebracht, so durchschneidet dieser den geschichteten Wasserkörper, wobei die aus der Nähe stammenden jüngeren Schichten oben angetroffen werden und die von weither transportierten älteren in der Tiefe liegen (Abb. 1). Bei Förderung des Wassers erfolgt eine intensive Durchmischung, so daß eine durchschnittliche Nitratkonzentration entsteht. Durch spezielle Entnahmetechniken können die einzelnen Schichten jedoch getrennt erfaßt und ihr Nitratgehalt bestimmt werden. Legt man in Richtung des anströmenden Grundwassers mehrere Brunnen an, so kann man die Herkunft der Nitratbelastungen verfolgen und schließlich den verursachenden Flurbereich

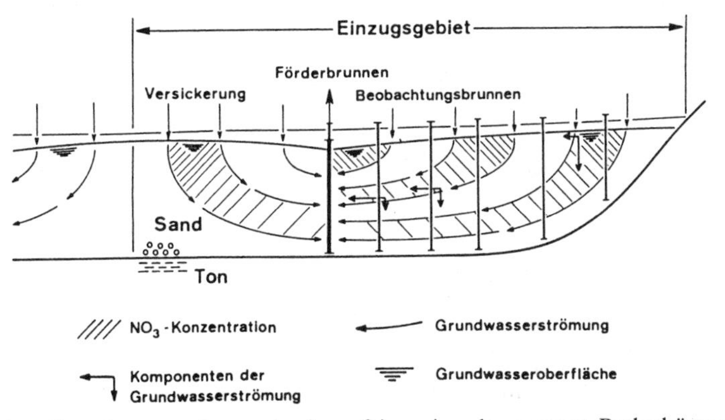

Abb. 1. Grundwasserströmung in einem feinporigen homogenen Bodenkörper [16].

erkennen. Das Verfahren ist aufwendig und bisher nur an wenigen Stellen realisiert worden (Obermann 1981) [16]. In der Regel liegt nitratreiches Wasser oben, nitratarmes Wasser unten. Dies mag in vielen Fällen auf einem stärkeren Eintrag von Nitrat in die jüngeren Schichten beruhen. Grundsätzlich sind die nach unten abnehmenden Nitratkonzentrationen aber auch durch die im Grundwasserleiter ablaufenden mikrobiologischen Abbauvorgänge zu erklären. Entgegen einer weit verbreiteten Ansicht ist der Untergrund selbst in großen Tiefen noch mit Mikroorganismen besiedelt [9]. Unsere Grundwässer wären wesentlich stärker nitrathaltig, wenn es diesen Abbau nicht gäbe. Die auch in tieferen Schichten aktiven Mikroorganismen reichen in der Regel aber nicht aus, um langfristig ein Ansteigen der Nitratkonzentrationen im Grundwasser zu verhindern.

In zahlreichen Wasserversorgungsanlagen, die in Gebieten mit intensiver landwirtschaftlicher Nutzung liegen, ist deshalb ein Ansteigen der Nitratkonzentrationen zu beobachten. Je nach Bodenbeschaffenheit, Düngungsverhalten und hydrogeologischer Situation kann dies zeitlich sehr unterschiedlich ablaufen (Abb. 2 und 3). Neben Wasservorkommen, die bereits in den sechziger Jahren hohe Nitratwerte erreicht hatten und nun stationär sind, findet man solche, die auch heute noch in stetigem Steigen begriffen sind [19]. Am Niederrhein wurde ein Anstieg von 2 bis

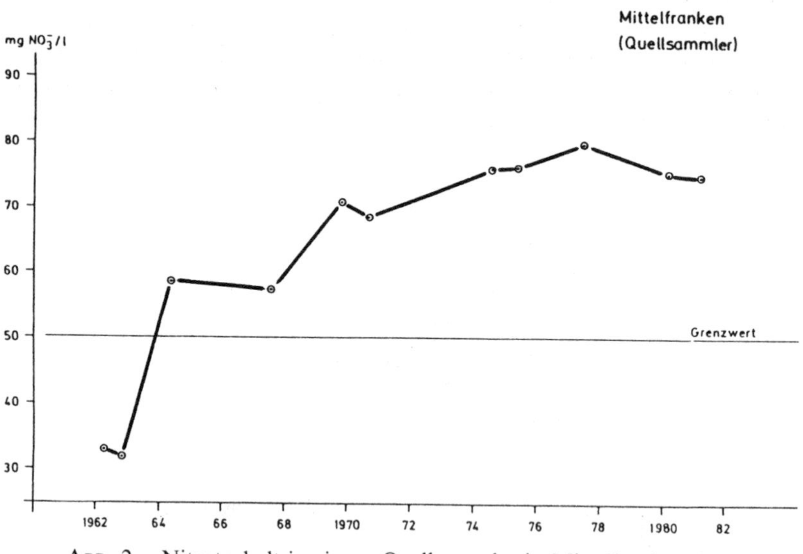

Abb. 2. Nitratgehalt in einem Quellsammler in Mittelfranken [19].

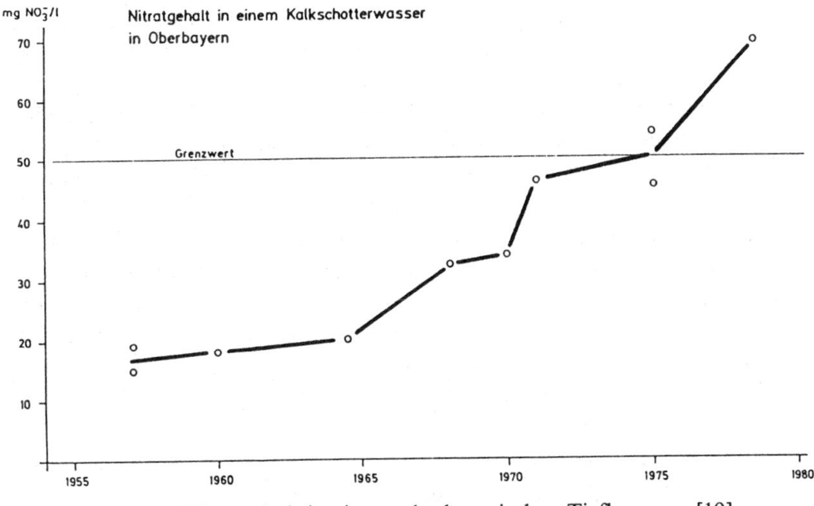

ABB. 3. Nitratgehalt in einem oberbayerischen Tiefbrunnen [19].

3 mg NO_3^- pro Jahr berechnet [8]. Es gibt aber auch Beispiele dafür, daß die Trinkwasser-Nitratwerte trotz Düngung gleichbleibend niedrig bleiben. Da das geförderte Wasser oft lange Fließstrecken im Untergrund hinter sich hat und damit in der Regel Jahre bis Jahrzehnte alt ist, dürfte der jetzige Zustand in einem Teil der Fälle erst den Beginn der Entwicklung darstellen. Von 870 Millionen m^3 Trinkwasser, das im Bundesland Bayern durch 4331 Wassergewinnungsanlagen gefördert und verteilt wurde, enthielten 73·5 % weniger als 25 mg/Liter NO_3^-, 5·5 % enthielten 50 bis 90 mg/Liter und 0·1 % mehr als 90 mg/Liter (Tab. 1).

Tabelle 1

Nitratgehalte im Trinkwasser Bayerischer Wassergewinnungsanlagen [19]

Nitratgehalt (mg/Liter)	*Wassergewinnungsanlagen*	*Trinkwassermenge (1 000 m³)*	*%*
unter 25	2 551	639 693	73·5
25–90	645	125 693	14·4
50–90	195	47 579	5·5
über 90	15	856	0·1
ohne Angabe	925	56 191	6·5
insgesamt	4 331	870 012	100·0

In der gesamten Bundesrepublik Deutschland werden 89·4% der Bevölkerung mit Trinkwasser versorgt, das weniger als 20 mg/Liter NO_3^- enthält, 5·7% trinken Wasser mit 50–90 mg/Liter und 0·9% über 90 mg/Liter NO_3^- [1]. Ein Problem stellen die besonders in ländlichen Gegenden vorhandenen privaten Einzelwasserversorgungen dar. Allein im Regierungsbezirk Münster gibt es 61 000, von denen 25 bis 30% mehr als 50 mg/Liter NO_3^- enthalten. Auch im Oberflächenwasser, das zur Trinkwassergewinning herangezogen wird, haben vielerorts die Nitratkonzentrationen durch die fast ausschließliche Anwendung oxidativ arbeitender Kläranlagen auf Werte um 20 mg/Liter NO_3^- zugenommen. Die über die Jahre 1963 bis 1979 erfolgende Abnahme der Ammoniumkonzentrationen und Zunahme der Nitratkonzentrationen ist sehr eindrucksvoll am Beispiel des Rheins zu erkennen [6].

3. AUFNAHME VON NITRAT UND NITRIT ÜBER WASSER UND LEBENSMITTEL

Die mit dem Trinkwasser aufgenommene Nitratmenge hängt von der Nitratkonzentration und dem Trinkwasserkonsum ab. In beiden Fällen ist eine große Varianzbreite gegeben. Auch wenn Analysen in engem zeitlichen Abstand durchgeführt werden, sind häufig beträchtliche Schwankungen der Nitratkonzentrationen festzustellen. Aus Sicherheitsgründen werden den Beurteilungen meist die Maximalwerte zugrundegelegt. Das bedeutet, daß die tatsächliche Langzeitbelastung in zahlreichen Fällen erheblich niedriger liegen kann. Die Probleme werden dadurch jedoch nicht geringer. Wie weiter unten gezeigt wird, sind aus toxikologischer Sicht sowohl die Maximalwerte als auch die langfristigen Durchschnittswerte von Bedeutung.

Über die tägliche Trinkwasseraufnahme sind divergierende Meinungen zu finden. Einerseits wird der absolute tägliche Wasserbedarf, der als Mittelwert für Sommer und Winter ca. 2·7 Liter beträgt, für Berechnungen zugrundegelegt. Da ein erheblicher Teil der Wasserzufuhr aber durch Obst, Gemüse, Fertiggetränke, Fleisch u.a. erfolgt, sollte von einem niedrigeren Wert, etwa 1·7 bis 2·0 Liter/Tag, ausgegangen werden. 1·2 Liter/Tag ist als Minimum zu betrachten. Diese Angaben berücksichtigen allerdings nicht, daß bestimmte Personengruppen (z.B. Schwerarbeiter, Hitzearbeiter, Zuckerkranke) einen darüberhinausgehenden Flüssigkeitsbedarf haben. Bei einem Trinkwasserkonsum von 1·7 Liter und einer Nitratkonzentration von 50 mg/Liter beträgt die tägliche Nitrataufnahme 85 mg, bei einer Nitratkonzentration von 90 mg/Liter 153 mg (Abb. 4).

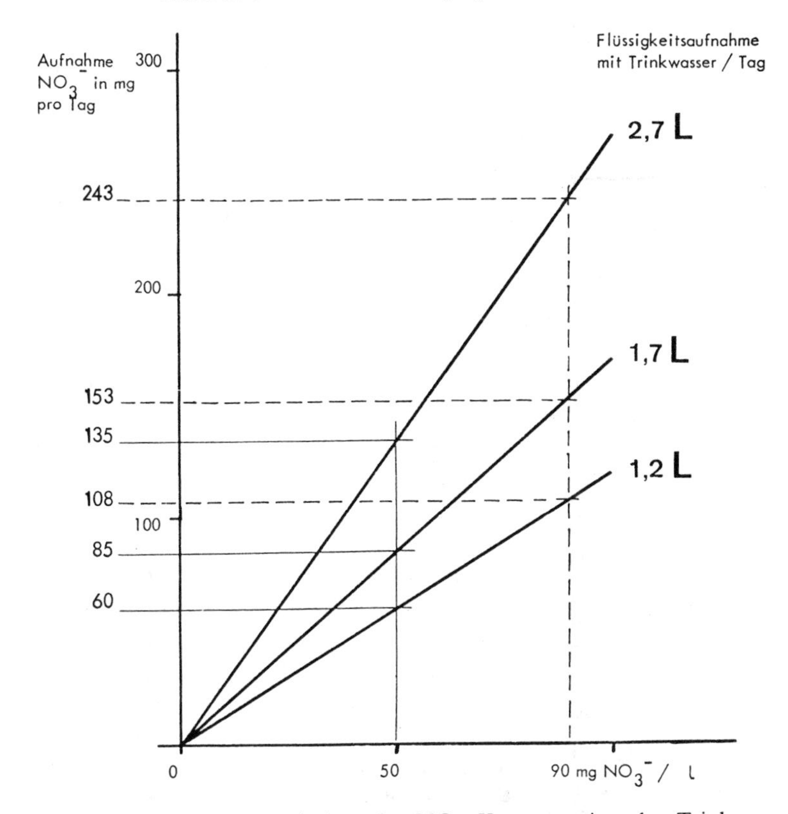

ABB. 4. Zusammenhang zwischen der NO_3^--Konzentration des Trinkwassers und der täglichen Nitrataufnahme [22].

Diese Mengen addieren sich zu den in der Nahrung enthaltenen. Den Berechnungen verschiedener Arbeitsgruppen zufolge werden in Europa durchschnittlich 36 bis 110 mg NO_3^- pro Tag aufgenommen (Tab. 2). Diesen Angaben entsprechend, sowie aufgrund eigener Untersuchungen gehen wir beim Erwachsenen im Mittel von einer täglichen Belastung von 75 mg NO_3^- über pflanzliche und tierische Lebensmittel aus. Vegetabilien sind für etwa 75 % des aufgenommenen Nitrats verantwortlich, Fleisch und Fleischwaren für etwa 19 % und Getreideprodukte für etwa 6 % [23].

Durch einseitige Kostformen, aber auch durch die Art der Zubereitung von Nahrungsmitteln können allerdings für das Individuum große Abweichungen von diesen Werten auftreten. Besonders hohe Nitrataufnahmen werden aus Japan berichtet. Sie erreichten in den Jahren 1972 bis

Tabelle 2
Tägliche Aufnahme von Nitrat über Lebensmittel in Westeuropa und in den USA. Analysenergebnisse (A) und nach Verzehrstatistiken (C) in den jahren 1970 bis 1978 (nach Ellen und Schuller 1983 [6] modifiziert)

Land (Zeitraum)	*Experimentelle Bedingungen*	*NO_3^- mg/Person/Tag Mittelwert*	*(Bereich)*	*Autoren*[b]
BR Deutschland (1971–73)	Gaststättengerichte (A), zusätzliche Mahlzeiten (C)	75	(55–95)	Selenka und Brand-Grimm (1976) [23]
BR Deutschland (1974–77)	24-Std. Krankenhaus-kost (A)	45	(12–114)	Möhler (1982) [13]
BR Deutschland (1978)	Vegetabilische Duplikatmahlzeiten (A), korrigiert mit 4/3	36	—	Krampe und Andre (1980)
Schweiz (1977)	NO_3 in Lebensmitteln nach Verzehrsstatistik (C)	81[a]	—	Tremp (1980)
Niederlande (1976–78)	24-Std.-Duplikatmahl-zeiten (A)	110[a]	(9–706)	Stephany und Schuller (1980)
Niederlande (1976–78)	Mahlzeiten (A) nach Marktkorbmethode	98	—	de Vos und v. Dokkum (1980)
Schweden —	24-Std.-Duplikatmahl-zeiten (A)	49[a]	(26–81)	Jagerstand *et al.* (1980)
Großbritannien (1970–71)	NO_3 in Lebensmitteln nach Verzehrsstatistik (C) ohne Getreide-produkte und Milch	76	—	Walker (1975)
USA (1971–73)	NO_3 in Lebensmitteln nach Verzehrsstatistik (C)	99	—	White (1975, 1976)

[a] Einschl. Getränke.
[b] Literatur siehe Ellen und Schuller, 1983 [6].

1977 das 3 bis 5fache der westeuropäischen Werte. Über eine Trendentwicklung der Nitratkonzentrationen von Lebensmitteln ist wenig bekannt. Ältere Analysen zeigen, daß bereits vor Jahrzehnten bei bestimmten Gemüsearten hohe Nitratbelastungen vorgelegen haben. So wurden in den USA im Jahre 1907 bei Salat 1700 (400–3500) mg NO_3^-/kg, bei Sellerie 1500 (800–2900) mg NO_3^-/kg und bei Rote Beete 2600 (900–8000) mg NO_3^-/kg nachgewiesen. Angaben aus den Niederlanden zeigen Werte, die gelegentlich doppelt so hoch liegen, z.t. aber auch deutlich niedriger [6]. Die zur Verfügung stehenden Angaben sind jedoch schwer zu vergleichen, da große Unterschiede im Nitratgehalt auf Pflanzensorten, Kulturbedingungen, Erntezeit u.a. beruhen können.

Nitrit kommt im Trinkwasser selten in höheren Konzentrationen als 1 bis 2 mg/Liter NO_2^- vor [21]. In Lebensmitteln werden durchschnittlich 2·6 bis 6·7 mg als tägliche Aufnahme errechnet [6]. Mehr als die Hälfte davon stammen aus Fleisch und Fleischwaren, die entsprechende Zusätze zur Verhinderung von Botulismus-Erkrankungen, aber auch zur Erzeugung von Geruch- und Geschmackstoffen und zur Erzielung einer gefälligen Farbe erhalten.

Von dem FAO/WHO Joint expert Committee on Food Additives (JECFA) sind folgende ADI-Werte für Nitrat und Nitrit aufgestellt worden:

$$3{\cdot}65\,\text{mg } NO_3^-/\text{kg Körpergewicht} = 219\,\text{mg } NO_3^-/60\,\text{kg und Tag}$$
$$0{\cdot}4\,\text{mg } NO_2^-/\text{kg Körpergewicht} = 8\,\text{mg } NO_2^-/60\,\text{kg und Tag}$$

Obgleich sich diese Werte nur auf künstliche Zusätze von NO_3^- und NO_2^- bei der Lebensmittelherstellung beziehen, werden sie häufig zur Beurteilung des Gesamtgehaltes von NO_3^- und NO_2^- in Lebensmitteln herangezogen.

4. VERHALTEN DES NITRATS IM KÖRPER

Das mit der Nahrung und dem Trinkwasser aufgenommene Nitrat wird in den oberen Darmabschnitten resorbiert und im Laufe von 4 bis 14 Stunden über die Nieren ausgeschieden. Im Mittel erscheinen nur 82% des aufgenommenen Nitrats im Harn wieder, der Rest wird anderweitig metabolisiert [26]. Etwa ein Drittel des resorbierten Nitrats gelangt in einen Nebenkreislauf zu den Speicheldrüsen, wird dort konzentriert und mit dem Speichel in die Mundhöhle ausgeschieden (Abb. 5). Beim Erwachsenen setzen die Bakterien der Mundhöhle etwa 20% des im Speichel enthaltenen Nitrats zu Nitrit um. Aus den Ergebnissen mehrerer Arbeitsgruppen kann geschlossen werden, daß auf diese Weise im Mittel $6{\cdot}3 \pm 0{\cdot}7\%$ des aufgenommenen Nitrats innerhalb von 24 Std. zu Nitrit umgewandelt werden [5]. Ein ähnlicher Vorgang findet auch im Magen statt, wobei die Nitritproduktion im Magen nach Müller u.a. [15] 6 mal größer als die der Mundhöhle sein kann. Die Umsetzungen erfolgen im Nüchternsekret, das vorwiegend während der Nachtruhe und morgens, aber auch untertags gebildet wird. Bei 130 gesunden, jungen Probanden fanden sich im Nüchternsekret überwiegend pH-Werte zwischen 6·0 und

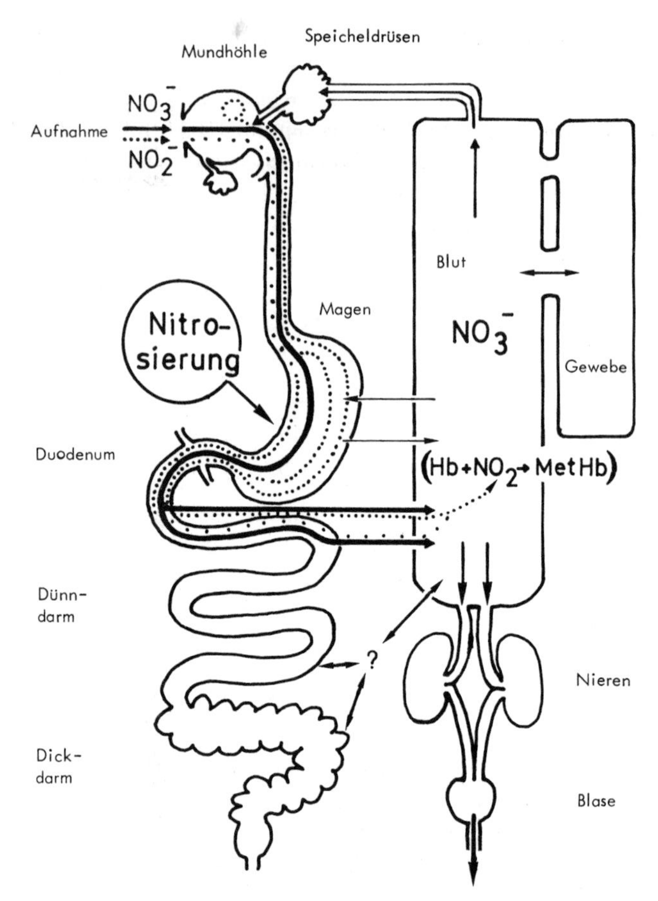

ABB. 5. Aufnahme und Verhalten von Nitrat im Organismus [21].

7·6 (70 Fälle). Die in der Größenordnung von 10^7/ml liegenden Koloniezahlen wiesen 35% bis 53% Nitritbildner auf, die sich etwa je zur Hälfte aus Sporenbildnern und Kokken zusammensetzten. Aber auch bei pH 4 und 5 (17 Fälle) lagen Koloniezahlen von 10^6 bis 10^7/ml und erhöhte Nitritkonzentrationen vor. Eine erhöhte endogene Nitritbildung ist ferner bei schwer verdaulicher Kost anzunehmen, da in diesen Falle der Speicheldrüsenkreislauf vom Nitrat über einen längeren Zeitraum auf einem höheren Konzentrationsniveau stattfindet [20, 26].

Sowohl die im Speichel enthaltenen Nitratkonzentrationen als auch die daraus entstehenden Nitritkonzentrationen verhalten sich direkt proportional zur Gesamtaufnahme von Nitrat [25]. Dies bedeutet, daß bei Anstieg der Nitratkonzentrationen in der Nahrung auch entsprechend erhöhte Nitritkonzentrationen im Speichel und Magensaft vorliegen. Die endogene Nitritbildung beim Erwachsenen übertrifft mit durchschnittlich 8–11 mg NO_2^-/Tag die mit Nahrungsmitteln erfolgende exogene Nitritzufuhr mindestens um den Faktor 3 bis 5.

Eine endogene Nitratbildung kann im Hungerzustand aus Eiweiß, Aminosäuren und anderen N-Verbindungen durch nitrifizierende Bakterien im Darm stattfinden [7]. Unter normalen Ernährungsbedingungen hat dieser Vorgang jedoch nur eine relativ geringe Bedeutung.

5. TOXIKOLOGISCHE BEDEUTUNG

Seit 40 Jahren ist bekannt, daß Nitrit beim Säugling das Krankheitsbild der Blausucht (Methämoglobinämie) auslöst [3]. Voraussetzung sind erhöhte Nitratkonzentrationen des Trinkwassers oder der vegetabilischen Beikost. Beim gesunden Säugling wird wenig endogenes Nitrit gebildet, da seine Mundhöhle, sein Magen und sein Dünndarm nur geringe Mengen nitratreduzierender Bakterien enthält. Beim kranken Säugling, vor allem im Falle von Darmerkrankungen, kann es jedoch zum Aufsteigen von nitratreduzierenden Keimen aus den tieferen Darmabschnitten und zu hohen Nitratumsetzungen im Magen und Dünndarm kommen. Zum Teil liegt eine quantitative Umwandlung von NO_3^- zu NO_2^- vor. Der gleiche Effekt findet statt, wenn stark keimhaltige, nitratreiche Nahrung verabreicht wird. Die Umwandlung kann auch außerhalb des Körpers während unsachgemäßer Bevorratung stattfinden. Das mit der Nahrung aufgenommene oder im Magen-DarmTrakt gebildete Nitrit wird wie Nitrat in den oberen Darmabschnitten rasch resorbiert und reagiert mit dem Blutfarbstoff. Das entstehende Methämoglobin (MetHB) ist nicht mehr in der Lage, Sauerstoff zu transportieren, so daß eine innere Erstickung eintritt. Bei 10 % MetHB stellen sich erste Krankheitszeichen ein, bei 40 bis 70 % erfolgt der Tod. Trotz hoher NO_3^--Gehalte des Trinkwassers ist die Krankheit heute selten geworden. Dies hat mehrere Gründe. Darmkranke Säuglinge, die in erster Linie gefährdet sind, kommen heute frühzeitig in ärztliche Behandlung, in schweren Fällen in die Klinik. Damit ist in der Regel eine Verringerung der Nitratzufuhr und der endogenen Nitratreduktion verbunden. Eine exogene Nitritzufuhr findet durch das üblich gewordene portionsweise Ansetzen von Pulvermilchnahrungen oder durch das Benutzen von Kühlschränken kaum noch statt. Auch eine endogene Nitratreduktion ist infolge der Begrenzung des

Keimgehalts von Milchpulvererzeugnissen, z.B. in der Bundesrepublik Deutschland auf 250 Kolonien/g, sowie durch verbesserte Hygiene bei der Nahrungsbereitung weniger wahrscheinlich. Außerdem wird von zahlreichen Eltern in Gebieten mit nitratbelastetem Trinkwasser nitratarmes Wasser für die Säuglingsnahrung benutzt.

Verläßliche Daten zeigen erst ab 10 bis 20 mg NO_2^-/kg Körpergewicht/Tag Methämoglobinämien. Dies entspricht 50–100 mg NO_2^-/Tag bei einem normal schweren Säugling oder 68 bis 136 mg vollständig reduziertes NO_3^-. Da ein Säugling pro Tag 850–1000 ml trinkt, kann ein Wasser ab 70 bis 140 mg/Liter NO_3^- krankheitsauslösend sein [22]. Eine leichte Erhöhung des Methämoglobinspiegels auf 2 bis 3 % ist bereits bei etwa 50 mg/Liter NO_3^- möglich. Dies besitzt jedoch noch keine gesundheitliche Relevanz. Eine symptomlose, lang anhaltende Methämoglobinämie scheint es nicht zu geben. Borneff [2] hat in Gebieten mit einem Trinkwassernitratgehalt von 110–140 mg NO_3^-/Liter Säuglinge untersucht und keine Unterschiede im MetHB-Gehalt und im Gedeihen gegenüber einer Kontrollgruppe gefunden.

Die meisten Methämoglobinämien wurden erst durch mehr als 100 mg NO_3^-/Liter ausgelöst. Andererseits liegen aber Berichte vor, wonach bereits bei 24 mg NO_3^-/Liter Erkrankungen auftraten [27].

Hier muß diskutiert werden, ob genetisch bedingte Stoffwechselstörungen oder andere Anomalien, wie z.B. die leicht oxidierbare Hämoglobinvariante Hb-M, vorlagen. Auch andere Erklärungen, wie Aufkonzentrierung durch Eindampfen des Wassers oder fehlerhafte Analytik, sollten bei den bereits 3 Jahrzehnte zurückliegenden Beobachtungen in Erwägung gezogen werden. Weitreichende Maßnahmen der Wasserversorgung und darüber hinaus der Landwirtschaft sind aufgrund derartiger Einzelberichte nicht gerechtfertigt.

Die Methämoglobinämie ist eine akute Erkrankung. Sie tritt innerhalb von Stunden bis höchstens Tagen nach einer erhöhten Nitrat- bzw. Nitritzufuhr auf. Aus diesem Grunde sind im Einzelfall auch kurzfristige Konzentrationsspitzen in der Lage, einen Gesundheitsschaden auszulösen. In Bezug auf die Säuglingsmethämoglobinämie sind daher nicht nur langfristige Durchschnittswerte, sondern auch die Maximalwerte von Bedeutung.

Das zweite mit Nitrit verbundene Risiko beruht auf einer Reaktion von NO_2^- mit sekundären und tertiären Aminen, wobei unter bestimmten Bedingungen krebserregende *N-Nitrosoverbindungen*, z.B. Nitrosamine, entstehen [10]. Die rein chemische Reaktion läuft bevorzugt bei sauren pH-Werten, wie sie z.B. im Magen vorliegen, ab (Abb. 6).

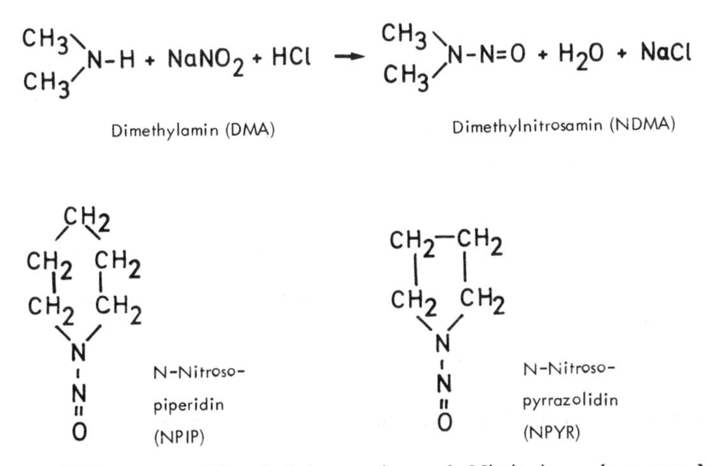

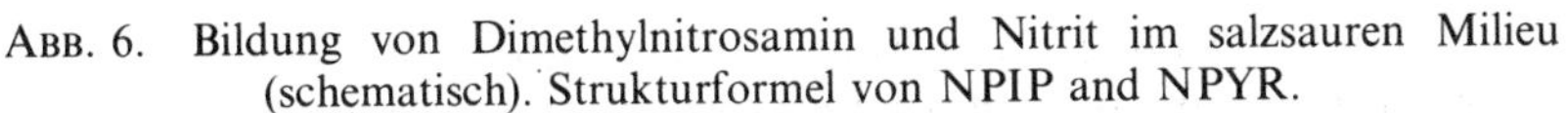

ABB. 6. Bildung von Dimethylnitrosamin und Nitrit im salzsauren Milieu (schematisch). Strukturformel von NPIP and NPYR.

Bei Anwesenheit von Katalysatoren, wie Formaldehyd, kommt die Reaktion aber auch bei neutralem pH zustande. Bestimmte Substanzen, wie Askorbinsäure oder SO_2, behindern die Reaktion. Da Nitrit bei Vorliegen von sekundären Aminen in die Reaktionsgleichung mit der 2. Potenz eingeht, wirkt sich eine Erhöhung der Nitritkonzentration in einer überproportionalen Nitrosaminausbeute aus. Die Nitrosierungsgeschwindigkeiten der verschiedenen Amine hängen u.a. von deren Basizität ab. Günstigerweise besitzen die hochcancerogenen Verbindungen Dimethylnitrosamin (NDMA) und Diethylnitrosamin (NDEA) eine starke Basizität und entstehen dadurch bedeutend langsamer als andere Nitrosamine mit mäßiger oder fehlender cancerogener Potenz [11]. Die durchschnittliche tägliche Nitrosaminaufnahme mit der Nahrung betrug 1981 in der Bundesrepublik Deutschland 0·5 μg rund 0·7 μg bei Männern und 0·5 μg bei Frauen [24]. Hauptkomponente ist das Dimethylnitrosamin (NDMA), das vor allem in Fleischwaren, aber auch in verschiedenen Bier- und Käsesorten und anderen Produkten gefunden wird. N-Nitroso-Pyrrolidin (NPYR) kommt überwiegend in Fleischwaren vor, wo es durch Erhitzungsprozesse erhebliche Konzentrationszunahmen erfährt. Sein Anteil an der täglichen Nitrosaminaufnahme beträgt durchschnittlich 23%. Die Aufnahme von N-Nitroso-Piperidin (NPIP) ist mit rund 0·01 μg/Tag demgegenüber zu vernachlässigen [24].

Die zur Auslösung von Tumoren bei der Ratte erforderliche minimale Dosis beträgt bei täglicher Gabe 10 μg NDMA bzw. 10 μg NDEA/kg

Tabelle 3
Minimal wirksame Dosen von NDEA, NDMA, NPYR und NPIP bei der Ratte [4]

Nitrosamin	*Dosis/kg Körpergewicht ($\mu g/kg$)*	*Applikation*	*Effekte*
NDEA	10	täglich	Lebercarcinom
NDMA	10	täglich	Lebercarcinom
NDMA	2	täglich	Hyperplasien
NPYR	290	täglich	Carcinom
NPIP	23	täglich	Carcinom

Körpergewicht (KG). Bei NPYR sind es 290 und bei NPIP 23 μg/kg KG (Tab. 3).

2 μg NDMA/kg KG führen während der Lebenszeit der Tiere zu keinem Tumor [4]. Allerdings entstehen Gewebsveränderungen, die als Vorstufen einer Krebsbildung bekannt sind. Beim Menschen läßt sich bei einer täglichen Aufnahme von rund 0·3 bis 0·5 μg NDMA auf eine durchschnittliche Belastung von 0·006 bis 0·009 μg/kg KG schließen. Zwischen der im Tierexperiment von Crampton [4] ermittelten minimalen Wirkungsdosis für präcanceröse Veränderungen und der beim Menschen beobachteten tatsächlichen Zufuhr besteht demnach nur ein Sicherheitsabstand von 1:200 bis 1:300. Dies erscheint relativ gering, zumal durchaus auch höhere Nitrosaminaufnahmen denkbar sind. Hier sind zunächst individuelle Belastungsunterschiede zu nennen, die auf einseitigen Verzehrsgewohnheiten, wie z.B. starken Fleisch-oder Bierkonsum, beruhen. Entscheidend aber ist die Frage, ob es eine zusätzliche endogene Nitrosaminentstehung gibt und ob diese einen nennenswerten Umfang erreicht. Durch Tierversuche ist die endogene Entstehung von N-Nitrosoverbindungen belegt. Daß auch beim Menschen Nitrosierungen ablaufen, wurde bereits 1967 von Sander [18] festgestellt. Eine Quantifizierung des Zusammenhanges zwischen Höhe der Nitrataufnahme und endogener nitrosierender Potenz des Menschen gelang jedoch erst kürzlich [17]. Die Autoren verwendeten die Aminosäure Prolin, die im Organismus mit Nitrit zu dem nicht cancerogenen N-Nitrosoprolin reagiert und in dieser Form mit dem Harn ausgeschieden wird. Dabei zeigte sich, daß bis zu einer täglichen Nitratbelastung von 195 mg NO_3^- eine nur mäßige Nitrosaminbildung erfolgt, daß aber darüberhinaus überproportionale Mengen entstehen. Bei 260 mg NO_3^- wird bereits das Zehnfache des Ausgangswertes erreicht (Abb. 7).

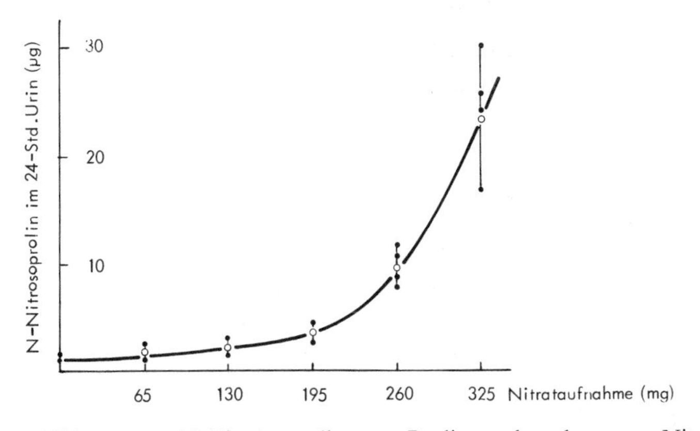

ABB. 7. Bildung von N-Nitrosoprolin aus Prolin und endogenem Nitrit im Menschen in Abhängigkeit von der Menge aufgenommenen Nitrats [17].

Dieses Ergebnis gilt nur für die angegebene Verbindung. Andere Nitrosamine können völlig andere Verhältnisse zeigen. Das Experiment zeigt aber, daß in Bereichen, die der täglichen Nitrataufnahme entsprechen, meßbare Nitrosierungen ablaufen.

Die Tatsache, daß die Magen-Carcinom-Rate seit Jahren rückläufig ist, während die Nitratgehalte vielerorts ansteigen oder zumindest gleich bleiben, spricht nicht gegen die Beteiligung von Nitrosaminen bei der allgemeinen Krebsgenese und insbesondere beim Magen-Carcinom. Anatomische und physiologische Besonderheiten des Magens, unterschiedliche Ernährungsbedingungen sowie andere Umweltfaktoren scheinen eine zusätzliche, im einzelnen noch unklare Rolle zu spielen. Interessant sind in diesem Zusammenhang Untersuchungen, die zeigen, daß der Magenschleim nicht nur ein Biotop für nitritbildende Bakterien darstellt, sondern auch Nitrit vorübergehend bindet, anreichert und zu gegebener Zeit freisetzt [14].

Ständig zugeführte minimale Nitrosaminmengen rufen keine akuten Schädigungen hervor. Ihre Wirkungen summieren sich aber und können nach Jahren bis Jahrzehnten einen Tumor entstehen lassen. Maßgeblich ist die Dauer und die durchschnittliche Höhe der Belastung. Dies gilt auch für die Vorläufer der Nitrosamine, das Nitrat und das Nitrit.

Die Europäische Gemeinschaft hat sich entschlossen, einen Grenzwert von 50 mg/Liter NO_3^- und einen Richtwert von 25 mg/Liter NO_3^- für Trinkwasser aufzustellen. Damit werden höhere Grenzwerte, wie in dem Vereinigten Königreich, der Bundesrepublik Deutschland oder Belgien, in

Kürze zurückgenommen werden müssen. Für die Wasserversorgungsunternehmen ergeben sich dabei zwangsläufig z.T. erhebliche Probleme. Wo es nicht gelingt, durch Zumischen nitratarmen Wassers aus dem eigenen Raum oder durch Fernleitungen den Anforderungen zu entsprechen, müssen Denitrifizierungsanlagen installiert werden. Von den bereits technisch realisierten Verfahren des Ionenaustausches, der Elektrodialyse, der Umkehrosmose und der biologischen Nitratreduktion bringt die Umkehrosmose die geringsten hygienischen Probleme. Erfolgversprechend sind auch Versuche, die Denitrifikation in den Untergrund zu verlegen. Ein wesentliches Problem stellt jedoch die Behandlung von hunderttausenden kleiner Eigenwasserversorgungen, meist in ländlichen Gebieten, dar. Für sie sind hygienisch sichere, wartungsarme Denitrifizierungsanlagen oder Anschlüsse an größere Versorgungseinheiten zu schaffen.

6. SCHLUSSFOLGERUNG

Alle Maßnahmen zur Verringerung von Nitrat in der Ernährung des Erwachsenen sind zum gegenwärtigen Zeitpunkt prophylaktischer Art. Der Beweis, daß dieses prophylaktische Handeln sinnvoll ist, steht noch aus. Es ist möglich, daß das Nitratproblem in einigen Jahren mit größerer Gelassenheit gesehen werden kann; wahrscheinlicher aber ist, daß die gegenwärtigen Restriktionen verschärft werden müssen. In dieser Situation ist dem hygienischen Grundsatz der Risikominimierung unbedingt zu folgen. Spitzenbelastungen in Wasser und Lebensmitteln sollten konsequent beschnitten werden.

LITERATUR

1. Aurand, K., Hässelbarth, V. and Wolter, R. (1982). Nitrat- und Nitratgehalte von Trinkwässern in der Bundesrepublik Deutschland. In: *DFG Deutsche Forschungsgemeinschaft: Nitrat-Nitrit-Nitrosamine in Gewässern*, Verlag Chemie Weinheim, s. 99–105.
2. Borneff, M. (1980). Untersuchungen an Säuglingen in Gegenden mit nitrathaltigem Trinkwasser. *Zbl. Bakt. Hyg. I. Abt. Orig. B.*, **172**, 59–66.
3. Comly, H. H. (1945). Cyanosis in infants. *J. Amer. med. Ass.*, **129**, 112.
4. Crampton, R. F. (1980). Carcinogenic dose-related response to nitrosamines. *Oncology*, **37**, 251–4.

5. Eisenbrand, G. (1981). N-Nitrosoverbindungen in Nahrung und Umwelt, Wissenschaftliche Verlagsgesellschaft mbH, Stuttgart, s. 20.
6. Ellen, G. und Schuller, P. L. (1983). Nitrate, origin of continuous anxiety. In: *DFG Deutsche Forschungsgemeinschaft: Das Nitrosamin-Problem*, Verlag Chemie Weinheim, s. 97–134.
7. Green, L., Tannenbaum, S. R. and Goldman, P. (1981). Nitrate Synthesis in the Germfree and Conventional Rat. *Science*, **212**, 56–8.
8. Hellekes, R. (1983). Langzeitmessungen des Nitratgehaltes im Grundwasser im Einzugsgebiet Mönchengladbach. DLG, DVGW, s. 134–9.
9. Hoos, E. und Schweisfurth, R. (1982). Untersuchungen über die Verteilung von Bakterien von 10 bis 90 Meter Unterbodenoberkante. *Vom Wasser*, **58**, 103–12.
10. Magee, P. N. und Barnes, J. M. (1956). The production of malignant primary hepatic tumors in the rat by feeding dimethylnitrosamine. *Brit. J. Cancer*, **10**, 114–22.
11. Mirvish, S. S. (1979). Formation of N-nitroso compounds. Chemistry, kinetics and in vivo occurrence. *Toxicol. appl. Pharmacol.*, **31**, 325–51.
12. Moeschlin, S. (1972). *Klinik und Therapie der Vergiftungen, 5. Aufl.* Georg Thieme Verlag, Stuttgart.
13. Möhler, K. (1982). Nitrat- und Nitritgehalt der Nahrungsmittel. In: *DFG Deutsche Forschungsgemeinschaft: Nitrat—Nitrit—Nitrosamine in Gewässern.* Verlag Chemie Weinheim, s. 106–14.
14. Mueller, R. L. und Henninger, H. (1984). Der Magenschleim—eine endogene Nitratquelle des Menschen. Vortrag, 10. Arbeitstagung der Deutschen Gesellschaft für Hygiene und Mikrobiologie in Mainz, 4.–6. Oktober 1984.
15. Mueller, R. L. u.a. (1983). Die endogene Synthese kanzerogener N-Nitrosoverbindungen: Bakterienflora und Nitritbildung im gesunden menschlichen Magen. *Zbl. Bakt. Hyg. I. Abt. Orig. B.*, **178**, 297–315.
16. Obermann, P. (1981). Hydrochemische/hydromechanische Untersuchungen zum Stoffgehalt von Grundwasser bei landwirtschaftlicher Nutzung. *Besondere Mitteilungen zum Deutschen Gewässerkundlichen Jahrbuch*, **42**, 1981.
17. Oshima, H. und Bartsch, H. (1981). Quantitative estimation of endogenous nitrosation in humans by monitoring N-Nitrosoprolin excreted in the urine. *Cancer Research*, **41**, 3658–62.
18. Sander, J. (1967). Kann Nitrit in der menschlichen Nahrung Ursache einer Krebsentstehung sein? *Arch. Hyg.*, **151**, 22–8.
19. Schmeing, F. (1983). Langzeitmessungen des Nitratgehaltes im Grundwasser im Einzugsgebiet Bayern. In: *DLG Deutsche Landwirtschafts-Gesellschaft: Nitrat—ein Problem für unsere Trinkwasserversorgung*, Arbeiten der DLG, Band 177, DLG-Verlag Frankfurt/Main, 140–6.
20. Selenka, F. (a) (1983). Nitrat und Nitrit in Wasser und Boden. In: *DFG Deutsche Forschungsgemeinschaft: Das Nitrosamin-Problem*, Verlag Chemie Weinheim, s. 135–44.
21. Selenka, F. (b) Gesundheitliche Bedeutung des Nitrats in der Nahrung. In: *DLG Deutsche Landwirtschafts-Gesellschaft: Nitrat—ein Problem für unsere Trinkwasserversorgung*, Arbeiten der DLG, Band 177, DLG-Verlag Frankfurt/Main, s. 7–24.

22. Selenka, F. (1982). Gesundheitliche Aspekte von Nitrat, Nitrit, Nitrosaminen. In: *DVGW Deutscher Verein des Gas- und Wasserfachs*, DVGW-Schriftenreihe, Wasser Nr. 31, Eschborn, GFR, 1982, s. 131–43.
23. Selenka, F. und Brand-Grimm (1976) Nitrat und Nitrit in der Ernährung des Menschen. Kalkulation der mittleren Tagesaufnahme und Abschätzung der Schwankungsbreite. *Zbl. Bakt. Hyg. I. Abt. Orig. B.*, **162**, 449–66.
24. Spiegelhalder, B. (1983). Vorkommen von Nitrosaminen in der Umwelt. In: *DFG Deutsche Forschungsgemeinschaft: Das Nitrosamin-Problem*, Verlag Chemie Weinheim, s. 35.
25. Spiegelhalder, B., Eisenbrand, G. und Preussmann, R. (1976). Influence of dietary nitrate on nitrite content of human saliva: Possible relevance to in vivo formation of N-nitroso compounds. *Fd. Cosmet. Toxicol.*, **14**, 545–8.
26. Suchomel, E. (1979). Nitratgehalte in Blut, Speichel und Harn des Menschen nach Genuß von Lebensmitteln unterschiedlicher Verdaulichkeit. Dissertation Universität Mainz, 1979.
27. Wedemeyer, F. W. (1956). Methämoglobinämie des jungen Säuglings durch nitrathaltiges Brunnenwasser. *Arch. f. Kinderheilkunde*, **152**, 267–75.

6

Effets Écologiques des Pesticides (et en Particulier au Niveau des Organismes Édaphiques)

PH. LEBRUN

Ecologie et Biogéographie, Université Catholique de Louvain, Belgique

SUMMARY

The ecological effects of pesticides are reviewed in the context of agricultural pest control. As a foreign substance which can affect any ecosystem (involving both target and non-target populations) the pesticide can cause both structural and functional changes. These changes may be reversible or irreversible. Examples are discussed with particular reference to soil biology. Of the major consequences of excessive pesticide use those of pest resurgence, pest substitution, loss of gene pool diversity, loss of ecological function, and food chain contamination are illustrated. The problem of resistance is emphasised and the rate of its extension with the recent use of organic insecticides is noted. With regard to structural and functional changes it appears that the complementary roles of soil micro-flora and -fauna in decomposition processes can be seriously affected by pesticide use.

1. INTRODUCTION

Développer le thème des effets écologiques des pesticides se heurte dès l'abord à de nombreuses difficultés tant le sujet est vaste et tant sont nombreux les points de vue auxquels on peut se situer pour évoquer les multiples aspects du devenir des pesticides dans l'environnement.

Il est cependant possible de classer les problèmes comme, par exemple, se limiter aux effets sur la vie sauvage et donc s'insérer dans le contexte de la conservation de la flore et de la faune. De même on peut s'attacher au devenir immédiat ou retardé des pesticides dans le milieu de vie et donc restreindre le problème sous l'angle de la toxicité aiguë ou chronique. Ceci revient à dissocier les effets directs, immédiatement perceptibles, des effets indirects, plus latents, n'exerçant leurs conséquences qu'à long terme.

Dans le même ordre d'idée, et du point de vue de l'écologiste, les effets des pesticides peuvent être analysés en fonction des différents systèmes (ou niveaux) écologiques: les effets sur une population (une seule espèce envisagée), les effets sur un réseau trophique (une relation entre deux espèces concurrentes et une plante-hôte, un prédateur et une proie ou entre un hôte et un parasite) ou enfin les effets sur une communauté d'organismes vivants, envisagée quant à sa structure ou à son fonctionnement.

Aucune de ces approches n'est pleinement satisfaisante et s'il est facile, sur le plan didactique, de séparer les effets directs des effets indirects, les effets à court terme des effets à long terme, les effets sur une seule espèce des effets sur un ensemble d'espèces, ces limites sont cependant très artificielles. Tôt ou tard on est amené à les aborder de manière simultanée car les unes et les autres sont très largement interdépendantes.

A cet égard la synthèse réalisée par Moore [15] pose très bien le problème dans les deux relations reprises à la Figure 1. Il y a lieu d'insister avec Moore [15] sur le fait que le concept initial de l'action d'un pesticide repris dans la première formule est entièrement erroné. L'action d'un pesticide n'est jamais exclusive; il agit toujours à tous les niveaux des composantes d'une biocénose et, les orateurs précédents l'ont bien montré, au niveau de tous les compartiments de l'écosystème.

Dès lors, dans cet exposé, nous nous situerons à un autre point de vue et comme le thème général de cette réunion est 'Chemicals in agriculture' l'accent sera mis sur un contexte agricole. En conséquence, on se placera au niveau même de la raison d'être d'un pesticide c'est-à-dire l'action sur les espèces-cibles d'une part, l'action sur les espèces non-cibles d'autre part. Il n'est peut-être pas inutile, en effet, de rappeler que l'usage des pesticides répond à un objectif précis: celui de limiter une espèce préjudiciable et non une utilisation purement gratuite.

(1) Pesticide $\rightarrow$ Ravageur $\pm$ Quelques 'effets secondaires'

(2) | Pesticide $\rightarrow$ Ensemble de l'écosystème |

(où $\rightarrow$ signifie: 'agit sur')

FIG. 1. Schéma de base des effets écologiques d'un pesticide. (1): effets escomptés par l'utilisateur. (2): effets réels (d'après Moore [15]).

D'autre part, toujours dans le contexte agricole de ce Symposium la majorité des exemples relèveront du compartiment sol des écosystèmes. Le sol est en effet le support de la productivité agricole et constitue le premier lieu où sont drainés les résidus des pesticides. Toutes les grandes écoles de pédologie s'accordent pour considérer le sol comme une sorte de superorganisme représentant à la fois la matrice où vivent une multitude d'organismes mais également le résultat de l'activité de ces organismes. Son fonctionnement, rien que sous l'angle du cycle des éléments minéraux, est extrêmement complexe et justifie pleinement l'étude approfondie des interférences des pesticides avec le rôle exercé par la flore et la faune édaphiques.

Enfin, il va sans dire que le problème qui nous préoccupe aujourd'hui a déjà fait l'objet de très nombreuses synthèses et réunions scientifiques [2, 7, 13, 14, 16, 18, 19, 20]... etc. En d'autres termes on ne mentionnera que pour mémoire les aspects que ces auteurs se sont déjà attachés à développer. Tous ces ouvrages sont très bien documentés quant aux aspects descriptifs du devenir des pesticides dans l'environnement et sur leurs effets à court terme. Dès lors, personnellement, je voudrais plutôt mettre l'accent sur les conséquences.

2. DEFINITION ECOLOGIQUE D'UN PESTICIDE

Ainsi que le schéma de la Figure 1 l'indiquait déjà un pesticide ne peut être défini, sous l'angle écologique, comme ayant une action ponctuelle et limitée.

Ainsi, on pourrait définir un pesticide sur le plan écologique comme étant un *facteur étranger* introduit dans un système écologique parvenu à un certain état susceptible d'induire des modifications structurelles suivies *ou non* de modifications fonctionnelles. D'autre part, la modification de structure (accompagnée ou non d'une modification de fonctionnement) s'effectue de manière *réversible* ou *irréversible*.

Prenons un premier exemple touchant la faune vivant dans les sols. Le graphique de la Figure 2 montre l'évolution d'une population d'un Collembole (*Folsomia candida*) espèce très abondante dans diverses terres de culture. Dans le contexte expérimental évoqué ici, un protocole réalisé en parcelles montre qu'à la dose de 2 kg de matière active à l'ha la population montre par rapport au témoin un début de restauration 24 semaines déjà après l'épandage, et si comme on pouvait s'y attendre l'effet de réhabilitation est plus lent à la dose de 10 kg de matière active à l'ha, il y a

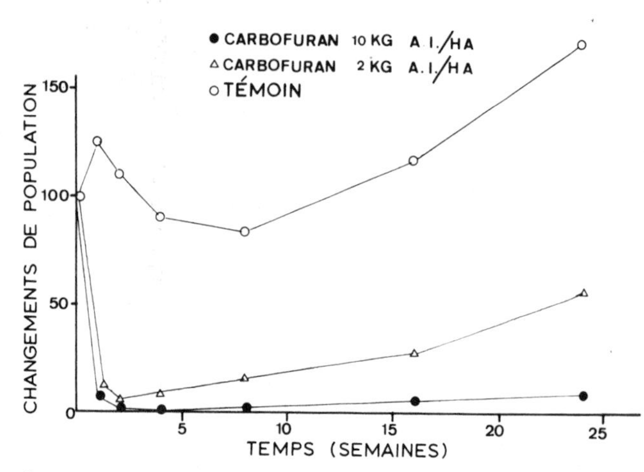

FIG. 2. Évolution relative de la densité d'une population d'un Collembole (*Folsomia candida*) soumise à un traitement insecticide: moyennes de 2 parcelles (Thirumurthi et Lebrun [24]).

indéniablement des signes de reconstitution [24]. Ceci dépend de la persistance de la toxicité de la molécule. Il y a donc réversibilité du phénomène et la modification induite dans cette population n'est donc que transitoire.

Par contre, dans l'exemple repris à la Figure 3, il est évident que la population de lombrics soumise à deux traitements annuels d'un insecticide (dose normale d'utilisation) accuse une décroissance progressive sans manifester la moindre tendance à la restauration d'un traitement à l'autre. Dans le cas présent la population de lombrics est exterminée de manière irréversible.

La différence entre les deux exemples tient tout simplement à l'écologie propre des taxons considérés. Les Collemboles ont des aptitudes considérables en matière de dispersion. Espèces relativement opportunistes elles recolonisent facilement tous les milieux au départ de petits foyers de survie. Par contre, il est bien connu que les lombrics sont des animaux à possibilités de dispersion médiocres et divers auteurs, dont notamment Bouche [1] ont montré que leurs déplacements de populations ne représentaient qu'un mètre par an. Cette faiblesse explique d'ailleurs pourquoi, en Moyenne Belgique, dans différents champs soumis à une culture betteravière intensive depuis plus de 20 ans, les populations de lombrics ont pratiquement disparues [5].

En ce qui touche les modifications de structure, suivies ou non d'effets sur

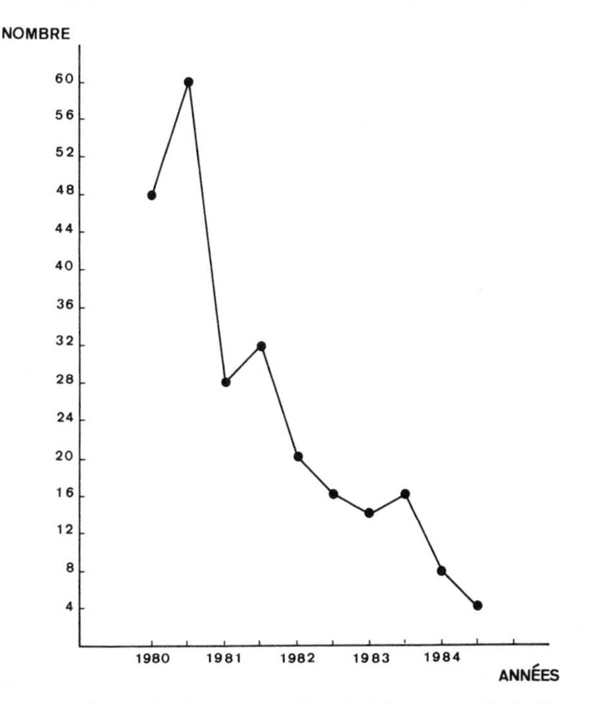

FIG. 3. Évolution d'un peuplement de Lombriciens soumis à 2 traitements insecticides annuels (aldicarbe 0·2 g m.a./m^2): moyennes de 5 quadrats de 0·25 m^2 (Lebrun, non publié).

le fonctionnement du système, on peut faire référence au modèle très classique et bien connu de la substitution d'organismes tel qu'il a été développé par Smith et Van den Bosch [23], dans le contexte particulier de substitution de pestes (Fig. 4). Ce modèle montre que la disparition d'une espèce dans un système écologique peut être suivie par le remplacement de cette espèce. Dans le cas présent, il s'agit d'une modification de structure qui n'influencera que très peu, si pas du tout, le fonctionnement du système en lui-même (un phytophage remplace un autre phytophage, un prédateur remplace un autre prédateur...). La fonction écologique est toujours assurée. C'est dans le domaine agricole que ces phénomènes sont les plus familiers et dans de nombreuses séquences culturales, on connaît ainsi des complexes d'insectes ravageurs qui se sont modifiés conjointement avec le développement de nouvelles molécules insecticides.

Par contre, la modification de structure peut affecter le fonctionnement lorsqu'il n'y a pas de compensation écologique. Si on revient à l'exemple des

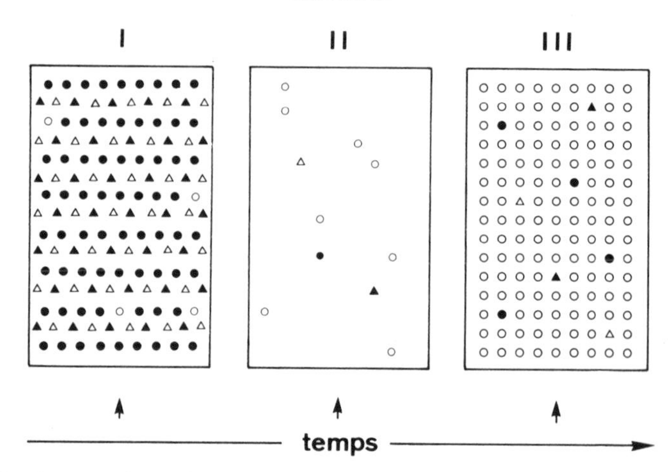

FIG. 4. Rupture d'équilibre biologique provoquée par des traitements insecticides (↑): le cas d'une substitution de peste (d'après Smith et Van den Bosch [23]). ●, Ravageur primaire et ses ennemis naturels (▲ et △); ○, ravageur secondaire.

lombrics évoqué il y a quelques instants, il faut noter que le rôle écologique exercé par ces animaux est tellement précis et tellement spécialisé qu'aucune compensation n'est possible et la disparition des vers revient à la disparition d'une fonction. La diversité des animaux est d'ailleurs tellement faible dans les terres de grande culture intensive qu'un remplacement des lombrics par un autre taxon est mathématiquement impossible. On y reviendra dans quelques instants mais dès à présent on peut signaler que la fonction des lombriciens réside principalement dans la régulation du cycle de l'azote.

3. VUE GÉNÉRALE DES EFFETS ÉCOLOGIQUES DES PESTICIDES AU NIVEAU D'UNE OU DE QUELQUES POPULATIONS INTERDÉPENDANTES

Dans le contexte que nous avons fixé en commençant l'exposé, il y a lieu, à présent, d'évoquer les principaux effets écologiques des pesticides en différenciant les effets observables sur les espèces-cibles (il s'agit bien sûr des pestes) et les espèces non-cibles. Pour la facilité, on séparera arbitrairement le niveau d'une seule (ou une association d'un nombre limité d'espèces) du niveau de plusieurs espèces (le niveau de la communauté).

En un premier temps, un pesticide peut accroître la mortalité de la population-cible aussi bien que celle de la population non-cible (Tab. 1).

Tableau 1
Effets écologiques des pesticides au niveau des populations (une seule espèce ou nombre limité d'espèces)

Espèces-cibles	*Espèces non-cibles*
Augmentation de la mortalité: —régression ou extinction de la population —possibilité de substitution de peste	Augmentation de la mortalité: —régression ou extinction de la population —*perte de diversité génétique* —*perte de fonction écologique*
Augmentation de la mortalité spécifique (sexe, âge, ...): —régression ou extinction de la population —*possibilité de substitution de peste*	Augmentation de la mortalité spécifique (sexe, âge, ...): —régression ou extinction de la population —*perte de diversité génétique* —*perte de fonction écologique*
Changements de fécondité: —régression ou extinction ou augmentation de la population —*possibilité de substitution de peste* —*possibilité de résurgence de peste*	Changements de fécondité: —régression ou extinction ou augmentation de la population —*perte de diversité génétique* —*perte de fonction écologique* —*banalisation des flores et des faunes*
Changements génotypiques: —*résistance* —*possibilité de résurgence de peste*	Changement génotypiques: —*résistance*
Bioconcentration: —contamination des chaînes trophiques	Bioconcentration —contamination des chaînes trophiques

Que cette mortalité soit spécifique ou non le résultat est le même: la régression, voire l'extinction, de la population. Les conséquences cependant sont différentes puisque pour les espèces-cibles la conséquence principale réside dans une possible substitution de pestes. Nous avons déjà évoqué ce phénomène, il n'est plus nécessaire d'y revenir. Pour les espèces non-cibles, un premier effet est la perte de diversité génétique, problème qui retient tant, et à juste titre, l'attention des protecteurs de la nature.

On ne peut, à cet égard, ignorer les lourdes menaces qui pèsent sur l'avifaune et nous en voulons comme preuve une récente enquête réalisée dans le sud de la Belgique qui fait apparaître que sur les 20 espèces protégées mentionées dans la directive No. 409/79 de la Commission des Communautés Européennes (Annexe I, et propositions à l'Annexe I) il y en a 9, soit pratiquement la moitié, qui sont indirectement menacées par les pesticides utilisés en agriculture [12].

En ce qui concerne la perte d'une fonction écologique consécutive à la régression ou à l'extinction d'une espèce on empruntera un exemple à l'écologie des lombrics encore. L'objectif expérimental de la recherche résumée ici consistait en une étude de la dynamique de l'azote (essentiellement la nitratation) dans des sols dépourvus de vers de terre par comparaison à des sols qui en sont pourvus. Cette situation correspond donc à l'extinction des populations de lombriciens à laquelle on a déjà fait allusion. La dynamique de l'azote est suivie après un épandage de matière organique sous forme de lisier, pratique très largement répandue dans les régions agricoles de l'Europe occidentale. Comme le montre le graphique de la Figure 5 la dynamique de l'azote se déroule de manière très contrastée

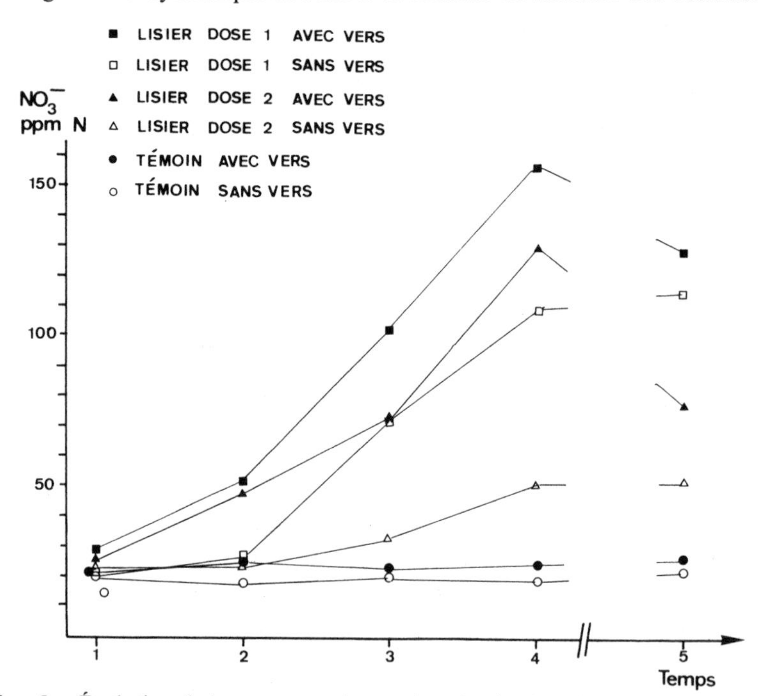

FIG. 5. Évolution de la teneur en nitrates (ppm) suite à un épandage de lisier de porcs (dose 1 = 200 m^3/ha; dose 2 = 100 m^3/ha) en présence ou en absence de Lombriciens: expérience réalisée dans des colonnes en laboratoire (d'Après Derby *et al.* [4]).

selon que les lombrics soient présents ou non. Dans les objets expérimentaux pourvus de vers de terre, en effet, la vitesse de nitratation est nettement plus rapide que dans les objets expérimentaux qui en sont dépourvus [4].

En d'autres termes, ceci est la démonstration *a contrario* de la plus importante des fonctions écologiques des lombrics: la minéralisation de l'azote et dans le contexte agricole préserver les lombrics revient à préserver le potentiel de fertilité d'un sol.

Il est clair que le 'génocide' des lombriciens consécutif à l'utilisation intensive de pesticides risque de se traduire, à terme, par un important déséquilibre dans le cycle de l'azote. De plus, comme la régression des lombrics est un phénomène récent (son point de départ coïncide avec l'avènement des insecticides carbamates), on n'a pas encore pleinement appréhendé ce type de conséquence.

Lorsqu'on aborde des effets écologiques moins évidents on doit évoquer les changements de fécondité observables aussi bien chez les populations-cibles que chez les populations non-cibles. Si la fécondité va dans un sens décroissant on rejoint le même type de conséquences comme la possibilité d'une substitution de peste d'un côté, perte de diversité génétique associée à des pertes de fonctions écologiques d'un autre côté. La diminution de fécondité constitue un des aspects parmi les plus pernicieux des conséquences écologiques des pesticides car elle survient sans que la survie des géniteurs ne soit affectée. On en connaît de nombreux exemples chez les Oiseaux [16], [20], mais les invertébrés n'échappent pas à la règle. Les populations de vers étudiées par van Rhee [25] représentent un bon exemple où les traitements pesticides n'affectent pas la survie mais l'avenir de la population est cependant compromis à cause de la baisse de fécondité (voir Fig. 6).

D'un autre côté, et la littérature en rapporte de nombreux cas [14], [19] etc...., le pesticide peut accroître la fécondité. Ceci se traduit par un phénomène de résurgence de peste (Fig. 7) pour les espèces-cibles, éventuellement par une prolifération des espèces non-cibles qui, inoffensives auparavant, peuvent devenir un réel problème pour l'agriculture. Si par contre, il s'agit d'espèces exerçant un rôle écologique non directement lié à la production agricole certaines fonctions écologiques pourraient être déséquilibrées par rapport à d'autres et changer notablement l'architecture d'origine subtilement mise en place par la nature. D'une manière générale, cette situation aboutit toujours à la banalisation des flores et des faunes.

C'est au niveau des changements génotypiques que s'inscrivent les

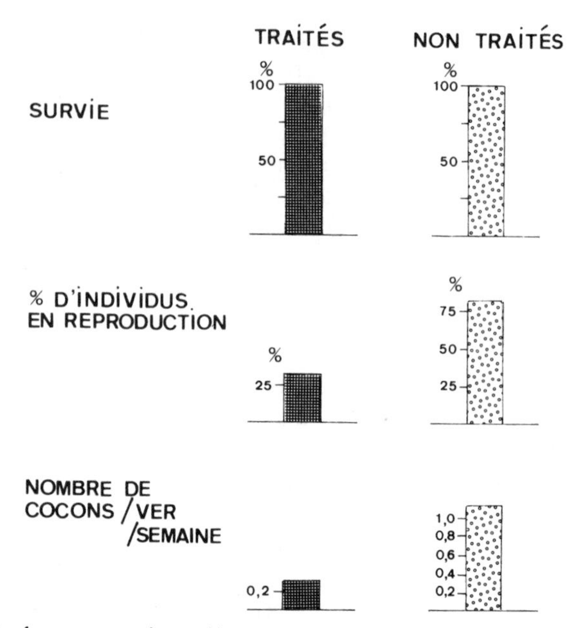

FIG. 6. Quelques paramètres démécologiques d'un Lombricien (*Allolobophora caliginosa*) soumis à un ensemble de traitements pesticides: expériences de laboratoire, test sur 11 semaines, traitements à raison d'$\frac{1}{2}$ dose normalement appliquée en vergers (captan, acricid, DDT, parathion, ométhoate, amitrol, grammoxone, simazine) (dessiné d'après les données chiffrées de Van Rhee [25]).

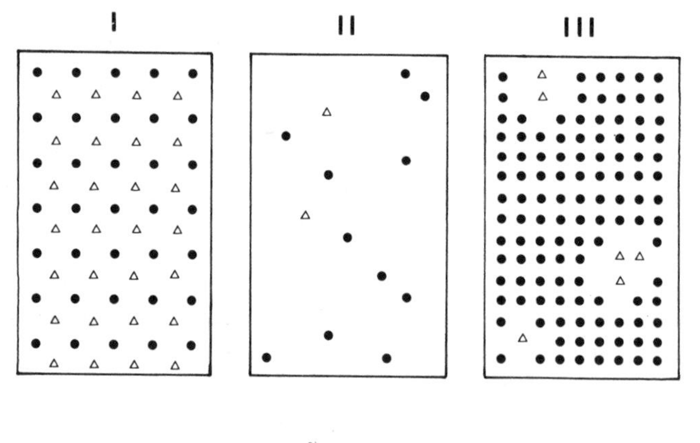

FIG. 7. Mécanisme schématique du phénomène de 'pest-resurgence' (d'après Smith et Van den Bosch [23]).

problèmes de résistance et de résurgence de peste. La Figure 7 représente le schéma classique du mécanisme de résurgence qui peut découler de l'acquisition d'une résistance ou de la disparition des ennemis naturels ou de la combinaison de ces deux phénomènes. Comme la résistance est un des problèmes les plus préoccupants parmi les effets écologiques des pesticides on en fera un paragraphe distinct (voir § 4).

Enfin, et pour mémoire, un effet écologique important réside dans le phénomène de bio-concentration. A titre d'exemple, et à nouveau repris dans le domaine de la biologie du sol, la Figure 8 montre une étroite corrélation entre les résidus des insecticides organochlorés du sol et des vers de terre et des limaces. Ces deux taxons ne sont pas pris au hasard car ils constituent, comme on le sait, un point de départ extrêmement important qui fait la liaison entre les chaînes trophiques des saprophages et les chaînes trophiques supérieures essentiellement constituées de Vertébrés (depuis les Amphibiens et les Reptiles jusqu'aux Mammifères, en passant par les Oiseaux). Si les organochlorés, d'une manière générale, sont relativement peu toxiques pour les lombrics il apparaît néanmoins que les concentrations observées chez ceux-ci constituent une menace réelle pour toutes les espèces qui s'en nourrissent. Un exemple tristement classique est

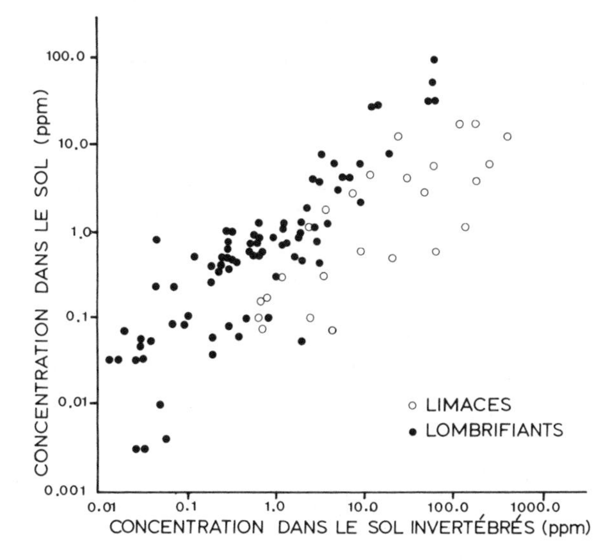

FIG. 8. Relation entre les résidus d'insecticides organochlorés du sol et des limaces et Lombriciens (d'après Edwards et Thompson [8]).

celui publié par Dimond *et al.* [6] qui mettent en évidence d'importantes mortalités (70 %) chez le merle migrateur suite aux traitements insecticides du printemps.

4. LE PROBLÈME DE LA RÉSISTANCE

Le problème de la résistance illustre bien un effet sur un système écologique affectant à la fois sa structure et son fonctionnement. Rappelons tout d'abord, que la résistance résulte de la pression de sélection exercée par une molécule chimique sur une population. Ceci veut donc dire que dans la diversité génétique de cette population un certain nombre d'individus sont en quelque sorte préadaptés à mieux supporter la toxicité d'une molécule. Comme ce sont ces individus dont la probabilité de survie est la plus élevée, la fréquence du gène (ou de la combinaison de gènes) associée à cette moindre sensibilité augmentera dans la population, génération après génération.

Il n'est pas dans l'intention de refaire tout un exposé sur la résistance, d'autres l'ont déjà fait antérieurement [3], [29], etc....

Nous voudrions simplement attirer l'attention sur quelques points. Le premier concerne le nombre d'espèces reconnues actuellement pour avoir développé un certain type de résistance vis-à-vis de certaines molécules insecticides. Que cette résistance soit du type simple, croisée ou généralisée importe peu. En valeur absolue le nombre d'espèces considérées est relativement faible: 300 à 350 connues tout au plus. Ce nombre est faible comparé au nombre important d'espèces d'Arthropodes existant dans le monde (plus d'un million deux cent mille). Mais, sans vouloir être alarmiste à l'excès, il est bon de préciser que ces 350 espèces *représentent déjà près de 10% des espèces nuisibles* sur lesquelles des traitement insecticides sont dirigés spécifiquement depuis de nombreuses années. Un autre point apparaît très clairement à la figure suivante (Fig. 9): c'est *la vitesse avec laquelle le phénomène de résistance s'est instauré depuis les années 1950* c'est-à-dire depuis l'avènement des insecticides organiques de synthèse. La courbe présente en effet une allure exponentielle. Ceci reflète l'extrême diversité et la multiplicité des aptitudes qu'ont les Arthropodes terrestres à mieux supporter les traitements dont ils sont l'objet, grâce à toute une série de mécanismes aussi bien enzymatiques (concentration anormalement croissante d'enzymes détoxifiant les molécules insecticides), que morphologiques (cuticule de plus en plus épaisse et à revêtement cireux de plus en plus important ralentissant la pénétration de la molécule)... etc.

Je ne voudrais pas anticiper sur ce que nos collègues exposeront demain lorsque les solutions seront envisagées. Je pense cependant devoir ajouter

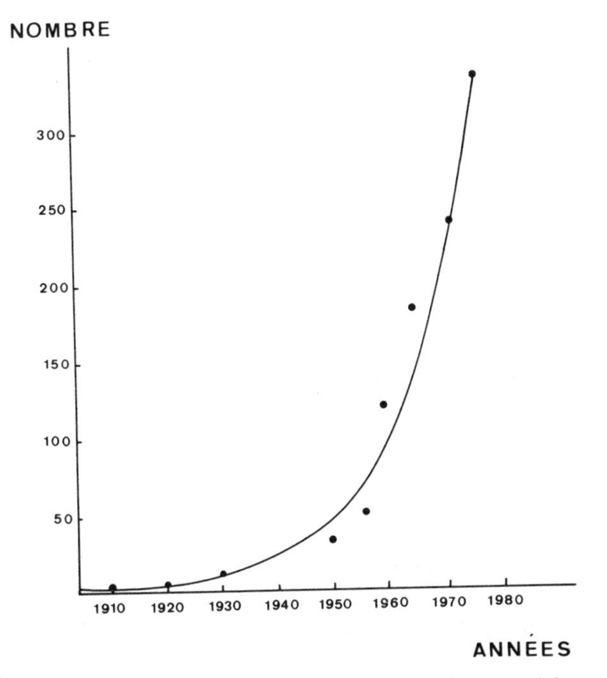

FIG. 9. Évolution du nombre d'espèces d'Arthropodes devenus résistants à un ou plusieurs groupes de pesticides (d'après les données de divers auteurs et en particulier Georghiou et Taylor [9]).

une troisième réflexion qui s'inscrit dans un contexte plus constructif. La résistance doit inciter fermement la mise en oeuvre simultanée de plusieurs méthodes de lutte pour assurer une meilleure couverture des traitements. Ainsi, par exemple, un traitement chimique peut être complété par une méthode de désorientation sexuelle par utilisation de phéromones en vue d'éliminer les individus résiduels qui sont précisément ceux porteurs des gènes de résistance. Plus que jamais une intégration de différentes méthodes de lutte doit être envisagée pour enrayer ce phénomène très préoccupant qu'est la résistance.

5. EFFETS ÉCOLOGIQUES DES PESTICIDES SUR LA STRUCTURE ET LE FONCTIONNEMENT DES BIOCÉNOSES DU SOL

Au niveau biocénotique ou si l'on préfère au niveau de la communauté il n'y a plus lieu de différencier entre espèces-cibles et espèces non-cibles

Tableau 2
Effets écologiques des pesticides au niveau de la communauté

Modifications dans la structure et la diversité des communautés
Changements dans les fonctions écologiques, par ex. dans le rôle joué par les organismes du sol

puisque les premières sont nécessairement insérées dans un ensemble interdépendant.

Le Tableau 2 présente les principaux effets écologiques des pesticides au niveau des communautés. On y a repris, quelques effets biocénotiques déjà mentionnés comme la substitution de peste, la résurgence de peste et la bioconcentration. On se situera ici à un niveau beaucoup plus global qui implique l'ensemble des espèces vivant dans un biotope.

On développera surtout les modifications de structure et de diversité des communautés et les changements dans les fonctions écologiques, en particulier le rôle exercé par les organismes édaphiques.

5.1 Modifications dans la structure et la diversité des communautés

Pour bien situer le débat au niveau des pratiques agricoles, nous avons retenu un exemple parmi les recherches entreprises dans le cadre du Comité de recherche sur l'utilisation des pesticides en agriculture (CRUPA) subsidié par l'Institut Belge pour l'Encouragement de la Recherche Scientifique dans l'Industrie et dans l'Agriculture (IRSIA).

Une vaste expérimentation a été menée en culture betteravière sur l'ensemble de l'époque allant de la mise en place de la culture jusqu'à un mois après l'émergence des plantules. Douze objets expérimentaux ont été retenus et répétés selon un dispositif classique par blocs aléatoires. Sans vouloir entrer dans les détails, précisons que chaque objet expérimental, mis à part le témoin (No. XII) comporte des traitements herbicides, des traitements insecticides, ou la combinaison des deux (voir Tab. 3). Les traitements peuvent être rassemblés en quatre groupes:

les objets I à IV comportent des traitements herbicides jusqu'en postémergence ainsi qu'un traitement insecticide commun à l'heptachlore;

les objets V et XI comportent un (ou 2) traitement(s) herbicide(s) précédent l'émergence;

les objets VI et VII comportent des traitements herbicides jusqu'en postémergence:

les objets VIII, IX et X comportent à la fois des traitements herbicides et un traitement insecticide par utilisation d'un carbamate.

Tableau 3
Caractéristiques du traitement (Grégoire-Wibo *et al.*, sous presse)

	Avant ensemencement	*Avant émergence*	*Après émergence*
I	Avadex 3·5 litres Heptachlore 12 litres	Pyramin 5 kg	Betanal 8 litres
II	Ro-neet 3·5 litres Venzar 0·5 kg Heptachlore 12 litres		Betanal 8 litres Venzar 1 kg
III	Avadex 3·5 litres	Pyramin 2 kg Heptachlore 5 litres	Betanal 8 litres
IV	Avadex 3·5 litres Heptachlore 12 litres		Betanal 8 litres
V	Avadex 3·5 litres	Pyramin 5 kg	
VI		Nortran 10 litres	Betanal 8 litres
VII		Nortran 20 litres	Betanal 8 litres
VIII	Avadex 3·5 litres Temik 10 kg		Betanal 8 litres
IX	Avadex 3·5 litres Temik 20 kg		Betanal 8 litres
X	Avadex 3·5 litres Carbofuran 15 kg		Betanal 8 litres
XI		Nortran 5 litres Pyramin 4 kg	
XII			

Tout au long de la période expérimentale des pièges d'activité ont été disposés dans chacune des parcelles de chaque objet expérimental. Ces pièges ont été relevés régulièrement et les animaux de la faune édaphique épigée ont été identifiés et dénombrés.

Les conclusions majeures de cette expérience apparaissent clairement (Tab. 4). Une réduction de la densité globale du peuplement tout d'abord, réduction d'autant plus accentuée que l'objet expérimental comporte un traitement insecticide soit par organochloré (objets I à IV) soit par carbamate (objets VIII à X). La réduction pour les objets comportant un traitement insecticide est au minimum de 70 % par rapport au témoin. La réduction observée avec les herbicides est nettement plus faible lorsque les herbicides ne sont pas appliqués après l'émergence (objet VI et XI).

Mais, beaucoup plus important est le profond bouleversement que l'on peut observer dans la structure de la biocénose des insectes épigés en culture betteravière. Comme le montre la Figure 10, il apparaît que le spectre d'abondance des principales familles d'insectes considérés change radicalement lorsqu'on les compare aux témoins. Ce sont surtout les fungivores

Tableau 4
Effets des pesticides sur les organismes utiles du sol (uniquement saprophages et prédateurs) (Grégoire-Wibo *et al.*, sous presse)

	N	%
I	483	−77
II	415	−81
III	676	−68
IV	448	−79
V	1 550	−28
VI	1 125	−47
VII	1 041	−51
VIII	320	−85
IX	597	−72
X	656	−69
XI	1 633	−24
XII	2 140	0

N = Nombre total d'individus (saprophages et prédateurs).
% = réduction par comparaison avec la parcelle non traités (XII):
I, II, III, IV, VIII, IX, X > −68 %
V, XI < −30 %
VI, VII ± −50 %.

comme les espèces de la famille des Sminthuridés qui abondent dans les parcelles témoins et dans les parcelles peu perturbées (à nouveau les objets V et XI caractérisés par un traitement herbicide uniquement en pré-émergence). Dans la plupart des autres objets expérimentaux la structure fait apparaître une dominance des phytophages comme les Aphides, les Erotylides (parmi lesquels l'Atomaire de la betterave—*Atomaria linearis*—espèce très préjudiciable aux jeunes semis betteraviers) et certains Isotomides. Ceci reflète la sensibilité différentielle des différentes espèces et familles selon leur position trophique. De telles conclusions sont désormais classiques et, à l'occasion d'une autre expérience le même phénomène a été mis en évidence: les animaux édaphiques saprophages et prédateurs sont nettement plus sensibles aux pesticides que les phytophages [10].

Un changement aussi radical dans la structure de la biocénose ne peut avoir que des conséquences importantes sur son fonctionnement. Ce point doit, dès lors, être développé; on se réfèrera à la faune édaphique endogée.

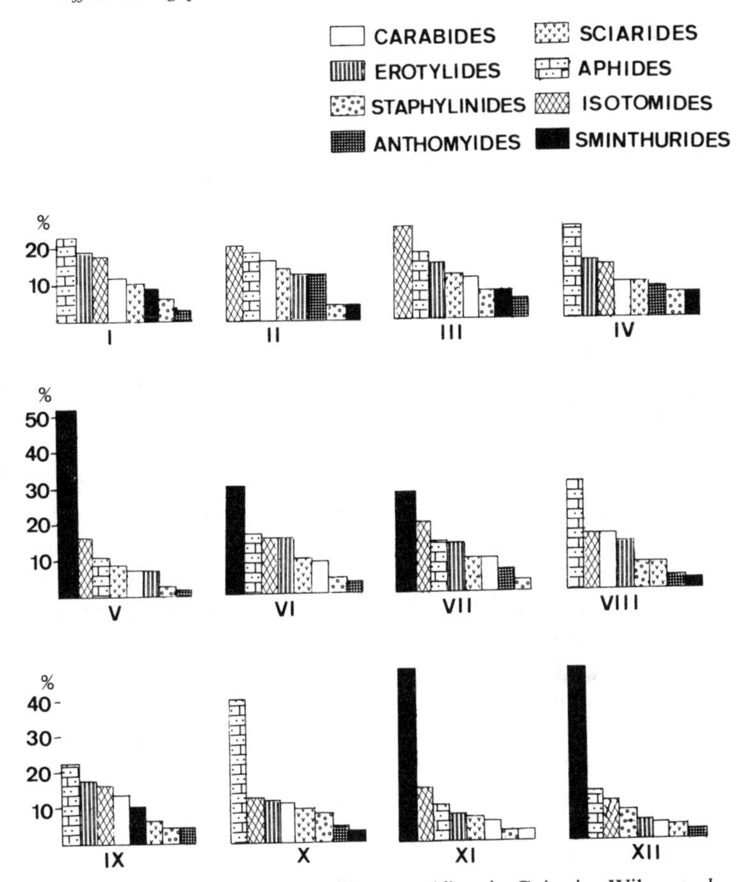

FIG. 10. Structure des biocénoses d'insectes (d'après Grégoire-Wibo *et al.*, sous presse).

5.2 Changements dans les fonctions écologiques et en particulier dans le rôle exercé par les animaux du sol

Il importe, tout d'abord, de bien définir le rôle exercé par les organismes du sol. Globalement, ce rôle peut être qualifié de 'fonction d'antiphotosynthèse' c'est-à-dire la dégradation de la matière organique et la libération des éléments biogènes (inorganiques). Il est connu depuis longtemps que certaines formes de vie dans le sol contribuent de manière importante à la fragmentation de la matière organique, son enfouissement, et à sa transformation biochimique par l'intermédiaire du transit à travers le tube digestif. C'est notamment le cas des vers de terre et d'animaux détritiphages de grande taille comme les larves d'insectes, les Diplopodes, certains Isopodes, et certains Acariens. Cependant, cette faune constitue une minorité quant à sa diversité en espèces. Il existe à côté, en effet, toute une multitude d'autres animaux représentés par des densités considérables (jusqu'à 300 000 individus/m^2 dans les sols forestiers) dont le rôle dans la fragmentation de la matière organique et dans les modifications biochimiques est relativement négligeable.

Quel est, par conséquent, la fonction de cette multitude d'organismes? Les données les plus récentes [11, 17, 22, 28] ont mis l'accent sur le rôle capital de ces organismes dans le cycle des éléments minéraux. La microflore, en effet, est un agent important de stockage et d'immobilisation des éléments minéraux. Cette immobilisation constitue donc une concurrence directe pour l'alimentation minérale au niveau des racines et donc la productivité des plantes supérieures. Or, les animaux interviennent tout d'abord par leur intense activité de broutage qui a pour résultat la mobilisation des éléments minéraux stockés dans la biomasse fongique. En plus, par leur activité alimentaire, ils dispersent les spores des microorganismes qui peuvent ainsi se développer et participer aux premières étapes de la dégradation de la matière organique dans toute l'étendue et la profondeur du substrat.

Il s'agit donc d'un véritable phénomène de synergie compétitive dont le résultat est clair. Comme il apparait à la Figure 11 la décomposition de la matière organique est fortement ralentie en l'absence de microorganismes et d'animaux; en présence de microorganismes isolés elle est loin d'atteindre la même vitesse que lorsque les animaux sont présents. C'est précisément ce que l'on observe lors de traitements insecticides qui éliminent les microarthropodes [26].

En outre, il existe de subtils rapports de force entre microorganismes et animaux du sol. Si les animaux mycophages et bactériophages sont trop abondants, la croissance microbienne est fortement ralentie par surbroutage et les premiers stades de la décomposition sont freinés. Lorsque les animaux sont trop peu nombreux la biomasse fongique croît à l'excès avec immobilisation des éléments minéraux. Il y a donc un équilibre précis entre les deux groupes d'organismes, équilibre où interviennent très largement toutes les espèces prédatrices. Ces dernières constituent par conséquent un important facteur de régulation qui optimalise les attaques de la matière organique par les microbes et la libération des éléments biogènes grâce au broutage des animaux. Ceci a notamment été démontré lors d'expériences réalisées en microcosmes dans des litières de créosote-bush (cf. Fig. 12). Les Acariens régulateurs des Nématodes bactériophages

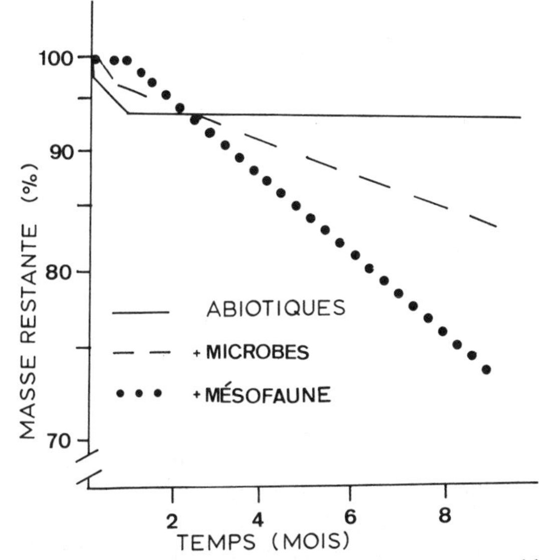

FIG. 11. Taux de décomposition de la feuille de bouteloue bleu (*Bouteloua gracilis*), faune et flore exclues (—, abiotiques), faune exclue (– –, + microbes) et microarthropodes inclus (. . . , + mésofaune) (d'après Vossbrinck *et al.* [26] et Seastedt [22]).

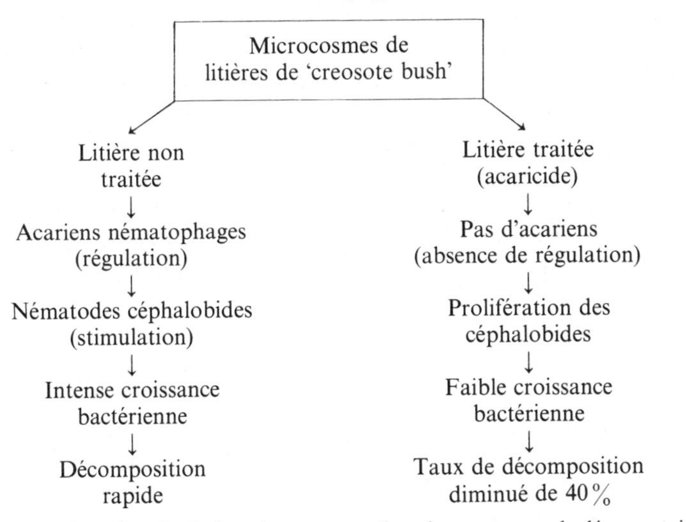

FIG. 12. Schéma des régulations intervenant dans les processus de décomposition de la matière organique: à gauche, conditions naturelles; à droite, avec l'interférence d'un pesticide (d'après Santos *et al.* [21], repris par Wallwork [27]).

Tableau 5
Decomposition de litière forestière sous l'action d'insecticide (Weary et Merriam [30])

Couverture morte d'une forêt d'erables rouges (Ontario, Canada)		
	Parcelle témoin	*Parcelle traitée (carbofuran à 285 g a.i./ha)*
Taux de décomposition	1·5–2·2 g m^{-2} $jour^{-1}$	1·0–1·3 g m^{-2} $jour^{-1}$
Quantite décomposée par année (Mai à Novembre)	315–460 g (100)	210–270 g (60)

$\Delta = 40\%$.

limitent le broutage et permettent une décomposition tout à fait normale de la matière organique par voie bactérienne. Par contre, lorsqu'un acaricide est introduit dans le système les Nématodes ne sont plus contrôlés, prolifèrent, exercent un surbroutage des bactéries et la décomposition est, en conséquence, fortement ralentie. Dans le cas présent, ce ralentissement représente plus de 40 % en valeur moyenne annuelle [21, 27].

D'une manière globale encore l'interférence des pesticides avec le système de décomposition organique et de libération des éléments biogènes peut être mis en évidence, *a contrario*, suite à un traitement insecticide généralisé. Au Tableau 5, en effet, sont reprises les données de Weary et Merriam [30] montrant que même à des doses extrêmement faibles l'insecticide affecte notablement la vitesse de décomposition des litières forestières. L'explication de ce phénomène a été donnée ci-dessus; ce type d'effet est probablement universel et droit se retrouver dans tous les types de sol. C'est, à notre sens, une des menaces les plus graves que l'abus de pesticides peut induire au niveau des biocénoses édaphiques.

6. RÉSUMÉ ET CONCLUSIONS

Les effets écologiques des pesticides considérés sous l'angle agronomique (c'est-à-cire comme agents de contrôle d'organismes-cibles nuisibles) ont été passés en revue.

Défini comme une substance étrangère, affectant un système écologique (population-cible ou non-cible, réseau trophique simplifié, ensemble de l'écosystème) un pesticide induit des modifications fonctionnelles aussi bien que structurelles dans ce système. Ces modifications sont en outre de type *réversible* ou *irréversible*. En cette matière divers exemples repris de la biologie des sols sont commentés.

L'attention a été attirée sur les problèmes directement liés à l'utilisation des pesticides dans les agrosystèmes. L'auteur rappelle, en les discutant quant à leurs mécanismes, les principales conséquences écologiques de l'abus des pesticides: la substitution et la résurgence de peste, la perte de diversité génétique, la perte de fonctions écologiques, la banalisation des flores et des faunes, la contamination des chaînes trophiques. Un accent tout particulier est mis sur le problème de la résistance des espèces-cibles aux pesticides et en particulier sur la vitesse avec laquelle ce phénomène s'est propagé depuis l'avènement des insecticides organiques de synthèse.

S'orientant vers les modifications de structure et de fonctionnement de l'ensemble de la biocénose soumise aux pesticides, l'exposé aborde ensuite le problème de la fertilité biologique des sols. Il apparaît que la synergie exercée par les microorganismes et les animaux édaphiques dans les processus de décomposition organique et surtout dans le cycle des éléments minéraux est gravement compromise par les pesticides. Ceci apparaît de divers exemples qui démontrent la complexité de ces processus et la nécessité de les préserver pour le bien de l'agriculture.

RÉFÉRENCES

1. Bouché, M. B. (1972). Lombriciens de France. *Écologie et Systématique*, INRA, Paris, Publ. 72-2, 671 pp.
2. Brown, A. W. A. (1978). *Ecology of pesticides*, John Wiley, New York, 525 pp.
3. Cremlyn, R. (1979). *Pesticides. Preparation and Mode of Action*, John Wiley, Chichester, 240 pp.
4. Debry, J. M., Houssiau, M., Lemasson-Florenville, M., Wauthy, G. et Lebrun, Ph. (1982). Impact de populations lombriciennes introduites sur le pH et sur la dynamique de l'azote dans un sol traité avec du lisier de porcs. *Pedobiologia*, **23**, 157–71.
5. De Medts, A. (1981). Effets de résidus de pesticides sur les Lombriciens en terre de culture. *Pedobiologia*, **21**, 439–45.
6. Dimond, J. B., Belyea, G. A., Kadunce, R. E., Getchell, A. S. et Blease, J. A. (1970). DDT residues in robins and earthworm associated with contaminated forest soils. *Canad. Entomol.*, **102**, 1122–30.
7. Edwards, C. A. (Ed.) (1973). *Environmental Pollution by Pesticides*, Plenum Press, New York et London, 542 pp.
8. Edwards, C. A. et Thompson, A. R. (1973). Pesticides and the soil fauna. *Residue Rev.*, **45**, 1–79.

9. Georghiou, G. P. et Taylor, C. E. (1977). Pesticide resistance as an evolutionary phenomenon. Proceedings of XV International Congress of Entomology, the Entomological Society of America, pp. 759–85.
10. Lebrun, Ph. (1977). Incidences écologiques des pesticides sur la faune du sol. *Pédologie*, **27**, 67–91.
11. Lebrun, Ph., André, H. M., De Medts, A., Grégoire-Wibo, C. et Wauthy, G. (Eds) (1983). New trends in soil biology. Proceedings of the VIIIth Int. Coll. Soil Zoology, Louvain-la-Neuve, 30th Aug.–2 Sept. 1982, Imp. Dieu Brichard, Ottignies, 711 pp.
12. Ledant, J.-P., Jacob, J-P. et Devillers, P. (1983). Animaux menacés en Wallonie. Protégeons nos oiseaux. Duculot. Région Wallonne, Gembloux, 325 pp.
13. Matsumura, F. (1975). *Toxicology of Insecticides*, Plenum Press, New York and London, 503 pp.
14. Moore, N. W. (Ed.) (1966). Pesticides in the environment and their effects on wildlife. Suppl. to vol. 3 of the *Journal of Applied Ecology*, Blackwell Scientific Publications, Oxford, 311 pp.
15. Moore, N. W. (1967). A synopsis of the pesticide problem. In: *Advances in Ecological Research, Vol. 4*, Academic Press, London, pp. 75–129.
16. Moriarty, F. (1983). *Ecotoxicology. The Study of Pollutants in Ecosystems*, Academic Press, London, 233 pp.
17. Parker, L. W., Santos, P. F., Phillips, J. et Whitford, W. G. (1984). Carbon and nitrogen dynamics during the decomposition of litter and roots of a Chihuahuan desert annual, *Lepidium lasiocarpum*. *Ecological Monographs*, Vol. 54, pp. 339–60.
18. Perring, F. H. et Mellanby, K. (Ed.) (1977). *Ecological Effects of Pesticides*, Linnean Society of London and Academic Press, London, 193 pp.
19. Ramade, F. (1974). *Eléments d'écologie appliquée*, Ediscience, McGraw-Hill, Paris, 522 pp.
20. Ramade, F. (1977). *Ecotoxicologie*, Masson & Cie, Paris, 205 pp.
21. Santos, P. F., Phillips, J. et Whitford, W. G. (1981). The role of mites and nematodes in early stages of buried litter decomposition in a desert. *Ecology*, **62**, 664–9.
22. Seastedt, T. R. (1984). The role of microarthropods in decomposition and mineralization processes. *Ann. Rev. Entomol.*, **29**, 25–46.
23. Smith, R. F. et Van Den Bosch, R. (1967). Integrated Control. In: *Pest Control* (Kilgore and Doutt Eds), Academic Press, London, pp. 295–340.
24. Thirumurthi, S. et Lebrun, Ph. (1977). Persistence and bioactivity of carbofuran in a grassland. *Med. Fac. Landbouww. Rijksuniv., Gent*, **42**, 1455–62.
25. Van Rhee, J. A. (1977). Effects of soil pollution on earthworms. *Pedobiologia*, **17**, 201–8.
26. Vossbrinck, C. R., Coleman, D. C. et Woolley, T. A. (1979). Abiotic and biotic factors in litter decomposition in a semi-arid grassland. *Ecology*, **60**, 265–71.
27. Wallwork, J. A. (1982). Mineral Cycling in Soil Systems. Conférence donnée à l'Université Catholique de Louvain, Louvain-la-Neuve, le 8 mai 1981 (non publié).

28. WALLWORK, J. A. (1983). Oribatids in forest ecosystems. *Ann. Rev. Entomol.*, **28**, 109–30.
29. WATSON, D. L. et BROWN, A. W. A. (1977). *Pesticide Management and Insecticide Resistance*, Academic Press, London, 638 pp.
30. WEARY, G. C. et MERRIAM, H. G. (1978). Litter decomposition in a red maple woodlot under natural conditions and under insecticide treatment. *Ecology*, **59**, 180–4.

SESSION II

Need for and Benefits from Fertilisers and Pesticides

Chairman: N. King
(*Department of the Environment, London, UK*)

7

Ökonomische Gesamtbewertung der Verwendung von Düngern und Pestiziden und Auswirkungen eines reduzierten Einsatzes

H.-J. GEBHARD und G. WEINSCHENCK

Institut für Landwirtschaftliche Betriebslehre,
Universität Hohenheim, Bundesrepublik Deutschland

SUMMARY

The continuously growing intensification of agricultural production in the countries of the European Community has, in the last few years, led not only to a burdening of public budgets, but also to negative environmental effects such as in raising levels of nitrate in foodstuffs and ground water or in endangering species diversification. In the current economic situation, however, the majority of farms have no alternative to the massive use of fertilisers and plant protection chemicals if they want to improve or even maintain their income situation. A desirable reduction, especially at top intensities, is only possible with a drastic shift in the price–cost relationship between products and production inputs, or with administrative regulations limiting quantitative factor input. The difficulty in this connection is in the large regional dissimilarities with respect to both natural and economical production conditions.

Farms without animals show a heavy dependence on purchases of mineral fertiliser. A price increase or quantitative restrictions, especially for nitrogen, under unfavourable production conditions, can threaten the existence of and lead to abandonment of these farms.

Farms with animals, on the other hand, have a number of advantages and substitution possibilities through improved handling and usage of animal excrements. Their income is therefore considerably less dependent on purchased fertiliser.

A necessary reduction in production intensity—whether due to alleviating markets or to reducing external effects—leads inexorably to negative income effects on all farms. If these are not desired, the negative income effects must be equalised with direct income payments having a neutral influence on production, i.e. intensity. The necessary instruments are known and applicable; corresponding political decisions and incentives are lacking.

1. EINLEITUNG

Die intensive Landbewirtschaftung in westlichen Industrieländern ist seit einiger Zeit wachsender Kritik aus allen Bevölkerungsschichten ausgesetzt.

So neigt die Mehrzahl der Ökologen bereits im Umweltgutachten der BR-Deutschland 1978 zu der Auffassung 'Die Landwirtschaft ist per Saldo ein Umweltverschmutzer'. Das Umweltgutachten sieht die wichtigsten Gründe in:

einer Vereinfachung der Fruchtfolgen;
der Spezialisierung auf ertragsreiche aber schädlingsanfällige Sorten;
einer erntemaximierenden Düngung;
den arbeitsextensiven Betriebsstrukturen; der Unkrautbekämpfung durch Herbizide;
einer Trennung von Viehhaltung und Pflanzenproduktion.

Diese Entwicklung in modern geführten landwirtschaftlichen Betrieben war möglich und auch ökonomisch sinnvoll, weil mineralische Dünge- und chemische Pflanzenbehandlungsmittel relativ preiswert und in beliebiger Menge zur Verfügung standen.

Im folgenden soll zunächst gezeigt werden, welche Entwicklungen in diesem Bereich in den vergangenen Jahrzehnten stattgefunden haben.

Nach der theoretischen Darstellung der ökonomischen Gegebenheiten beim Einsatz von Dünge- und Pflanzenbehandlungsmitteln werden dann die Wirkungen politischer Handlungsalternativen zur Senkung des ertragssteigernden und ertragssichernden Aufwands am Beispiel unterschiedlicher landwirtschaftlicher Betriebe überprüft. Den Abschluß bildet eine gesamtwirtschaftliche Einordnung des vorliegenden Problems.

2. ENTWICKLUNG DES DÜNGE- UND PFLANZENBEHANDLUNGSMITTELEINSATZES SOWIE DER GETREIDEERTRÄGE IN DEN LÄNDERN DER EG-9

In den vergangenen 2 Jahrzehnten stieg der Verbrauch an mineralischen Düngemitteln—insbesondere Stickstoff—in einzelnen Ländern der EG-9

Tabelle 1
Entwicklung des Handelsdüngerverbrauchs in den Ländern der EG-9 (kg Reinnährstoff je ha LF und Jahr im 3-jährigen Mittel)

	B/L	*NL*	*F*	*BRD*	*I*	*GB*	*IRL*	*DK*
				Stickstoff				
1959/60 –61/62	55·3	97·7	16·4	43·5	16·7	23·4	5·3	39·5
1969/70 –71/72	105·7	178·6	42·9	81·9	30·7	42·5	18·0	97·3
1979/80 –81/82	124·8	238·3	68·3	118·4	58·8	71·3	45·8	131·2
				Phosphor				
1959/60 –61/62	51·4	46·9	27·6	47·3	18·6	21·6	17·1	36·9
1969/70 –71/72	88·0	48·6	55·2	66·2	27·4	26·6	36·4	43·4
1979/80 –81/82	63·2	40·8	57·3	68·1	42·9	23·7	25·4	40·0
				Kalium				
1959/60 –61/62	87·1	60·0	22·0	72·2	5·5	22·8	14·4	57·5
1969/70 –71/72	108·2	58·7	47·6	86·5	11·7	26·1	30·7	62·2
1979/80 –81/82	92·0	56·5	54·6	92·6	21·7	24·3	31·3	51·3

Quelle: United Nations, *Statistical Yearbook*, verschiedene Jahrgänge.

um das 3- und 4-fache seines Ausgangsniveaus Anfang der 60er Jahre (Tab. 1).

Obwohl die Entwicklung in den betrachteten Ländern im Prinzip ähnlich verlief, lassen sich für einzelne Parameter doch erhebliche Unterschiede nachweisen.

Während das Verbrauchsniveau z.B. von N zu Beginn des Betrachtungszeitraumes in Irland bei gerade 5 kg/ha LF lag, betrug es in den Niederlanden fast das 20-fache. Auch die Wachstumsraten in den einzelnen Ländern gehen weit auseinander.

So fand in Frankreich innerhalb 20 Jahren eine Vervierfachung des Stickstoffverbrauchs statt, während er sich in Belgien/Luxemburg 'nur' verdoppelte.

Schließlich fand eine spürbare Veränderung im Verhältnis der ausgebrachten Nährstoffe statt. Die ursprünglich enge Relation zwischen

Stickstoff, Phosphor und Kali erweiterte sich deutlich zu gunsten des Stickstoffs, der heute in allen EG-Ländern mengenmäßig die dominierende Rolle spielt.

Betrachtet man die Entwicklung in den vergangenen Jahren, so ist eine deutliche Verlangsamung der Zuwachsraten zu erkennen. Das gilt insbesondere für die Nährstoffe Phosphor und Kalium, bei denen schon seit Anfang bis Mitte der 70er Jahre eine gewisse Stagnation im Verbrauch sichtbar wird. Beim Stickstoffeinsatz ist diese Tendenz erst seit 3 bis 4 Jahren erkennbar.

Es ist jedoch nicht auszuschließen, daß im Zuge des Züchtungsfortschritts und daraus resultierender Ertragsverbesserungen auch der Nährstoffeinsatz wieder ansteigen wird.

Auf der anderen Seite sind gerade aus Umwelt- und Kostengesichtspunkten in jüngster Zeit Verfahren in der Entwicklung bzw. praktischen Erprobung, die auf eine gezieltere Nährstoffversorgung landwirtschaftlicher Nutzpflanzen und damit eine in der Regel sparsamere Nährstoffausbringung ausgerichtet sind.

Parallel zur Entwicklung des Stickstoffeinsatzes stiegen auch die Erträge aller wichtigen Kulturen an. Bei Getreide insgesamt erhöhte sich das Ertragsniveau innerhalb von 20 Jahren um 50 bis 100 %. Die Unterschiede zwischen den einzelnen EG-Ländern sind absolut gesehen dabei eher größer geworden (Tab. 2).

Mit Sicherheit wären diese Ertragssteigerungen nicht allein durch eine erhöhte Nährstoffzufuhr zu erzielen gewesen. Es ist allerdings kaum möglich, den Einfluß der ertragsverbessernden Züchtungskomponenten wie verbesserte Bodenbearbeitung, Dünge- und Pflanzenschutzmitteleinsatz, etc., im einzelnen zu quantifizieren.

Die Optimierung der allgemeinen Wachstumsbedingungen und die ständig steigenden Erträge erhöhten jedoch auch die Konkurrenzfähigkeit

Tabelle 2

Entwicklung der Getreideerträge in den Ländern der EG-9 (dt je ha LF und Jahr im 3-jährigen Mittel)

	B	*L*	*NL*	*F*	*BRD*	*I*	*GB*	*IRL*	*DK*
				Getreide insgesamt					
1960–62	35·2	23·6	36·7	24·9	29·4	20·1	31·6	31·4	35·3
1970–72	39·2	28·8	40·2	38·1	37·2	28·4	38·4	38·5	38·6
1980–82	51·7	51·9	61·5	48·0	45·8	35·3	51·0	45·0	41·7

Quelle: United Nations, *Statistical Yearbook*, verschiedene Jahrgänge.

von Unkräutern und den Befalls- und Krankheitsdruck durch Schädlinge und Pilze. Eine Realisierung des wachsenden Ertragspotentials war und ist daher nur durch einen ständig sich erweiternden Einsatz von Pflanzenbehandlungsmitteln möglich.

Allerdings ist gerade die kaum noch überschaubare und sich weiter vergrößernde Palette von Präparaten Anlaß zu einer wachsenden Kritik aus der Bevölkerung.

Insgesamt gesehen, ist der internationale Markt für Pflanzenbehandlungsmittel einer eingehenderen Betrachtung nur sehr schwer zugänglich. Man weiß auch wenig über die Zahl der Behandlungen und angewendeten Wirkstoffmengen in einzelnen Kulturen bzw. Regionen.

Die folgenden Ausführungen beschränken sich daher auf die amtliche Statistik der BR-Deutschland (Tab. 3). Demnach hat hier die abgesetzte Wirkstoffmenge insgesamt von 1970 bis 1979 ganz beträchtlich zugenommen und lag im Jahr 1979 fast 80 % über dem Ausgangswert. Den größten Anteil machen nach wie vor die Herbizide mit gut 60 % aus.

Seit 1980 ist nun ein ständiger Rückgang der abgesetzten Wirkstoffmenge insgesamt zu beobachten, wobei die Fungizide eher eine steigende Tendenz aufweisen. Es ist allerdings verfrüht, diese Entwicklung als Trendwende im flächenbezogenen Pflanzenschutzmitteleinsatz zu interpretieren, da neue

Tabelle 3

Abgesetzte Wirkstoffmengen an Pflanzenbehandlungsmitteln in der BR-Deutschland 1970 bis 1982 in t

Jahr	*Herbizide*	*Fungizide*	*Insektizide*	*And. Mittel*	*Summe*
1970	10629	4745	1519	2088	18981
1971	11063	4897	1637	2078	16675
1972	12744	4526	1579	2130	20979
1973	14918	5133	2098	2266	24415
1974	16894	6144	1615	2070	26723
1975	15700	5291	1648	2342	24981
1976	14906	5400	2073	2597	24976
1977	16876	5706	2143	2839	27564
1978	18234	6918	2175	3056	30383
1979	20510	7112	2341	3687	33650
1980	20857	6549	2341	3183	32930
1981	19507	7012	2405	2871	31795
1982	17776	7211	1948	2429	29364

Quelle: *Statistisches Jahrbuch über Ernährung, Landwirtschaft und Forsten*, verschiedene Jahrgänge (BR-Deutschland).

Verfahren und Mittel in der Regel mit geringeren Wirkstoffmengen auskommen.

Es ist in den meisten Fällen nämlich unerläßlich, bei steigenden Düngergaben und Erträgen aus Gründen der Risikominderung begleitende Pflanzenbehandlungsmaßnahmen zu ergreifen. Zwischen Dünge- und Pflanzenschutzmitteleinsatz bestehen somit in weiten Bereichen komplementäre Beziehungen.

3. PROBLEMATIK DER BESTIMMUNG OPTIMALER AUFWANDMENGEN AN ERTRAGSSTEIGERNDEN PRODUKTIONSMITTELN

Geht man davon aus, daß Dünge- und Pflanzenbehandlungsmittel in beliebiger Menge zur Verfügung stehen, so ist ihr Einsatz solange ökonomisch sinnvoll, wie die Kosten der letzten Aufwandseinheit durch den erzielten Mehrerlös noch gedeckt werden. Bei einem gegebenen Preis-Kosten-Verhältnis hängt das optimale Faktoreinsatzniveau daher letztendlich vom Verlauf der jeweiligen Produktionsfunktion ab. Aus einer Vielzahl von Versuchen geht hervor, daß die Ertragsverläufe landwirtschaftlicher Kulturen mit zunehmendem Dünger- insbesondere Stickstoffeinsatz—fallende Grenzerträge aufweisen.

Würden die Düngemittel nichts kosten, so wäre ihr Einsatz bis zur Erreichung des Maximalertrages sinnvoll. Der Einsatz von Pflanzenbehandlungsmitteln hat komplementären Charakter und dient der Sicherung des angestrebten Ertrages. Je höher die Preise für die genannten Produktionsmittel allerdings liegen, desto mehr weichen Maximal- und ökonomischer Optimalertrag in der Regel voneinander ab.

Das Hauptproblem für nahezu alle Landwirte besteht darin, daß sie die spezifischen Ertragsfunktionen für ihren Standort nicht kennen. Somit basiert die Bemessung des jeweiligen Faktoreinsatzes fast ausschließlich auf Erfahrungswerten und orientiert sich meist unbewußt am Maximalertrag. Zusätzlich ergibt sich die Schwierigkeit, daß im Zuge des technischen Fortschritts (Züchtung, Pflanzenschutz, Bodenbearbeitung etc.) die Produktionsfunktionen einer ständigen Veränderung in der Weise unterliegen, daß sich der Bereich stark sinkender Grenzerträge in Richtung auf ein immer höheres Ertragsniveau verschoben hat.

Nur so ist es zu erklären, daß trotz andauernder Verschlechterung der Faktor- Produktpreis-Relationen eine Steigerung des Dünge- und Pflanzenschutzmitteleinsatzes dem Einzelbetrieb ökonomische Vorteile brachte.

Das Ausrichten der Bewirtschaftungsintensität am theoretischen Maximalertrag bedeutet zumindest bei Getreide unter den gegenwärtigen Preis-Kosten-Verhältnissen keinen allzugroßen Gewinnverzicht, da Optimal- und Maximalertrag nicht gravierend voneinander abweichen. Die hohe Rentabilität des Handelsdüngereinsatzes läßt sich nicht zuletzt damit erklären, daß die staatlich festgesetzten Produktpreise sich an den Durchschnittskosten der Produktion und nicht an ihren Grenzkosten orientieren. Die sehr geringe Elastizität der Nachfrage nach Handelsdüngern ist nur eine Folge dieser Gegebenheiten.

Auf der anderen Seite treten in neuerer Zeit auch in der Landwirtschaft die Auswirkungen des oft kurzfristigen ökonomischen Denkens und Handelns auf die Umwelt verstärkt zu Tage. Das einzelbetrieblich optimale Verhalten, bei dem externe Effekte z.B. auf Boden und Wasser entstehen, entfernt sich immer weiter vom gesamtwirtschaftlichen Optimum. Dies gilt umso mehr dann, wenn Produkte erzeugt werden, die am Markt keinen Nachfrager mehr finden.

Alles deutet darauf hin, daß eine Begrenzung der Intensität des Dünge- und Pflanzenschutzmitteleinsatzes durch marktbedingte Verschiebungen der Preisrelationen kurzfristig kaum zu erwarten ist.

Will man—aus welchen Gründen auch immer—die Landwirte dazu bewegen, weniger Chemie einzusetzen, so muß man entweder das Preis-Kosten-Verhältnis und damit das ökonomisch sinnvolle Aufwandsniveau verändern, oder durch eine Kontingentierung auf der Faktor—bzw. Produktseite die Intensität bzw. Produktion verringern.

4. AUSWIRKUNGEN EINER BEGRENZUNG DES EINSATZES VON DÜNGE- UND PFLANZENBEHANDLUNGSMITTELN AUF UNTERSCHIEDLICHE LANDWIRTSCHAFTLICHE BETRIEBE

Aufgrund der weiter oben beschriebenen Rahmenbedingungen der landwirtschaftlichen Produktion ist in absehbarer Zeit kaum damit zu rechnen, daß die Vorteilhaftigkeit der bisher praktizierten und teilweise weiter wachsenden Bewirtschaftungsintensität durch marktbedingte Änderungen relevanter Parameter spürbar geschmälert wird.

Gleichzeitig nützen Appelle an den einzelnen Landwirt solange wenig, wie mit der Verminderung der speziellen Intensität Einkommenseinbußen verbunden sind.

Eine Senkung des hohen Faktoreinsatzes kann daher nur auf administrative Weise erfolgen. Eingriffsmöglichkeiten bestehen im Prinzip sowohl auf der Produkt- als auch auf der Faktorseite und zwar durch eine Veränderung der Faktor-Produktpreis-Verhältnisse oder eine direkte Mengensteuerung der Produktion bzw. des Faktoreinsatzes.

Es würde an dieser Stelle zu weit führen, alle genannten Alternativen ausführlicher zu behandeln. Darüber hinaus würde bei einer Mengenbegrenzung auf der Produktseite eher eine Änderung der Anbaustruktur als eine Reduzierung der speziellen Intensität zu erwarten sein. Letzteres könnte man zweifellos durch eine spürbare Absenkung des nominalen Agrarpreisniveaus erreichen. Die Folge wäre jedoch ein mehr oder weniger großer Einkommensdruck in der Landwirtschaft, was politisch zur Zeit kaum vertretbar erscheint.

Die folgenden Modellrechnungen beschränken sich deshalb auf denkbare Veränderungen der Rahmenbedingungen im Bereich des Faktoreinsatzes—insbesondere Dünge- und Pflanzenbehandlungsmittel—bei gegebenem Agrarpreisniveau.

Im Prinzip gibt es zwei Möglichkeiten auf den Faktoreinsatz—in unserem Zusammenhang speziell den Stickstoffeinsatz—Einfluß zu nehmen:

1. Durch Anhebung des Stickstoffpreises, etwa in Form einer Umweltabgabe.
2. Durch eine an die Fläche gebundene, jährlich in gleicher Höhe zur Verfügung stehende Höchstmenge.

Ausgehend von einer mittleren Standortqualität, die gekennzeichnet ist durch maximale Erträge von ca. 57 dt/ha Winterweizen bzw. 53 dt/ha Wintergerste und 530 dt/ha Zuckerrüben bzw. 430 dt/ha Silomais werden für einen Marktfrucht- und einen Futterbau-Betrieb die Wirkungen der

Tabelle 4
Faktorausstattung der ausgewählten Modellbetriebe

	Kenngrößen	*Marktfrücht-Betrieb*	*Futterbau-Betrieb*
Landwirtschaftlich genutzte Fläche (LF)	ha	50	30
Ackerfläche	ha	50	9
Grünland	ha	—	21
Arbeitskräfte	AK	1·5	1·8
Stallplätze			
Milchkühe	Stck.	—	25
Rinder	Stck.	—	30

vorgenannten Maßnahmen auf Produktionsstruktur und Einkommen überprüft.

Die Faktorausstattung der ausgewählten Betriebe ist in Tabelle 4 dargestellt. Die nicht ausgewiesenen Kapazitäten, wie verfügbare Maschinen, Siloraum etc. entsprechen dem Durchschnitt des jeweiligen Betriebstyps (siehe Tab. 4). Mit Hilfe linearer Optimierungsrechnungen wurden die Ertragsverläufe relevanter Kulturen in Abhängigkeit von Stickstoff- und Pflanzenbehandlungsmitteleinsatz sowie die Bereiche Nährstoffbilanzierung und Humusersatzwirtschaft in die Betrachtung einbezogen.

4.1. Auswirkungen einer Stickstoffpreiserhöhung auf Organisation und Einkommen landwirtschaftlicher Betriebe

Eine spürbare Verteuerung des zugekauften mineralischen Stickstoffs wirkt theoretisch auf 2 Ebenen: Einmal wird die im Vergleich zur Ausgangsorganisation optimale Einsatzmenge reduziert und zum anderen ergeben sich Verschiebungen in der Organisations- bzw. Anbaustruktur. Beim Marktfruchtbetrieb wird dies besonders deutlich (Tab. 5).

Eine Verdopplung des Stickstoffpreises von 1·50 DM/kg auf 3.-DM/kg N hat eine Reduzierung der insgesamt im Betrieb eingesetzten N-Menge um fast 50 % zur Folge. Der Raps wird durch den Anbau von Ackerbohnen vollständig verdrängt. Gleichzeitig verschlechtert sich der Gewinn des Betriebes um fast ein Viertel. Bei einer weiteren Verteuerung wird der Stickstoffeinsatz zu Zuckerrüben sehr stark reduziert. Hier kommt die relativ flache Ertragsfunktion zum Ausdruck, die aus einer gezwungenermaßen kurzfristigen Betrachtungsweise resultiert. Andererseits wächst die Vorteilhaftigkeit stickstoffakkumulierender Pflanzen, wie z.B. der Ackerbohnen, die jetzt bis zur Fruchtfolgegrenze ausgedehnt werden.

Das Betriebsergebnis des ausgewählten Futterbaubetriebes wird von einer Stickstoffpreiserhöhung wesentlich weniger tangiert als das des Marktfruchtbaubetriebes (Tab. 6). Dies hängt in erster Linie mit dem Anfall tierischer Exkremente und den darin enthaltenen Nährstoffen zusammen. So bewirkt eine Preissteigerung von 0·50 DM je kg Stickstoff zwar eine Reduzierung der Spitzenintensitäten, der Gewinn des Betriebes wird jedoch nur unwesentlich geschmälert. Auffallend ist die Änderung der Organisation der Grünlandwirtschaft zugunsten der Wiesennutzung. Die Erklärung dafür liegt in einer suboptimalen Verwertung der auf der Weide anfallenden Nährstoffe bei steigenden Stickstoffpreisen. Die Tierhaltung bleibt dagegen über den gesamten Bereich der untersuchten Preisanhebung stabil.

Tabelle 5
Auswirkungen einer Stickstoffpreiserhöhung in einem Marktfruchtbetrieb

	Kenngrößen	*Ausgangsorganisation (N-Preis 1·50 DM/kg)*	*Erhöhung des Stickstoffpreises um ... DM/kg*					
			0·50	*1·00*	*1·50*	*2·00*	*3·00*	*4·50*
1. Anbauumfang und Stickstoffintensität								
Winterweizen	in % AF	33·3 (185)	33·3 (175)	33·3 (165)	33·3 (120)	33·3 (115)	33·3 (100)	14 (88)
Sommergerste	bzw.	33·3 (65)	33·3 (65)	33·3 (65)	33·3 (60)	33·3 (60)	33·3 (55)	33·3 (40)
Raps	(kg N/ha)	21·3 (170)	21·2 (165)	21 (155)	—	—	—	—
Zuckerrüben		12·1 (150)	12·2 (140)	12·3 (125)	12·4 (120)	19·3 (26)	19·3 (25)	19·4 (25)
Ackerbohnen		—	—	—	21 (60)	14 (55)	14 (55)	33·3 (55)
2. Zugekaufte min. N-Menge	kg/ha LF	129	123	116	68	54	47	23
Verfügbarer Stickstoff insg.	kg/ha LF	137	132	125	87	71	64	49
Pflanzenschutz-intensität	%	100	100	100	79	66	66	54
3. Betriebsdeckungs-beitrag	DM	92 050	88 920	85 945	83 435	82 033	79 490	76 818
4. Gewinn	DM	37 050	33 920	30 945	28 435	27 033	24 490	21 828
Gewinnänderung	%		−8·5	−16·5	−23·3	−27·0	−33·9	−41·1

Tabelle 6
Auswirkungen einer Stickstoffpreiserhöhung in einem Futterbaubetrieb

	Kenngrößen	*Ausgangsorganisation (N-Preis 1·50 DM/kg)*	*Erhöhung des Stickstoffpreises um . . . DM/kg*					
			0·50	*1·00*	*1·50*	*2·00*	*3·00*	*4·5*
1. Anbauumfang und Stickstoffintensität								
Winterweizen	% AF/	33·3 (190)	33·3 (180)	33·3 (130)	33·3 (130)	33·3 (125)	33·3 (120)	31 (105)
Sommergerste	GF	16·7 (70)	16·7 (65)	16·7 (65)	16·7 (65)	16·7 (60)	16·7 (50)	9·7 (45)
Körnermais	bzw.	8 (195)	—	—	—	—	—	—
Silomais	(kgN/ha)	42 (220)	50 (210)	50 (200)	50 (145)	50 (140)	50 (135)	50 (122)
Kleegras		8·3 (165)	8·3 (165)	8·3 (165)	8·3 (165)	8·3 (165)	8·3 (165)	4·3 (130)
Wiese	% GF	77 (85)	97 (73)	92 (81)	100 (75)	100 (68)	100 (68)	100 (57)
Weide		23 (185)	3 (170)	8 (0)	—	—	—	—
2. Zugekaufte min. N-Menge	kgN/ha LF	71	36	27	17	9	8	0
Verfügbarer Stickstoff insg.	kgN/ha LF	131	106	98	91	84	82	73
Pflanzenschutz-intensität	%	100	100	90	76	76	76	69
3. Betriebsdeckungs-beitrag	DM	89 146	88 519	88 033	87 707	87 526	87 396	87 167
4. Gewinn	DM	44 146	43 519	43 033	42 707	42 526	42 396	42 167
Gewinnänderung	%		−1·4	−2·5	−3·3	−3·8	−4·0	−4·5

Tabelle 7
Gewinnreduzierung mit und ohne Anpassung der Organisation bzw. Intensität in unterschiedlichen landwirtschaftlichen Betrieben

	Kenngrößen	*Ausgangsorganisation (N-Preis 1·50 DM/kg)*	*Erhöhung des Stickstoffpreises um ... DM/kg*					
			0·50	*1·00*	*1·50*	*2·00*	*3·00*	*4·5*
1. Marktfruchtbaubetrieb								
(a) mit Anpassung								
Gewinn	DM	37 050	33 920	30 945	28 435	27 033	24 490	21 818
Gewinnänderung	%		−8·5	−16·5	−23·3	−27·0	−33·9	−41·1
	DM/ha LF		−63	−122	−172	−200	−251	−304
(b) ohne Anpassung								
Gewinn	DM	37 050	33 825	30 600	27 375	24 150	17 700	8 025
Gewinnänderung	%		−8·7	−17·4	−26·1	−34·8	−52·2	−78·3
	DM/ha LF		−65	−129	−194	−258	−387	−581
(c) Differenz mit/ohne Anpassung	DM/ha LF		−2	−7	−22	−58	−136	−277
2. Futterbaubetrieb								
(a) mit Anpassung								
Gewinn	DM	44 146	43 519	43 033	42 707	42 526	42 396	42 167
Gewinnänderung	%		−1·4	−2·5	−3·3	−3·8	−4·0	−4·5
	DM/ha LF		−21	−37	−48	−54	−58	−66
(b) ohne Anpassung								
Gewinn	DM	44 146	43 081	42 016	40 951	39 946	37 756	34 561
Gewinnänderung	%		−2·4	−4·8	−7·2	−9·5	−14·5	−22·0
	DM/ha LF		−36	−72	−107	−140	−213	−320
(c) Differenz mit/ohne Anpassung	DM/ha LF		−15	−35	−59	−86	−155	−254

Da gegenwärtig der Stickstoffpreis wieder näher an 2·00 DM je kg als an 1·50 DM liegt, würde eine Preissteigerung um 0·50 DM je kg im dargestellten Futterbaubetrieb auch nur eine sehr viel geringere Auswirkung auf die zugekaufte Stickstoffmenge hervorrufen (Reduzierung von 36 kg N/ha auf 27 kg N/ha). Selbst bei einer Vervierfachung des Stickstoffpreises auf 6·00 DM je kg Stickstoff und einem generellen Verzicht auf den Einsatz zugekaufter Stickstoffdünger würde der Gewinn des Betriebes um nur 4·5 % sinken. Beim Marktfruchtbaubetrieb beträgt die Gewinneinbuße bei diesem Preis dagegen mehr als 40 %. Als zusätzliche Fruchtart werden dann Leguminosen (Kleegras) in die Organisation aufgenommen. Es ist zu vermuten, daß Futterbaubetriebe mit einem geringeren Grünlandanteil als hier unterstellt, bereits bei geringeren Preisanhebungen mit dem Anbau derartiger Feldfutterpflanzen beginnen.

Die Wirkung einer Verteuerung mineralischer Stickstoffdüngemittel auf den Gewinn landwirtschaftlicher Betriebe ist also ganz entscheidend abhängig vom Betriebssystem und den bestehenden Substitutionsmöglichkeiten. Betriebe mit einem hohen Viehbesatz und großem Nährstoffanfall aus tierischen Exkrementen können durch eine Verbesserung von Aufbereitung und Ausbringung besagter Stoffe den Einkauf teurer mineralischer Düngemittel umgehen. Im Gegensatz dazu besitzt ein reiner Marktfruchtbaubetrieb nur wenige ackerbauliche Möglichkeiten, teuren Mineraldünger zu substituieren. Bei dieser Betriebsform schlägt sich eine Stickstoffpreisanhebung daher sehr gravierend im Gewinn nieder, obwohl durch die Anpassung von Intensität und Organisation noch stärkere Gewinneinbußen vermieden werden. Dies kommt zum Vorschein, wenn man ein Verharren der Landwirte auf der Ausgangsorganisation mit ihren relativ hohen Intensitäten unterstellt. Dabei spielt die Betriebsform und die Höhe der angenommenen Preissteigerung eine entscheidende Rolle (Tab. 7). Wollte man die Intensitätsreduzierung möglichst einkommensneutral gestalten—und nur so hat sie überhaupt eine Chance der politischen Realisierbarkeit—so wäre lediglich die Gewinnminderung bei Anpassung der Organisation und Intensitäten als direkte Transferzahlung an die Landwirte zu leisten. Bei einer Anhebung des Stickstoffpreises auf 3·00 DM/kg N würde dieser Betrag zwischen 100 und 200 DM/ha LF liegen. Die entsprechende Intensitätsreduktion beim mineralischen Stickstoffeinsatz würde zwischen 50 und 80 % ausmachen. Ein guter Teil des notwendigen Einkommensausgleichs könnte dabei über die Einnahmen aus dem erhöhten Stickstoffpreis erfolgen und somit eine Entlastung des öffentlichen Haushaltes ermöglichen.

Tabelle 8
Auswirkungen einer Stickstoffkontingentierung in einem Marktfruchtbaubetrieb

	Kenngrößen	*Ausgangsorganisation (ohne Konting.)*	*Reduzierung der eingesetzten mineralischen N-Menge um ... %*				
			15	*30*	*45*	*60*	*100*
1. Anbauumfang und Stickstoffintensität							
Winterweizen	% AF/	33·3 (185)	33·3 (160)	33·3 (160)	33·3 (126)	33·3 (107)	9·9 (0)
Sommergerste	GF	33·3 (65)	33·3 (60)	33·3 (60)	33·3 (60)	33·3 (55)	33·3 (31)
Raps	bzw.	21·3 (170)	20·5 (150)	6·3 (150)	—	—	—
Zuckerrüben	(kgN/ha)	12·1 (150)	12·4 (120)	12·4 (120)	12·4 (120)	19·3 (26)	23·5 (0)
Ackerbohnen		—	0·5 (60)	14·7 (60)	21 (60)	14 (55)	33·3 (35)
2. Verfügbare min. N-Menge	kg/ha LF	129	110	90	71	52	0
Verfügbarer Stickstoff insg.	kg/ha LF	137	119	106	90	67	22
Pflanzenschutz-intensität	%	100	100	92	81	66	36
3. Betriebsdeckungs-beitrag	DM	92 050	91 401	90·055	88 675	86 971	74 620
4. Gewinn	DM	37 050	36 401	35 055	33 675	31 971	19 620
Gewinnänderung	%		−1·8	−5·4	−9·1	−13·7	−53·0

Vergleichbare Intensitätssenkungen mit wesentlich geringeren negativen Einkommenswirkungen sind aber auch durch eine Vergabe von z.B. flächengebundenen N-Mengen erzielbar. Allerdings wäre hierbei ein erhöhter Verwaltungs- und Überwachungsaufwand nötig.

4.2. Auswirkungen einer Stickstoffkontingentierung auf Organisation und Einkommen unterschiedlicher landwirtschaftlicher Betriebe

Betrachten wir zunächst wieder den viehlosen Marktfruchtbaubetrieb (Tab. 8). Die Ausgangsorganisation bei freier Verfügbarkeit mineralischer Stickstoffdüngemittel weist ein Düngungsniveau von ca. 130 kg N/ha LF auf, wobei Betriebe mit einem größeren Anteil düngungsintensiver Kulturen, wie Winterweizen und Raps, auf ertragreichen Standorten wesentlich höhere Mengen einsetzen.

Reduziert man die verfügbare Stickstoffmenge in dem dargestellten Modellbetrieb um 15 %, d.h. auf 110 kg N/ha LF, so ergibt sich eine Gewinnminderung um nur 1·8 %. Will man eine vergleichbare Intensitätsreduzierung mit einer Verteuerung des Stickstoffs erreichen, so würde der Gewinn um etwa das Zehnfache sinken (vgl. Tab. 5). Der Anbau von Ackerbohnen mit ihrer Fähigkeit Stickstoff aus der Luft zu binden, gewinnt sehr schnell an Vorteilhaftigkeit und nimmt mit weiterer Verknappung des mineralischen Stickstoffs einen wachsenden Anteil in der Fruchtfolge ein. Diese Entwicklung wirkt sich auch in Richtung auf eine Verminderung der Pflanzenschutzintensität aus, die zunächst langsam, dann aber sehr deutlich zurückgenommen wird.

Der Futterbaubetrieb hat wie bereits bei der Stickstoffpreiserhöhung so auch im Falle einer Kontingentierung erhebliche Vorteile gegenüber dem betrachteten Marktfruchtbaubetrieb (Tab. 9). Eine spürbare Reduzierung der Düngungs- und Pflanzenbehandlungsintensität, bei nahezu unverändertem Gewinn, tritt erst bei einer Reduzierung der Stickstoffausgangsmenge um 60 % ein. Die Umstellung der Organisation, insbesondere der Grünlandbewirtschaftung läuft ähnlich wie bei einer Verteuerung des Stickstoffs.

Zusammenfassend können die hier angestellten einzelbetrieblichen Modellrechnungen wie folgt beurteilt werden: Der ökonomische Nutzen des Einsatzes ertragssteigernder und ertragssichernder Produktionsmittel muß unter den gegebenen Preis-Kosten-Relationen für den Einzelbetrieb als sehr bedeutend eingestuft werden. Bei viehlosen Betrieben ist die freie Verfügbarkeit dieser Produktionsmittel gegenwärtig in der Regel sogar Voraussetzung zur Existenzerhaltung.

Die exakte Quantifizierung der Auswirkungen veränderter Rahmenbedingungen sowohl auf die Bewirtschaftungsintensität bzw. das

Tabelle 9

Auswirkungen einer Stickstoffkontingentierung in einem Futterbaubetrieb

	Kenngrößen	*Ausgangsorganisation (ohne Konting.)*	*Reduzierung der verfügbaren mineralischen N-Menge um ... %*				
			15	*30*	*45*	*60*	*100*
1. Anbauumfang und Stickstoffintensität							
Winterweizen	% AF/	33·3 (185)	33·3 (185)	33·3 (185)	33·3 (185)	33·3 (136)	31 (105)
Sommergerste	GF	8·3 (150)	8·3 (150)	8·3 (150)	16·7 (65)	16·7 (65)	9·7 (45)
Körnermais	bzw.	8 (195)	8 (190)	8 (190)	7 (190)	—	—
Silomais	(kgN/ha)	42 (220)	42 (220)	42 (220)	43 (215)	50 (205)	50 (122)
Kleegras		8·3 (165)	8·3 (165)	8·3 (165)	8·3 (165)	8·3 (165)	4·3 (130)
Wiese	% GF	76 (86)	83 (83)	90 (81)	97 (79)	97 (73)	100 (57)
Weide		24 (180)	17 (180)	10 (180)	3 (180)	3 (165)	—
2. Verfügbare min. N-Menge	kg/ha LF	2 238	61	50	39	29	0
Verfügbarer Stickstoff insg.	kg/ha LF	1 807	124	117	110	100	73
Pflanzenschutz-intensität	%	100	100	100	100	90	69
3. Betriebsdeckungs-beitrag	DM	89 146	89 125	89 101	89 077	88 885	87 167
4. Gewinn	DM	44 146	44 125	44 101	44 077	43 885	42 167
Gewinnänderung	%		−0·05	−0·01	−0·16	−0·6	−4·5

Anbauverhältnis als auch auf den Gewinn scheitert heute in der Hauptsache noch daran, daß regions-, kulturart- oder sortenspezifische Kenntnisse über die technischen Beziehungen zwischen Aufwand und Ertrag fehlen. In diesem Sinne können die hier dargestellten Ergebnisse auch nur Modellcharakter haben und die Tendenz der Wirkung unterschiedlicher administrativer Maßnahmen zur Begrenzung der Intensität aufzeigen.

5. GESAMTWIRTSCHAFTLICHE BEURTEILUNG DER HOHEN PRODUKTIONSINTENSITÄT

Bisher haben wir uns ausschließlich mit dem Einzelbetrieb, seinen Produktionsbedingungen und seinen Schwierigkeiten befaßt, das einzelbetriebliche Optimum, insbesondere hinsichtlich der speziellen Intensität möglichst genau zu treffen.

Nun kann aber gerade in Ländern mit administrativ festgesetzten Preisen und Intervention der Produktionsüberschüsse das einzelbetriebliche Optimum mehr oder weniger stark vom gesamtwirtschaftlichen Optimum abweichen.

In den Ländern der EG ist dieser Trend seit einiger Zeit sichtbar. Er verstärkt sich nicht zuletzt durch die wachsende Intensität des Einsatzes ertragssteigernder und ertragssichernder Produktionsmittel und die damit verbundene zunehmende Realisierung vorhandener Produktionsreserven.

Als Gründe für diese Entwicklung lassen sich nennen:

(1) Die im Zuge der gemeinsamen Agrarmarkt- und Preispolitik für alle wichtigen landwirtschaftlichen Produkte festgelegten Mindestpreise, die nicht den Marktgleichgewichtspreisen entsprechen und sich aus einkommenspolitischen Gründen mehr an den Durchschnittskosten der Produktion als an ihren Grenzkosten orientieren.

(2) Eine zunehmende Zahl externer Effekte und daraus resultierende volkswirtschaftliche Kosten, die sich aus einer wachsenden Intensivierung der landwirtschaftlichen Produktion ergeben und für die—bei Anwendung des Verursacherprinzips—eigentlich der Einzelbetrieb aufkommen müßte. Hierunter fallen insbesondere die Nitratanreicherung der Böden und des Grundwassers durch zu hohe organische und mineralische Nährstoffgaben, Erosionsgefahr durch eine Monotonisierung der Anbauverhältnisse bzw. der Landschaft und im Zusammenhang mit dem Einsatz von Pflanzenschutzmitteln, eine Verarmung der Flora und Fauna.

Die Hinnahme einer weiteren Verstärkung dieser externen Effekte und die teilweise irreversible Schädigung natürlicher Ressourcen muß insbesondere dann als äußerst fragwürdig erscheinen, wenn die Produktion ständig über der Nachfrage liegt.

Auch aus Gründen der Versorgungssicherheit ist eine Produktion über den Eigenbedarf hinaus nur dann gerechtfertigt, wenn sie zumindest ohne Verlust abgesetzt werden kann. Eine Berücksichtigung des krisenanfälligen Vorleistungseinsatzes, wie Dieselöl und Düngemittel, entkräftet ohnehin dieses von Politikern gerne gebrauchte Argument zur Rechtfertigung wachsender Überschüsse.

Aus gesamtwirtschaftlicher Sicht ist daher entweder der Umfang oder die Intensität der landwirtschaftlichen Produktion zu begrenzen.

Eine Belebung des Strukturwandels in der Landwirtschaft und somit das Ausscheiden von Arbeitskräften und Fläche aus der Produktion ist in der gegenwärtigen gesamtwirtschaftlichen Situation kaum zu erwarten und würde, außer durch eine Verringerung der Überschußproduktion kaum zu Wohlfahrtsgewinnen führen, da zumindest die freiwerdende Arbeit nicht produktiv eingesetzt werden könnte.

Eine Verringerung des überhöhten Einsatzes ertragssteigernder bzw. ertragssichernder Produktionsmittel könnte dagegen sehr wohl einen volkswirtschaftlichen Gewinn bedeuten, da für ihre Produktion alternativ verwendbare Ressourcen benötigt werden und möglicherweise negative externe Effekte verhindert oder verringert werden könnten. Die theoretischen Möglichkeiten hierfür sind bekannt. Theoretisch müßte einer Senkung des Agrarpreisniveaus aus gesamtwirtschaftlichen Kostengesichtspunkten der Vorzug eingeräumt werden. Dies scheint aus politischen Gründen kaum realistisch. Eine Kontingentierung der gesamten Agrarproduktion ist aus verwaltungstechnischen Gründen nicht machbar und hätte aufgrund mangelnder Konsensfähigkeit auch wenig Erfolg. Aus Gründen optimaler Faktorallokation wäre sie aus volkswirtschaftlicher Sicht neben der Produktpreissenkung nur die nächst beste Lösung.

Auf der Faktoreinsatzseite würde eine Verteuerung z.B. des Stickstoffs die Intensität der Produktion in Abhängigkeit vom Standort mehr oder weniger reduzieren. Ihr Vorteil gegenüber einer Kontingentierung der einzelbetrieblichen Stickstoffmengen läge—ähnlich wie bei einer Agrarpreissenkung—darin, daß keine negativen Allokationseffekte auftreten.

Insgesamt muß der volkswirtschaftliche Nutzen der intensiven landwirtschaftlichen Produktion gegenwärtig eher negativ eingeschätzt werden. Die wachsende Produktionsintensität, die aus einem überhöhten Agrarpreisniveau resultiert und einzelbetrieblich durchaus ökonomische

Vorteile bringt, führt zu steigenden volkswirtschaftlichen Kosten, die auf die Gesamtbevölkerung überwälzt werden. Bei fehlendem Wirtschaftswachstum sinkt jedoch deren Zahlungswilligkeit und Zahlungsfähigkeit zumal für Produktmengen, die niemand haben will.

Eine Änderung der politischen Strategie des Abwartens scheint unvermeidbar.

8

Improvement of Agricultural Production by Pesticides

C. AHRENS and H. H. CRAMER
Agrochemicals Division, Bayer AG, Leverkusen, Federal Republic of Germany

SUMMARY

Crop protection with pesticides has improved, and continues to improve, agricultural production from several points of view. One of the most important aspects is certainly the effort to reduce pre-harvest losses, both in quantity and quality, thus increasing crop yields and nutritional value of agricultural produce. The assessment of post-harvest losses, i.e. during storage or transport, is difficult, but the control of loss factors with pesticides is nonetheless very important, again with regard to quantity and quality of stored foodstuffs. Pesticides can also serve to improve agricultural production systems against constraints of climate or soil. Within the concept of integrated crop production, pesticides are an essential element, and play a decisive role in environmentally adapted and technologically advanced farming systems.

1. INTRODUCTION

Two main factors can be considered responsible for great fluctuations of production in agriculture: inclement weather, and pests. The first factor is beyond human control; Man has been fighting against the second factor since the beginnings of agriculture. It must be admitted that in only about the last 50 years, with modern plant protection technologies, have we been successful in bringing the greatest problems under control. There is no question that, in the areas where plant protection measures are sensibly applied to intensive production systems, extreme variations in production are avoided and the level of yield is increased.

However, the employment of production factors such as high-yielding varieties, mineral fertilisers and plant protection for increasing and protecting yields require a considerable input. Therefore, the question must be repeatedly asked whether there is an actual improvement in agricultural production with respect to economic and environmental consequences.

1.1. Assessment of potential pre-harvest crop losses

The losses in yields caused by plant pests are quantitative and qualitative in nature. An initial global estimate of the quantitative losses for 60 important crops was made in 1967 by Cramer [5] who came to the conclusion that about 35% of the potential yield was lost through plant pests, such as insects, fungi and weeds. Various authors later made estimates for particular crops or areas. A new method was employed in the estimate of losses in which a computer evaluation was made, based on a large number of trial reports [1]. From this evaluation, it was found that, in the cultivation of rice, approx. 24% of the potential harvest was lost through insects, and that 10% of the loss in wheat yield could be attributed to fungal diseases. These data are in accordance with the earlier estimates [5].

Global or even regional estimates of loss are exceptionally problematic and burdened with many possible sources of error. On the other hand, increases in yield are quite easily seen in retrospect over a definite period of time, for example, through the data in Table 1 based on FAO statistics [6a]. These indicate what increases in yield could be achieved, even in the

Table 1
Increase in global cereal production (1970–1981)

Region	*Total production Mt*			*Cultivated area Mt*			*Yield t ha^{-1}*		
	1970	*1981*	*% Increase*	*1970*	*1981*	*% Increase*	*1970*	*1981*	*% Increase*
Developing countries	362	502	38·6	291	314	7·9	1·24	1·60	29
Developed countries	422	605	43·4	147	174	18·3	2·88	3·47	20·5
Centr. planned countries	447	557	24·6	244	251	2·8	1·83	2·22	20·1
World	1 232	1 664	35·1	682	740	8·5	1·81	2·25	24·3

Source: FAO [6a].

Table 2
Production increase in cereals through area expansion and yield intensification (1970–1981)

Increase in global cereal production through	*Mt*	%
Expansion of cultivated area	+131	30
Yield increase	+301	70
Total	+432	100

Source: FAO [6b].

developing countries, through the application of modern agrarian technology, including new varieties, mineral fertilisers and plant protection.

It is especially important, from an economic and ecological point of view, to be able to understand this impressive, world-wide increase in yield for the most important basic food and feed grains that by far the greatest degree of growth was achieved from an increase in intensity and a much smaller degree through an increase in the acreage cultivated (Table 2).

It is equally important to emphasise the fact that the expansion of production in the developing countries in particular was achieved primarily by an intensification of cultivation rather than an expansion of land under cultivation (Table 3).

Almost three-quarters of the increase in grain production realised in the developing countries was achieved by intensive farming, that is, by the increased use of new, high-yielding varieties, mineral fertilisers and plant protection as well as by improved cultivation and partial irrigation. To assign a quantifiable, separate value to these main elements of the 'Green

Table 3
Cereal production increase in developing countries through area expansion and intensification (1970–1981)

Increase in cereal production in developing countries	*Mt*	%
Expansion of cultivated area	+37	26
Yield intensification	+103	74
Total	+140	100

Source: FAO [6b].

Revolution', as some authors have tried, seems superfluous since all components play an essential part in present-day agrarian technology and are mutually interdependent.

If we assume that the world-wide pre-harvest losses caused by animal pests, plant diseases and weeds stand on average at about 30% for all kinds of cultivated plants, a relatively modest prevention of these losses could considerably increase effective yields. Thus, if there were a 30% reduction of these losses in the developing- and centrally planned countries alone, 100 million tons of additional grain would become available [4].

The estimates of FAO [2] that, intermediately, the greatest agricultural production reserves are provided by intensive farming, must be looked at realistically. If we were to succeed in solving the world-wide problem of food production in this way, the alarming loss of natural- and quasi-natural areas such as, for example, the tropical rain forests, could be checked (Table 4).

1.2. Improvement of quality by plant protection

In principle, it is a general responsibility of agriculture to produce food of the highest possible quality. For a number of basic foods, however, for example, grain, subsequent handling and processing have a greater influence on the end-product than has the raw material itself.

Of course there is a danger of pollution with noxious or other disagreeable substances, when inadequate plant protection measures lead to an infestation by moulds, insect pests, etc. This in turn can promote the appearance of, for example, carcinogenic mycotoxins, which can be a risk to health. On the other hand, residue tolerances for some of these toxins have been established [9] as illustrated in Table 5.

Moreover, most mycotoxins are not affected by heat and may not be destroyed by boiling [9]. Finally, it should not be overlooked that the seeds,

Table 4
Decrease of forest and woodland (Mha)

	⌀ *1969–71*	*1981*	*Difference*
World	4 209	4 090	−119
Developed countries	912	879	−33
Developing countries	2 191	2 090	−101
Centrally planned	1 106	1 121	+15

Compared: Total forest area EC of 9 = 32 Mha.

Table 5
Acute toxicity and residue tolerances of some mycotoxins

Mycotoxin	*LD_{50} oral (mg/kg)*	*Residue tolerance (g/kg)*
Aflatoxin B_1	7·2 (rat)	0·000 005
T_2-Toxin	3·8 (rat)	n.d.
Fusarenon X	4·4 (rat)	n.d.
Malformin	4·0 (duck)	n.d.
Ochratoxin A	20 (rat)	n.d.
Patulin	30 (mouse)	0·000 05
Citrinin	50 (rat)	n.d.

Source: Heinze [9].
n.d. = not determined.

of some weeds, which, under certain conditions, are able to pollute grain, can be toxic for warm-blooded organisms.

But there is still another quality aspect that should not be overlooked in this connection since it can be of great importance for the agicultural producer. It is a matter of marketable quality, chiefly of fresh produce like fruit and vegetables, but also of other products such as coffee, tea, cacao, potatoes destined for industrial use, etc. The consumer or the processor is very particular about the quality of the goods he buys, which, as a rule, is evaluated with respect to size, perfection and cleanliness of surface or substances contained (starch, oil, sugar, aroma, etc.). The majority of quality criteria can be realised only through the use of plant protection chemicals such as insecticides and fungicides, but also molluscicides, nematicides and herbicides for the prevention of direct and indirect damage (e.g. reduction of storage life).

An example from the fruit sector should make this connection clear. In long-term testing over a period of 15 years on the experimental farm at Laacherhof, run by Bayer AG, the findings showed very distinctly that the deficit or even the insufficiency of the crop protection measures in apple orchards had a negative influence on the yield generally, and to a quite particular degree on the percentage of marketable fruit [10]. Where no plant protection measures were carried out, the yield fell to 57% of that which would have been realised with a full-season program. The percentage of fruit which, on the basis of quality criteria, could be graded as marketable, sank to only 19% of the possible total yield (Fig. 1).

Crop protection measures have a similarly decisive influence on quality

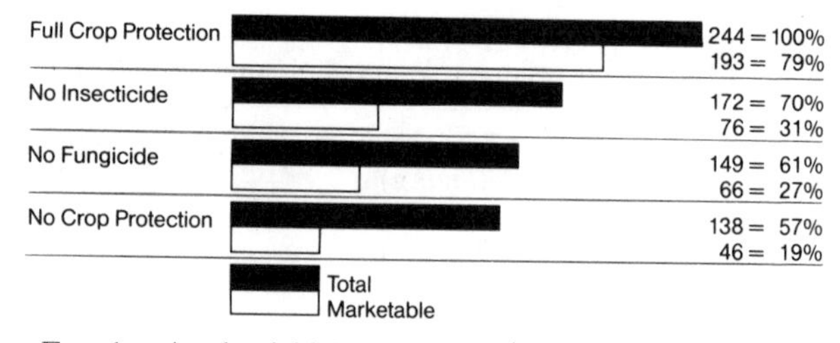

FIG. 1. Apple yield Laacherhof ⌀ 1967 to 1981 (dt/ha).

in vegetable cultivation and the cultures for export of developing countries, which must prevail in an exacting market.

1.3. Post-harvest losses

Subsequent to harvesting, potential foodstuffs suffer further considerable losses during transport, storage and processing. Only scant information exists concerning the magnitude of these losses. Statistics from FAO show that for 'durables', that is, primarily cereals and legumes, the average lower limit of losses lies at about 10 %, the average upper limit at 26 %. These are figures calculated from estimates which, for the lower limit, span the range from 1 to 40 % and for the upper limit, from 4 to 100 %. In other words, exact data are missing. The significance of this problem is illustrated by the following historical example.

In 1917, the British Government stored 3·5 million tons of cereals in Australia which was not shipped out during the U-boat war. In 1919, when they wanted to pull down their reserves, 2·5 million tons had been destroyed by grain weevils [12].

If such losses can gravely affect state budgets, then for the farmer, the economic significance of safe storage is, if possible, even more serious.

For him, in consideration of the seasonal price of grain, it is important to be able to determine his sale date according to the market situation and not from the aspect of possible infestation with pest organisms. In this context, it must be taken into consideration that quantitative losses alone occur insofar as a certain percentage of stocks are eaten by insects or rodents. A considerably greater percentage is made unusable by contamination. In the USA, flour with more than 40 insect particles per lb is not approved for human nutrition; in Germany, from a large material sample examined in 1964, 450 insect particles were found in 500 g of flour [7]. Rats and mice, with their excrement, can make unusable many times the volume of stored goods which they consume.

All these factors are difficult to quantify statistically. This may explain in part the lack of data concerning the actual magnitudes of loss. As a basis, experts in the economics of storage often quote an average loss of about 20 % in the world-wide warehousing of foodstuffs.

The world cereal production in 1982 amounted to 1695 million tons. Before processing or consumption, this grain is always stored, somehow, over a longer or shorter period of time. If 20 % of this yield is lost, the result corresponds to the tremendous quantity of 339 million tons, or precisely the total cereal production in the USA for the year 1982.

It should be pointed out however, that when dealing with percentages for estimates of loss, we must consider the reference magnitudes. Occasionally the statement can be read that, of the total possible foodstuff production, over 30 % is lost up to the time of harvest, an additional 20 % during transport and storage as well as 20 % with processing. This may not simply be added up to 70 %, so that—if we include even the losses in the pantry and the kitchen—finally there is hardly anything left. These percentages are each related to other initial magnitudes as shown in Fig. 2.

Nevertheless, it can, therefore, be seen as realistic that, of the potential harvest, scarcely half is, in fact, utilised, and of the harvest actually realised, about 25 % is lost. Just as in the energy sector, therefore, we have a considerable opportunity for improved utilisation of resources by a reduction of losses.

In this context, the technical possibilities for protecting stores have been considerably improved in the last few decades. It must not be overlooked, however, that an increase in potential dangers has come into being as well. Among these there is the considerably intensified and diversified world trade, with its ever shorter transport times. This enormously increases the danger of spreading warehouse pests and that of the repeated infestation of storage facilities. One of these dangers is also the change in production and storage conditions which has led to the fact that warehouse pests, formerly known only from imported cereals, now have the possibility to establish themselves and multiply even in domestic stores. There are basically 4

Potential Harvest		100%
Until Harvest	30% Loss	70%
Storage	20% Loss	56%
Processing	20% Loss	45%

FIG. 2. Reduction of potential harvest until consumption (schematic).

reasons for this: (1) As the cereals taken in with the combine harvester can have a moisture content of over 20 %, artificial drying is necessary so that it will keep. The dried grain has a temperature of between 25 and 35 °C and cools off only very slowly if it is not ventilated. (2) The storage of greater quantities of grain than in former times also delays the cooling-off process. For example, 50 tons of grain in a container 6 m high requires from 10 to 13 weeks until the initial storage temperature of from 26 to 32 °C drops to 20 °C. In early winter it finally reaches a temperature of 15 °C. At these high initial temperatures, warehouse pests can multiply rapidly and in great numbers and their metabolic heat hinders further cooling. (3) During the storage of grain with a moisture content of over 16 % and through the development of moulds and bacteria, as well as from the exposure of the storage spaces to direct sunlight, spontaneous heating occurs. (4) Finally the presence of a great many broken and damaged grains strongly promotes the development of certain weevils, and thus once again the development of spontaneous heating [13].

This is a very impressive example, demonstrating that a change in procedure can result in a shifting of problems, which makes it necessary to look beyond the individual components and consider the total context.

1.4. Improvement of crop production systems

Pesticides can serve as a means to improve existing crop production systems, and even to create, under particular circumstances, new crop production techniques. Two examples make this clear.

(a) Sugar beet used to be a highly labour-intensive crop because weed competition has to be eliminated over a long period between seed germination and complete leaf-coverage of the soil. Without chemical weed control, of the 130 man-hours per hectare from drilling to harvest, more than 100 were required for weeding. Today, less than 30 h are needed to grow a hectare of sugar beet. The introduction of new techniques such as the use of monogerm seed and precision drilling were only possible with the development of weed control with herbicides (Fig. 3).

In this context, another point is of importance: in a crop such as sugar beet, the number of plants per hectare is of paramount importance, since with the introduction of monogerm seed precision-drilling to the final stand is more and more carried out under practical farming conditions. Under these circumstances, any weeding program must be absolutely selective to the crop, because any plant loss or damage would be detrimental to the final yield.

Trial results over a 10 year period indicate that sugar beet yield in plots

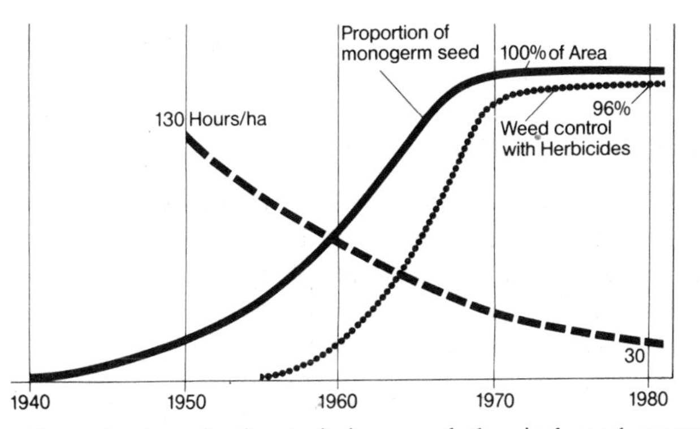

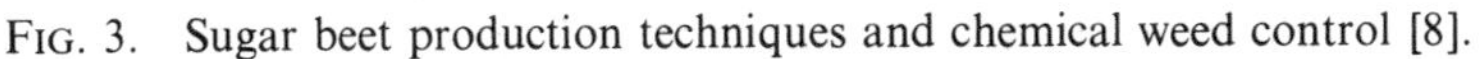

FIG. 3. Sugar beet production techniques and chemical weed control [8].

with herbicide weed control are superior to those with mechanical weed control (Table 6).

These figures show that all yield components of sugar beet are positively influenced by chemical weed control in comparison to mechanical control, so that the final sugar yield in the herbicide plots was on average 27% higher.

(b) In many parts of the world, traditional soil tillage systems are challenged by machinery innovations and new agricultural chemicals.

The steel mouldboard plough introduced in 1830 is still today the farmer's basic tool for the preparation of the seedbed aiming mainly at improving the physical condition and aeration of the soil, managing

Table 6

Comparison of weed control with herbicides and mechanical weed control in sugar beet (1974–1983)

Weed control programs	*Number of plants (rel.)*	*Beet yield (rel.)*	*Sugar concentration (rel.)*	*Sugar yield*	
				dt/ha	*rel.*
Mechanical weed control	100	100	100	72·7	100
Chemical weed control (metamitron):					
1. Presowing incorporated	137	124	102	91·6	126
2. Preemergence	139	131	105	99·8	137
3. Postemergence	127	116	102	85·6	118

Source: Kolbe [11].

surface trash and crop residues, controlling weeds, plant diseases and insect pests, incorporating fertiliser and levelling the field.

But in certain climates and with particular soil types ploughing has some serious disadvantages:

A turned and open soil surface is exposed to wind and water erosion.
The soil water complex can be disturbed and disrupted.
Wrong ploughing can lead to hard plough pans.
Ploughing can preserve weed seeds for later germination.
In general, ploughing consumes a lot of energy.

Reduced soil tillage leads to minimum-tillage and ultimately to no-tillage cropping systems, often also referred to as zero-tillage, direct-drilling or chemical tillage systems. Simply described, no-tillage is drilling or planting crops in previously unprepared soil by opening a narrow trench or slot only sufficiently wide and deep to obtain proper seed coverage. No other soil cultivation is done. Unwanted weeds and grasses are controlled by herbicides, an eventual build-up of soil pests can be kept under control by soil insecticides and nematicides, thus substituting much of a farmer's tractor power with chemical energy.

Such a cropping system can offer significant advantages of better soil moisture retention, reduced rainfall runoff, less water and wind erosion, less soil damage from machinery, better timing in planting or drilling, reduced energy input and lower production costs.

No-tillage, based on the use of herbicides, offers a valuable soil management option for areas where farming meets serious constraints, such as drought- or erosion-prone areas, areas with unstable soils or unfavourable climatic conditions.

2. DISCUSSION AND CONCLUSION

The examples, limited and selective as they may be, have shown that pesticides can improve agricultural production in many different ways. They reduce pre- and post-harvest crop-losses, they improve the quality, nutritional value and durability of harvested crops, and they constitute essential factors or even basic elements of crop production systems [3].

The concept of 'Integrated Crop Production' can be briefly defined as an agrosystem which is optimally suited for the prevailing environmental conditions. The basic farm management decisions, as well as the day-to-day management, should be based on the consideration of environmental

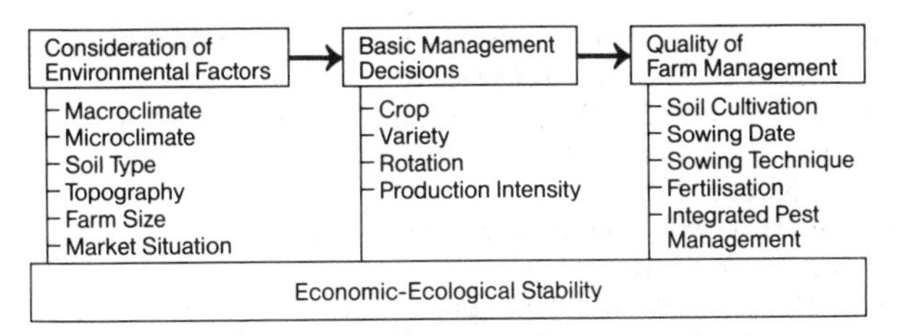

FIG. 4. Elements of integrated crop production.

factors in order to achieve economic success and ecological stability (Fig. 4).

Referring to the two examples, herbicides in sugar beet greatly reduced labour input, thus saving a crop which otherwise would have disappeared in industrialised countries because of rising labour costs. So herbicides saved an important and valuable crop—valuable in terms as a break crop in a crop rotation, and important in terms of income for many farming enterprises—and at the same time contributed to considerable increases in yield and quality.

In the case of no-tillage systems, herbicides, soil insecticides and nematicides are the very foundation on which the system rests. This technique renders farming possible in regions which must be considered as ecologically unstable with regard to climatic conditions, soil quality or topography.

Pesticides are essential elements of integrated crop production systems which aim at increasing yield and quality of crops within the framework of prevailing environmental conditions. These systems are continually improved by the development of new active substances, like the new systemic fungicides of the triazole-group, low dose herbicides such as the sulphonylureas, or the last-generation insecticides such as pyrethroids and growth inhibitors, in short—the more target-oriented pesticides. Additionally, progress in pest forecasting, in application techniques, in the formulation of the active ingredients and in the educational level of farmers is promising. We are, therefore, convinced that the ecological and economic aims in agriculture become more and more compatible, not conflicting.

REFERENCES

1. AHRENS, C., CRAMER, H. H., MOGK, M. and PESCHEL, H. (1983). Economic impact of crop losses. Proc. of 10th Internat. Congr. of Plant Protection, 1983, Brighton, pp. 65–73.

2. BOMMER, D. F. R. (1982). Landwirtschaft zwischen Mangel und Überfluß-Analysen und Perspektiven. *BASF-Mitt. f.d. Landbau*, 5/82.
3. BÜCHEL, K. H. (1982). Political, economic and philosophical aspects of pesticide use for human welfare. Fifth International Congress of Pesticide Chemistry, Kyoto, 1982.
4. BÜCHEL, K. H. (1983). Die Zukunft der Chemie in der Landwirtschaft. *Ber. über Landwirtschaft*, **61**, 382–99.
5. CRAMER, H. H. (1967). Plant protection and world crop production. *Pflanzenschutz-Nachr. Bayer* (*Engl. Ed.*), **20**, 1–523.
6a. FAO (1982). *Production Yearbook 1981*, **35**, Rome.
6b. FAO (1983). *Production Yearbook 1982*, **36**, Rome.
7. FREY, W. (1964). Der gegenwärtige Stand und aktuelle Probleme der Schädlingsbekämpfung im Vorratsschutz. *Gesunde Pflanzen*, **5**, 81–91.
8. HANF, M. (1975). The history of chemical weed control in beet growing. *BASF Mitt. f.d. Landbau*, 3/75.
9. HEINZE, K. (1983). *Leitfaden der Schädlingsbekämpfung*. Bd. IV, Vorrats- und Materialschädlinge (Vorratsschutz) Stuttgart.
10. KOLBE, W. (1982). Effect of different crop protection programmes on yield and quality of apples. II (1967–1981). *Pflanzenschutz-Nachr. Bayer* (*Engl. Ed.*), **35**, 125–33.
11. KOLBE, W. (1984). Zehn Jahre Versuche mit [R] Goltix zur Unkrautbekämpfung in Zucker- und Futterrüben (1974–1984) mit einem Rückblick auf die verschiedenen Bekämpfungsverfahren seit 200 Jahren. *Pflanzenschutz-Nachr. Bayer*, **37** (in print; English Edition in preparation).
12. LING, L. (1961). Ein Fünftel aller Ernten geht verloren. *Atlantis*, **33**, 427–8.
13. WEIDNER, H. (1983). Vorratsschädlinge. In: *Leitfaden der Schädlingsbekämpfung*, **Bd.** IV (Ed. Heinze, K.), Stuttgart.

9

The Present Use and Efficiency of Fertilisers and Their Future Potential in Agricultural Production Systems

G. W. COOKE

Rothamsted Experimental Station, Harpenden, UK

SUMMARY

The use of fertilisers, their efficiencies, and their place in agricultural systems are reviewed. The needs for further research and development are identified.

Fertilisers are essential for building soil fertility to enhance crop productivity. The extra plant nutrients that they supply eliminate nutritional constraints and make it possible to achieve required levels of yield because they interact with inputs used to control other constraints to crop growth. The aim must be 100% efficiency in fertiliser use, the whole being used to promote extra crop growth and none being lost to the nutrients in soil/plant systems based on long-term field experiments and on associated multidisciplinary research. The evaluation of current efficiency, and plans for improvement, also require the construction of the balance between inputs and outputs of nutrients on both local and national scales; the latter should lead to national fertiliser policies. The development of these models and balance sheets will indicate the future research that is essential. This must concentrate on the role of nitrogen and its pathways in the farming systems where large losses of this very mobile nutrient now occur. No losses of phosphorus supplied by fertilisers to the environment need occur if soil erosion is prevented.

Inputs used efficiently increase production and profit; reducing inputs, and particularly lessening fertiliser use, is a sure way to lower yields, lower margins, and to more expensive food. The aims of those concerned with research on fertiliser efficiency in agricultural systems, and of those working on the abatement of environmental problems associated with fertilisers, are identical; the two kinds of research should be closely coordinated.

1. INTRODUCTION: WHY FERTILISERS ARE NEEDED

Constraints that prevent plants achieving the potential, otherwise provided for by their genetic make-up and available solar radiation, limit growth and final yield. The constraints may be overcome by inputs to the agricultural production system; these inputs may be chemical, physical or biological in nature, or they may be improved practices for the management of soil and/or crop. The discovery in the last century of the principles of plant nutrition, and the development of fertilisers which followed, constitutes the greatest breakthrough in man's ability to ensure his supplies of food from plants grown in soil. When applied in correct forms and amounts fertilisers ensure that crops are not limited in their growth by shortages of nutrients. Fertilisers are now an important input to agriculture in most developed countries, for example they are responsible for one-seventh of the total expenditure by farmers in UK [1]. It is clear that our scientific objective must be to ensure that the whole of the amounts of plant nutrients applied serves its purpose of promoting plant growth, and that none escapes directly into the environment. Here I must say that those concerned with environmental research, and those concerned with research on the efficient use of fertilisers, have identical interests which should be closely coordinated. I emphasise that the farmers' and consumers' interests are best served when all losses of plant nutrients from agricultural systems are prevented. This Symposium has an important part to play in attaining the objective of research and development. Inputs to production systems interact to build up yields and investigations on their interactions which are intended to discipline working together on common projects. When this research has been successful, and it has been applied in practice by skilled farmers, the benefits will be of two kinds; all pollution will be avoided, and the amounts of food that we need will be produced at minimum cost per unit of produce.

1.1. The nutrients needed by crops

In assessing the need for fertilisers we require information on the amounts of nutrients which crops must have to achieve a given level of growth. Then we must assess the capability of soil and atmosphere to supply these nutrients; the difference represents the need for fertilisers to supply the extra nutrients.

Table 1
Amounts of nutrients in total crops at harvest and in the produce that is normally sold

	In total crops at harvest			*In grain or roots sold*		
	N	*P*	*K*	*N*	*P*	*K*
	(*kg per tonne of product sold*)					
Barley	20	3·0	18	15	2·7	4·6
Wheat	19	3·4	17	14	2·8	4·0
Potatoes[a]				3·2	0·5	4·5
Sugar beet	6·2	0·7	7·3	2·3	0·3	1·7
Oilseed rape	100	25	80	33	7	9

[a] Potato tops contain nutrients, but they are always returned to soil and analyses of the material are not available.

The composition of crops varies with variety, soil, climate, and cultural practice. In calculations made in this paper I have used the results of long-term Reference Experiments made on clay-loam soil at Rothamsted [2], and on sandy-loam soil at Woburn in Bedfordshire [3]; these sites are broadly representative of soils and climates in south-east England. The average contents of N, P and K (the major nutrients supplied by fertilisers) are given in Table 1 for cereals, potatoes and sugar beet from these experiments; data for oilseed rape are selected from that published by Holmes [4]. In a study of nutrient balances in European agriculture Hébert [5] used the following data expressed as kg of N, P and K per tonne of product: cereals 17N–3·5P–4·2K; potatoes 3·2N–0·7P–5·0K; sugar beet 2·5N–0·5P–2·4K; oilseeds 37N–6P–8K. For cereals, potatoes, and oilseeds these figures are similar to those in Table 1, indicating that my data are applicable to much of western Europe.

Table 2 gives the national average yields of five important crops in UK in 1982 (from [1]) and the total amounts of nutrients in the harvested crops, and in produce sold using the data in Table 1 on crop composition.

Differences between the two sets of data for each crop in Table 2 draw our attention to the important question of the effect of the disposal of crop by-products on fertiliser needs. If cereal straw or sugar beet tops are ploughed in, the nutrients they contain are returned to the system; but if removed for sale or for use elsewhere on the farm the 'export', particularly of N and K, requires increased use of fertilisers to balance the system and prevent the depletion of reserves of nutrients in the soil. Advisory recommendations normally take account of this point.

Table 2
Average yields of crops in UK in 1982 and their nutrient contents

	Yield of product ($t\ ha^{-1}$)	*In total crops harvested*			*In product sold*		
		N	*P*	*K* ($kg\ ha^{-1}$)	*N*	*P*	*K*
Wheat (grain)	6·2	118	21	105	87	17	25
Barley (grain)	4·9	98	15	88	74	13	23
Potatoes (tubers)	35·8				115	18	161
Sugar beet (roots)	49·8	309	35	364	115	15	85
Oilseed rape (grain)	3·3	330	82	264	109	23	30

The yields in Table 2 are recent national averages, but if we look further ahead we must consider the improvements in crop production that will occur as we move nearer to the proved potential of our crops. For example 14 t ha^{-1} of wheat grain and 90 t ha^{-1} of potatoes have been proved to be possible because they have been grown by farmers in Britain. These yields would remove the following amounts of nutrients from the field (in kg ha^{-1}):

	N	*P*	*K*
Wheat (grain plus straw)	266	48	238
Potatoes (tubers)	288	45	405

Another important point is that the amounts of nutrients in harvested produce are less than amounts needed and contained in the crop at the peak of active growth. Some of the nutrients contained in the whole plant will be returned to the soil, perhaps by translocation, certainly in fallen leaves and by leaching of the mature plant by rain, and some N may be lost by volatilisation from leaves by the process described by Wetselaar and Farquhar [6]. These losses are particularly important in assessing the need of cereals (and crops such as oilseed rape) for potassium. When large yields are achieved, for example 10 t ha^{-1} of wheat grain, the crop will have taken up around 200–250 kg ha^{-1} of K by flowering time. In temperate climates half of this amount is often lost before harvest by the processes mentioned above. Nevertheless we must recognise that the large quantity must be available in the soil for uptake during the growing season up to the time that the grain is formed. An example of the changes in K contents of wheat as the crop matures was provided by Prew *et al.* [7]; the crop received a total of 250 kg ha^{-1} of N and yielded about 10 t ha^{-1} of grain; amounts of N and K in the whole crop on three occasions were (in kg ha^{-1}):

	2 July (anthesis)	*6 August*	*28 August (harvest)*
N	221	222	217
K	276	236	142

1.2. The capacity of soils to supply plant nutrients

Soils vary greatly in the amounts of nutrients that they can provide. In 'natural' unfarmed situations these amounts depend on the parent materials from which the soil was formed and on the weathering and soil-forming processes as affected by climate and natural vegetation. When land is farmed the nutrient reserves are affected by farming system and by inputs of manures and fertilisers. The ability of average farmed soils in England to supply nutrients may be gauged from the results of the long-term Reference Experiments at the Rothamsted and Woburn Experimental Stations. The Rothamsted experiments on clay-loam soil were begun in 1956 on land which had long been under-grazed grassland farmed at low intensity and receiving no fertiliser; results for the 1971–75 period were given by Widdowson *et al.* [2]. The Woburn experiment began in 1960 on old arable land which had never been heavily fertilised; results for the 1975–79 period were given by Widdowson *et al.* [3]. In both experiments a five-year arable rotation was used; spring barley, potatoes and a clover-grass ley were grown on both sites; the other crops were winter wheat and kale at Rothamsted, and sugar beet and winter oats at Woburn. At Rothamsted a strip of the original permanent pasture was retained and received the experimental treatments; at Woburn a strip of long-term grass-clover ley was included throughout the period of the experiments. The fertiliser treatments used are outlined at the bottom of Table 3.

The amounts of N, P and K supplied by soil alone to crops receiving adequate amounts of the other two nutrients for the last 5-year periods of the experiments, which had then continued for 20 years, were:

	N[a]	*P*	*K*
	mean amounts supplied annually (kg ha^{-1})		
Rothamsted (1971–75)	65	9	34
Woburn (1975–79)	39	10	29

[a] The data for N exclude amounts harvested in clover-grass leys.

Table 3

The effects of fertilisers and farmyard manure on yields of crops in the Reference Experiments and the responses to nutrients tested in these Experiments

	Yields of agricultural produce (t ha^{-1})			*Yields of dry matter (t ha^{-1})*				
	No fertil.	*With N_2 PK*	*With FYM*	*Without fertil.*	*Response to*			
					N_1	N_2–N_1	*P*	*K*
Rothamsted 1971–75								
Barley grain	2·8	5·9	4·8	2·4	1·1	0·9	0·9	1·7
straw				2·1	1·4	1·1	0·7	1·3
Wheat grain	3·5	7·8	5·7	3·0	1·7	0·9	0·6	4·2
straw				3·8	2·7	1·4	1·2	4·2
Potato tubers	8·8	47·6	44·2	2·1	2·5	0·8	2·0	8·9
Kale, fresh crop	13·3	54·8	34·4	3·0	3·9	1·7	7·2	2·1
Grass-clover ley (in the crop rotation)				4·3	0·5	−0·3	3·0	4·5
Permanent grass				2·5	2·2	3·0	−0·2	0·9
Woburn 1975–79								
Barley grain	1·6	5·0	2·6	1·3	2·0	0·7	0·6	1·8
straw				1·2	2·2	1·2	0·4	1·3
Oats grain	1·7	4·5	2·6	1·5	1·7	0·6	0·4	0·1
straw				1·7	2·7	1·2	1·0	1·3
Potato tubers	6·6	28·1	25·2	1·7	2·1	1·0	1·3	4·2
Sugar beet	10·4	30·6	26·4	2·4	3·5	0·8	0·3	3·5
tops	8·0	29·0	17·8	1·7	2·1	1·1	0·2	1·0
sugar	1·6	5·0	4·5					
Grass-clover ley (in the crop rotation)				2·7	1·6	0·2	1·3	3·4
Long ley (established in 1960)				3·6	2·1	1·1	0·3	1·0

Notes:

Yields of grain are at 85% dry matter.

Fertilisers applied: 27·4 kg ha^{-1} of P and 208·5 kg ha^{-1} of K to all crops in both experiments.

Nitrogen: N_1 At Rothamsted 57 kg ha^{-1} of N to barley, 75 kg ha^{-1} to wheat and potatoes, 126 kg ha^{-1} to kale, 19 kg ha^{-1} to rotation ley, and 126 kg ha^{-1} to permanent grass. At Woburn 63 kg ha^{-1} of N to barley and oats, 126 kg ha^{-1} to potatoes and sugar beet, 31 kg ha^{-1} to the rotation ley and 188 kg ha^{-1} to the long ley.

N_2 At both centres was twice N_1.

Farmyard manure (FYM) applied at 50 t ha^{-1} for potatoes and kale only, this supplied 380 kg N, 64 kg P, and 504 kg K at Rothamsted; at Woburn it was applied at 50 t ha^{-1} for potatoes and sugar beet only and this dressing supplied 359 kg of N, 58 kg of P and 510 kg of K.

Experimental methods full details are given in references ([2] for Rothamsted and [3] for Woburn); the Standard 8-plot factorial combination of none versus N_1, P, and K, was tested on all crops. N_2 was tested with P and K. FYM was tested alone and with fertilisers.

The amounts of each nutrient extracted from soil varied from crop to crop, and the amounts of N secured by the potato crop immediately following the clover-grass ley were not materially greater than those secured by later crops in the rotation. The mean annual amounts of N, P and K extracted from Rothamsted clay soil declined only slightly during the 20 years. The corresponding diminution in supplies from the sandy soil at Woburn was more marked for all three nutrients, and particularly for K. In the first two years (1960, 1961) Woburn soil supplied 120 kg ha^{-1} of K, by 1964 the release had fallen to 62 kg ha^{-1} of K, and finally to only 28 kg ha^{-1} of K by 1979.

1.3. The effects of fertilisers and manures on crop yields

When the amounts of nutrients supplied by soil as indicated above, are compared with the amounts needed by crops to achieve even the average yields as indicated in Table 2, the need for extra nutrients to be supplied by fertilisers or other sources is immediately clear. The benefits of fertiliser inputs to arable cropping systems have been reviewed very recently by Lidgate [8] and the value of fertilisers for grassland was discussed by Ryden [9]. The Reference Experiments done by Rothamsted provide good examples of the returns in increased crop yields obtained by supplying the needed nutrients by fertilisers or by farmyard manure (FYM). The results published for the 5-year period which concluded the first 20 years of these experiments are summarised in Table 3. The responses to N which are shown are similar to those which may be expected under average conditions in southeast England; responses to P and K are typical of those experienced where good reserves of these nutrients have not accumulated in the soils as a result of long periods of manuring or fertilising. The increases in yields clearly justify the use of fertilisers to produce more food for man or animals; the question of economic benefits is discussed later.

1.4. Nutrient balances in fertiliser experiments

Having established that the use of fertilisers is justified plans must be made for their rational use based on scientific evidence. The ideal is that the extra nutrients supplied should exactly meet the needs of the crops grown to achieve the required yields, and the nutrients should be completely taken up by the crops leaving no residue in the soil. The Reference Experiments provide an opportunity for determining whether this ideal is achieved in practice. Table 4 gives the amounts of N, P and K applied for, and removed by the five crops grown in the rotation during the last 5 years for which results are available, and for the whole 20-year period of the experiments.

Table 4

The amounts of nutrients supplied for, and removed by five crops grown in rotation in the Reference Experiments, for the last five years and the whole twenty-year period

(Amounts of nutrients in kg/ha in each 5-year period, added and removed)

	N_1PK-expts			*N_2PK-expts*			*Farmyard manure-expts*		
	Added	*Removed*	*Difference*	*Added*	*Removed*	*Difference*	*Added*	*Removed*	*Difference*
Rothamsted, period 1971–75									
N	352	625	−273	703	745	−42	760	564	+196
P	137	98	+39	137	107	+30	128	97	+31
K	1 042	783	+259	1 042	867	+175	1 008	748	+260
Woburn, period 1975–79									
N	408	463	−55	816	605	+211	718	392	+326
P	137	68	+69	137	78	+59	116	64	+52
K	1 042	620	+422	1 042	734	+308	1 020	560	+460
Rothamsted, whole period 1956–75									
N	352	689	−337	703	802	−99	763	652	+111
P	137	96	+41	137	102	+35	151	93	+58
K	860	687	+173	860	750	+110	1 013	717	+296
Woburn, whole period 1960–79									
N	369	567	−198	738	700	+38	699	508	+191
P	137	81	+56	137	88	+49	174	74	+96
K	964	681	+283	964	157	+207	960	634	+326

The single rate of N-fertiliser was not equal to the amounts removed by the arable crops at both sites. The double dressing of N was less than the crops removed at Rothamsted, but it left a surplus at Woburn. The amounts of P and K supplied were greater than the amounts removed by crops at both sites. The amounts of N, P and K supplied by farmyard manure (FYM) were similar to those supplied by the double dressing of N plus P and K fertilisers. There was a considerable surplus of N supplied by FYM; this suggests that the N supplied by FYM is much over that removed in crops; this suggests that the N supplied by FYM is much less efficient than fertiliser N. On the whole the surplus of P supplied by FYM tended to be a little more than the surplus from the fertiliser. The surplus of K from FYM was notably greater than the surplus from fertiliser. Table 3 has shown that the yields of agricultural produce given by FYM are less than the yields from fertiliser dressings that supply similar amounts of nutrients;

therefore we must conclude that FYM contains nutrients in forms that are less efficient than those supplied by fertilisers.

1.5. Nutrients other than N, P and K

This paper is mainly concerned with the nutrients N, P and K as these are the main components of fertilisers; in addition most discussions on environmental problems are concerned with the roles of N and P. In concentrating on NPK we should not ignore the other three major nutrients (S, Ca and Mg) or the six micronutrients (B, Cu, Fe, Mn, Mo and Zn). There are interactions between all nutrients and the full benefit cannot be obtained from any one major nutrient or micronutrient unless all other nutrients are available in adequate amounts. In all research which is planned to improve the efficiency of any one nutrient, the supplies of other nutrients in relation to crop needs should be carefully assessed. Such information will usually be available in areas of well-developed agriculture served by efficient advisory services; but problems may be expected with novel crops, with land newly reclaimed for agriculture, or when soil or climatic conditions change.

Sulphur is an example of a nutrient which must receive more attention in future. Supplies of this nutrient were adequate when single superphosphate and ammonium sulphate supplied much of the P and N in fertilisers; these materials are now largely replaced by more concentrated sources of N and P which contain no S. Over most of western Europe sufficient S for our crops is supplied by the atmosphere in rain or by dry deposition. If the present campaign to investigate 'acid rain' results in regulations to lessen the emissions of sulphur dioxide to the atmosphere then sulphur deficiency will be increasingly recognised as a constraint to crop production in Europe. Serious deficiencies now occur in parts of Ireland and in northeast Scotland. When sulphur is required we will have to decide on the best way of adding it to fertilisers to secure maximum efficiency in the use of the nutrient.

Oilseed rape is an example of a crop which is occupying an increasing area of cropland in Europe and on which we require more information to ensure its efficient nutrition. Rape takes up large amounts of N, P and K, but only small proportions are removed in the seed. Good crops also require up to 100 kg ha^{-1} of sulphur; of this total a 3 t ha^{-1} crop of seed removes about 30 kg of S from the field. The crop is also sensitive to acidity in soil though no examples appear to have been reported where calcium itself was a nutrient that limited growth. A seed yield of 3 t ha^{-1} removes about 10–12 kg ha^{-1} of Mg and Ca. Holmes [4] reports that rape is sensitive to deficiencies of boron and manganese but that deficiencies of other micronutrients are rarely reported in Europe though they occur in other regions.

2. THE EFFICIENCIES OF NUTRIENTS SUPPLIED BY FERTILISERS AND MANURES

The data in Table 4 indicate that there is a serious need to examine the overall efficiencies of the nutrients we supply by means of fertilisers and organic manures.

2.1. Recoveries of nutrients from fertilisers

The apparent percentage recoveries by crops grown in the Reference Experiments at Rothamsted and Woburn from the fertilisers that were applied are shown in Table 5.

Nitrogen. The arable crops recovered no more than about half of the N applied; much of the remainder is likely to have been lost by leaching or denitrification. The data reported are for the lower rate–N_1; the percentages of the N_2 dressings recovered by the arable crops were similar. By contrast the permanent grass at Rothamsted recovered only a third of the N_1 dressing, but half of the N_2 application.

Table 5

Apparent recoveries of nutrients supplied by fertilisers in the Rothamsted and Woburn Reference Experiments

	Apparent percentage recoveries from		
	N_1	*P*	*K*
Rothamsted, 1971–75 period			
Wheat (grain + straw)	51	30	40
Barley (grain + straw)	47	27	36
Potatoes (tubers)	53	37	84
Kale (whole crop)	40	49	46
Ley (clover-ryegrass)	—	44	89
Permanent grass	34	18	46
Woburn, 1975–79 period			
Barley (grain + straw)	60	12	22
Oats (grain + straw)	62	7	33
Potatoes (tubers)	32	8	48
Sugar beet (roots + tops)	54	14	66
Ley (clover-ryegrass)	—	16	60
Long ley (grass + clovers)	—	11	43

Phosphorus. Recoveries of P from the clay soil at Rothamsted were much greater than from the sandy soil at Woburn. The cereals at Rothamsted gave relatively smaller responses to fertiliser P than the other crops did (Table 3), but they recovered more than a quarter of the P applied. Kale gave a large response to P and recovered half of the dressing. The responses to P by barley, potatoes and ley were smaller at Woburn than at Rothamsted.

Potassium. At Rothamsted wheat gave a large response to K and at harvest contained 40 % of that applied (earlier in the season it would have contained a larger percentage, as was indicated in Section 1.1). Potatoes and ley gave large responses to K and recovered most of the dressing. Kale gave a proportionally smaller response to K than the other crops did, but it recovered 46 % of the dressing; this indicates 'luxury' uptake which might have repercussions in animal feeding. At Woburn all the crops gave good

Table 6
Apparent recoveries of nutrients supplied by farmyard manure in the Reference Experiments

	Percentage recovery of applied		
	N	*P*	*K*
Rothamsted, period 1971–75			
FYM applied directly for			
potatoes	21	28	43
kale	10	17	20
permanent grass	18	20	29
FYM applied for roots 1 year previously			
wheat (total crop)	7	14	11
barley (total crop)	7	15	10
FYM for potatoes 2 years previously			
grass-clover ley	—	20	33
Woburn, period 1975–79			
FYM applied directly for			
potatoes	16	14	23
sugar beet (roots + tops)	16	17	24
long ley	—	11	16
FYM applied for roots 1 year previously			
barley (total crop)	4	6	4
oats (total crop)	4	7	10
FYM applied for potatoes 2 years previously			
grass-clover ley	—	16	23

responses to K-fertiliser and sugar beet and the rotation ley recovered about two-thirds of that applied. Recoveries of N, P and K by the permanent grass at Rothamsted, and of P and K by the long ley at Woburn, were smaller than the best recoveries by the arable crops, and much smaller than the recoveries of P and K by the one-year rotation ley.

2.2. Recoveries of nutrients supplied by farmyard manure

The apparent recoveries of the N, P and K supplied by FYM at the two sites are shown in Table 6.

Comparisons of the data in Tables 5 and 6 show that organic manuring does not increase the efficiency of nutrients. The crops receiving FYM directly recovered less of the N and K applied in FYM than they did from fertilisers. The organic manure was applied in autumn and much of the mobile N that it contained was no doubt lost by leaching in winter. These differences suggest that systems based on organic manuring may cause more pollution of water with nitrate than is caused by fertilisers used in the same cropping systems.

3. LOSSES OF NUTRIENTS FROM AGRICULTURAL SYSTEMS

Losses of plant nutrients from farming systems are the cause of serious loss to farmers and one cause of environmental pollution, therefore they are reviewed here. The emphasis is on N and P as these nutrients are involved in environmental problems.

3.1. Nitrogen

This nutrient is lost from all systems by leaching of nitrate, denitrification, volatilisation of ammonia from ammonium compounds and urea applied to soil, and by loss of ammonia from the leaves of plants. In humid climates the processes of loss from soil make it impossible to accumulate long-lasting reserves of inorganic nitrogen supplied by fertilisers. Therefore it is essential that the fertiliser be applied in amounts and forms, and at times and in places, that facilitate immediate uptake by plants. This is the objective of much research work and good progress is being made. Excellent work at the Letcombe Laboratory relates to conditions in Britain, and is relevant to much of the agriculture of north-western Europe. In our conditions nearly all of the loss is from nitrate which is leached from soil by drainage water, or is denitrified microbiologically to release nitrogen and its oxides as gases. The Letcombe work shows that under favourable conditions half of the applied N is recovered in the harvested crop, a quarter enters soil organic matter (by microbial intervention) but may be released for use by later crops, and up to a quarter may be lost by leaching and denitrification processes [10]. Parallel work by Powlson *et al.* [11] at Rothamsted shows that the recovery of applied nitrogen depends on the size of the crop grown; the large yields of the modern varieties now grown are much more efficient and recover twice as much of the applied nitrogen as did the lower yields of 30 years ago. The recovery of applied nitrogen has been shown to depend on the time of the application in relation to rainfall. With careful timing 80–90% of the applied N can be accounted for in crop and soil. My view is that with the aid of the models now being developed on nitrate relationships in soils, good scientific information on the soils, and better weather forecasts, losses of nitrogen from ammonium and nitrate salts used as fertilisers will be minimised and these fertilisers will become highly efficient in British agriculture.

One other pathway for loss of nitrogen does require further work; this is the loss of ammonia as a gas to the atmosphere that may occur when urea hydrolyses on or near the soil surface or in irrigation water. This is a particularly serious matter because urea now supplies 30% of the world's fertiliser-N, and it is the main form used in some developing countries—for example 62% of the N applied in Asia is in the form of urea. In arable cultivation urea can be as efficient as other N-fertilisers if it is placed well below the soil surface so that ammonia released by hydrolysis does not escape to the air. But much is applied as top-dressings to the soil surface for established crops and serious losses of ammonia may then occur.

Other work planned to make N-fertilisers more efficient has involved the use of slow-acting materials. These have ranged from early tests of urea–formaldehyde resins, to materials like isobutylidene-diurea (IBDU) and to oxamide (which is probably the most promising as the solubility can be controlled by varying the granule size). At present all these materials appear to be too costly for use on ordinary farm crops. Sulphur-coated urea has been widely tested; sometimes this material has been more efficient than untreated urea. Although the idea of a fertiliser that releases its nitrogen at a rate which matches uptake by the crop is attractive in theory, the prospects in practice are not promising. Probably the greatest weakness is that the rate of action of such fertilisers is not under the farmer's control because it depends on the weather and the biological conditions in the soil; there is always a risk that release may be completed after the crop is mature at the end of the growing season and then nitrate will be lost by leaching in winter.

Another approach has been to inhibit, or to delay, the processes that may lead to loss of nitrogen. Nitrification inhibitors have been developed and widely tested; their potential has been reviewed by Slangen and Kerkhoff [12] two well-known examples are N-Serve (2-chloro-6-trichloromethyl-pyridine) and DCD (dicyandiamide). Although they are effective in delaying the conversion of ammonium-N to nitrate-N these inhibitors are not widely applied.

Another important initiative has been the development of substances which inhibit the hydrolysis of urea by the enzyme urease. It is hoped that by using inhibitors urea broadcast on the soil surface will remain unchanged until it can be washed well below the soil surface by rainfall or by irrigation, thereby avoiding the loss of ammonia which occurs when urea hydrolyses on or near the surface. Some successes have been achieved with metal salts and with organic substances (phenyl phosphorodiamidate (PPD) is an example) as inhibitors. This is a promising approach to a problem that will inevitably become more important since the proportion of the total N-fertiliser supplied by urea is likely to increase.

A much closer approach to full efficiency of the N applied is a complete answer to pressures against the use of fertilisers. It must however be recognised that losses from arable farming systems cannot be completely prevented, even when correct timing and placement have resulted in 100% utilisation of the fertiliser-N, most being taken up by the crop and a small proportion retained by combination with the soil. Nitrate is released by mineralisation of soil organic matter, of crop residues, and of organic manures residues, in autumn after crops have ceased to grow and have been harvested. This nitrate, which is inevitably produced, is liable to be lost by leaching or denitrification during autumn, winter and early spring. While such losses cannot be entirely prevented, they are minimised by early sowing of cereals in autumn, or by growing catch crops for fodder, to take up the nitrate as it is released.

Grassland systems involve the same considerations regarding the efficiency of applied N as were discussed above for arable crops. As a long duration grass sward can grow for much of the year loss of nitrate by leaching is a much less serious problem than it is for arable land. Much experimental work on grassland that is fertilised and cut regularly has shown that much less nitrate is leached than is lost from arable land. The situation with grazed grassland is very different; although the direct losses of fertiliser-N are very small, losses of N from the system as a whole are large, and with present management they are unavoidable. Ryden *et al.* [13] found that the amount of nitrate leached below a grass sward grazed by cattle was 5·6 times greater than that leached from a comparable cut sward and exceeded the losses normally measured from arable land. The average amounts leached annually were 29 kg ha^{-1} of N from the cut sward, but 162 kg ha^{-1} of N from the grazed sward; both swards received 420 kg ha^{-1} of N annually. The concentrations of nitrate-N in drainage water were 5·4–13 mg N $litre^{-1}$ from the cut sward; concentrations in drainage from grazed swards, and from areas to which slurry or FYM had been applied, greatly exceeded the upper limit (22·6 mg N $litre^{-1}$) set by WHO as 'acceptable' for water for drinking. The reason for these large losses is that up to 90% of the N in herbage eaten by cattle is returned to the soil in excreta which falls in concentrated patches which cover only a small fraction of the surface area of the field. Much of this N is quickly lost by volatilisation of ammonia, but much is nitrified and the nitrate produced greatly exceeds the capacity for uptake by the plants growing near the patch and so it is liable to be leached. The actual size of these losses is related to the intensity of animal production which is reflected in stocking rate which is, of course, increased by the use of nitrogen fertiliser. However, it must be recognised that large losses of nitrate may also be expected where cattle graze well-managed grass-legume associations.

3.2. Phosphorus

Inorganic phosphate added to soil either in fertiliser or in organic residues is retained by most soils in 'fixed' forms and very little P is lost by leaching. Measurements made in England and in other countries indicate that between 0·05 and 0·25 kg ha^{-1} of P is lost annually by leaching from agricultural soils; the largest losses occur from sandy soils. Phosphate becomes more soluble in soils which are rich in organic matter and in those which are water-logged; large amounts may be lost from land that is used for flooded rice.

Losses of P may be greater from animal farming systems where surface run-off carries animal excreta, or applied wastes, down crevices in the soil to the drains, or directly into streams by flow over the surface of the soil.

The only other way in which P originating in fertilisers may reach natural waters is through the erosion of soil by air or water. Estimates of 6–12 kg ha^{-1} of P being lost annually by erosion have been made for USA as a whole. The relevance of such figures for European conditions is not known. While every effort should be made to prevent erosion by suitable soil management, some losses from steep slopes are unavoidable and these may remove much P; thus if 1 mm of soil is lost from a hectare of arable land this soil may remove 10 kg of P (together with 10–20 kg of N and 100–200 kg of K).

3.3. Potassium

K is much more liable to be lost by leaching than is P, particularly from sandy soils and from those devoid of the 2:1 clay minerals which 'fix' K in insoluble forms. Measurements on normal agricultural soils in England indicate that only about 2 kg ha^{-1} of K is lost annually by leaching, but estimates of losses in some other countries are much larger. Where sizeable losses of K appear likely they should be minimised by careful attention to the timing of applications of fertiliser-K in relation to crop growth and rainfall patterns.

3.4. Environmental implications

Our research target must be to develop systems of cropping and fertiliser use which permit no direct escape of the N and P supplied into the environment. It does not appear that any losses of K create any environmental problems. When our target is achieved we will be able to say that fertilisers have no direct effect on the environment. But because fertilisers increase soil fertility they make it possible to grow larger crops and these crops contain more nutrients. Therefore the nutrient cycle involving soil reserves, crop roots and residues, and return of organic manures and of animal excreta, is enlarged. Consequently the inevitable losses of nitrate by leaching in winter, or of denitrification in wet soils, and losses from the animal excreta cycle, must all be greater where fertilisers have been used to increase the nutrients in the soil/crop cycle and to increase stock carrying capacity. No P should escape from arable land if soil erosion is prevented. Careful management should aim to avoid all movement of P from animal wastes into surface waters, either by percolation through soil or surface run-off.

4. THE EVALUATION OF THE RESIDUES OF FERTILISERS IN SOILS

The effect of modern farming systems involving the use of fertilisers at recommended rates is to build up the reserves of nutrients in the soils. These reserves have a very important place in crop nutrition. It is essential to evaluate the fertiliser residues in field experiments so that the over-all long-term efficiency (that is immediate effect plus residual effects) of a fertiliser dressing can be calculated.

4.1. Nitrogen

The processes of loss which have been described above prevent an accumulation of mobile inorganic nitrogen in the forms supplied by fertilisers. Some ammonium-N may combine with clay minerals, but this fraction is not mobile and, if released, will be slow in action; it can only be valued by using fertilisers labelled with ^{15}N. The large reserves of N which build up in soil are in combination with soil organic matter. They result from crop waste and organic manures returned to the soil and from root residues which remain in the soil. These reserves are slowly mineralised to release nitrate-N at rates which depend on weather and microbial activity in the soil. This nitrate can be used by crops growing at the time and does make a considerable contribution to nitrogen nutrition, but it is liable to be lost by leaching or denitrification if released at times or in zones of the soil when and where plant roots are not active. No chemical technique has been developed which provides an entirely satisfactory measurement of the value of the nitrogenous status of soils which can be used as a guide to the total levels of nitrogen fertilising of the land. A good example is the system used in England and Wales by the Agricultural Development and Advisory Service [14].

4.2. Phosphorus

Engelstad and Terman [15] stated that soil-test laboratories in USA find that a half or more of all the soils tested are 'high' in soluble P. Ansiaux [16] reported on an enquiry in Europe in 1976 which confirmed the results of others mentioned below in this section—that a high level of soluble P enables larger yields to be obtained than can be got from poorer soils, however much fresh phosphate fertiliser is applied. This is because in the rich soil all the mature root system is in contact with soil solution rich in phosphate; this state cannot be achieved with a fresh fertiliser dressing because this cannot be mixed with all the soil and because the water-soluble phosphate added quickly reacts with the soil and the compounds formed have only a slight solubility. Ansiaux [16] reported the percentages of soils testing 'high' in soluble P in the countries surveyed; the percentages for the EEC countries surveyed were:

Belgium	68	Germany	38	Italy	18
Denmark	45	Greece	35	Netherlands	51
France	20	Ireland	4	UK	17

This building up of the levels of soluble P in soils is an inevitable result of the use of phosphate fertilisers. The ultimate process of the diffusion of phosphate ions to the root surface is so slow that only a limited proportion of the P applied will ever be taken up by the first crop for which it is applied.

The remainder of the fertiliser-P remains combined with soil and, as it has a slight solubility in water, it benefits later crops. A striking example of the long-term benefits of phosphate residues was described by Johnston and Poulton [17]; residues of P-fertilisers applied at Rothamsted between 1852 and 1901 (with no P applied since 1901) were still being recovered in the 1970s so that barley grown on plots with these old residues of P-fertilisers produces larger yields than can be obtained from poorer soils, however much fresh phosphate is applied to the poorer soil [18]. Vetter [19] and Gachon [20] have also provided evidence of the benefits resulting from maintaining the soil phosphate status at good levels. It must be emphasised that by building up the levels of available phosphate in soils we raise the fertility of the soil so that it has a greater potential for crop production (the same is true for building up of potassium reserves).

4.3. Potassium

The residual effect of a moderate dressing of K-fertiliser is usually too small to be measured accurately but when many dressings have been applied the residues accumulate, being held by the clay minerals present in most European soils so that they are not lost by leaching. These residues are slowly recovered by later crops and the long term experiments made by Rothamsted indicate that all the K applied by fertilisers or organic manures may eventually be recovered by crops. Johnston and Poulton [17] also reported on the investigation of these reserves of K. Residues accumulated from dressings of fertilisers and farmyard manure applied between 1852 and 1901 were still effective 70 years after the last application and in the 1970s they increased barley yields by $0{\cdot}5\ \mathrm{t\ ha^{-1}}$. As mentioned above, soils containing residues of K-fertilisers have often produced larger yields, irrespective of the amount of fresh K-fertiliser that is applied. These residues in the soil raise the yield potential of the site [18]. These results with residues of P and K in soils justify policies of using fertilisers and manures to build up reserves of P and K when poor soils are being improved; this is in contrast to the traditional policy of working only for immediate effects.

5. MANAGING THE LEVELS OF PHOSPHORUS AND POTASSIUM IN SOILS

Having accepted that residues of P and K fertilisers in soils will be used by later crops, and that they enhance the productivity of the soils which contain them, we require good scientific methods of managing these residues. When the levels in soils are assessed plans can be made to apply sufficient amounts of fresh nutrients as fertilisers or in manures, so that the required satisfactory status is achieved and maintained while avoiding excessive applications. Obviously soil analysis will be an important tool for providing this guidance, but it will need to be supplemented by calculations of the balance between inputs of P and K in fertilisers, manures, and crop wastes, and outputs in crops which are removed from each field. In this way efficient use of P and K will be achieved. (Nutrient balances are discussed in Section 9.)

5.1. Soil analysis

Long-term field experiments must be established to provide data on the relationship between crop responses to P and K fertilisers and the levels of soluble P and K in the soils. Using 'low', 'medium' and 'high' levels of analysis ascertained in this way, standardised recommendations will be developed which will be modified to adapt them to soil type, cropping system, and climate. Adjustments will also be required to take account of the yield levels expected; amounts of soluble P and K in the soils that are sufficient to produce average yields may not be sufficient to achieve yields that are near to the maximum for the crop. This fact is illustrated by work reported by Mattingly [21]: while 20 ppm of $NaHCO_3$-soluble P was sufficient for the maximum yield of sugar beet that was grown, wheat and barley yields increased steadily as soluble P increased from 20 to 50 ppm. Such research is particularly important when systems are being intensified to provide the conditions which permit crops to approach their maximum potential yield. There is a need for work on this topic to be done with potassium, which tends to have been neglected in the past in spite of the large amounts of K which are taken up by large yields. There is also a need for an inter-disciplinary approach involving the use of agronomic and plant physiological data on crop growth, rooting patterns and nutrient requirements. The purpose of such work will be to develop a model or models of the soil/plant system as it affects supply of P and K from the soil and uptake by the crop. Future systems for giving advisory recommendations to farmers will be based on these models which will ensure high efficiency in the use of P and K inputs.

6. THE AMOUNTS OF FERTILISERS USED IN EEC COUNTRIES

The amounts of fertilisers used in all countries are published annually by FAO [22], and here used to provide the data given in Table 7 which shows the amounts used in EEC countries.

Table 7
Total quantities of fertilisers used in EEC countries in 1982 and the mean rates of application to agricultural land in 1981

	Total use in 1982			*Applied to agricultural area in 1981*		
	N	*P* (*kt*)	*K*	*N*	*P* ($kg\ ha^{-1}$)	*K*
The world	60 443	134	19 862	13	3	4
Western Europe	9 760	2 315	4 350	59	14	26
Belgium/Luxembourg	197	41	116	125	26	74
Denmark	376	46	113	130	16	39
France	2 193	732	1 411	70	23	45
Germany	1 323	329	876	108	27	71
Greece	335	68	28	36	7	3
Ireland	275	62	146	47	11	25
Italy	981	303	292	56	17	17
Netherlands	477	35	88	237	17	44
UK	1 386	194	390	76	11	21

Comparison between the rates applied in different countries is difficult because of differences in land use and cropping, and particularly the relative proportions of grassland and arable land, and the way that grassland is fertilised and used. The highest rate of N is used in the Netherlands where much is used on grassland; more than 100 kg ha^{-1} of N is applied in Belgium/Luxembourg, Denmark and Germany. The combined use of P and K fertilisers exceeds 60 kg ha^{-1} in Belgium/Luxembourg, France, Germany and the Netherlands. Climate has a large effect on crop production potential and therefore on the use of fertilisers. The smallest rate of use of N + P + K is in Greece where the Mediterranean climate limits the yields of non-irrigated crops; a similar limitation applies in the southern part of Italy.

6.1. Changes in fertiliser use in the last 30 years

In considering the present use of fertilisers in various countries it is useful to consider the changes in application rates that have occurred in the past. Table 8 gives the rates of application to agricultural land published for a few countries by FAO in their Annual Fertilizer Review for 1953. The rates applied in 1952 may be compared with the amounts used 30 years later in 1982 which are given in Table 7 above.

Table 8

	Applied to agricultural land in 1951/2				*Applied in 1981/2*
	N	*P*	*K*	*N + P + K* (*kg ha*$^{-1}$)	*N + P + K*
Netherlands	67	18	57	142	299
Belgium	46	20	68	134	225
Germany	27	15	42	84	206
Denmark	23	13	35	71	185
United Kingdom	15	10	11	36	109
France	9	7	11	27	138

In 1952 the Netherlands used the largest rate of N + P + K as they do today; but the great increase in that country has been in the amount of N applied, the P and K applied has changed little. Much the same changes have occurred in Belgium and Denmark as in Netherlands, in other countries listed the amounts of P and/or K have increased as well as the amounts of N. The order of the countries for total use of N + P + K remains the same in 1982 as in 1952, except that the rates of application in France have climbed above those in UK—which is now at the bottom of this short list of countries. In the 30 years the total of N + P + K has increased by the following factors:

Netherlands	2·1	Germany	2·4	Denmark	2·6
Belgium	1·7	France	5·1	UK	3·0

7. FERTILISERS APPLIED TO ARABLE LAND IN RELATION TO THE NEEDS OF CROPS

The question of how well current fertiliser use matches the needs of crops may be assessed in two ways. Actual rates used may be compared with local advisory recommendations, the other method is by considering the national nutrient balance sheets.

7.1. Nitrogen: recommendations and use compared

In most countries advice on the use of N for particular crops takes account of the previous cropping and fertilising of the land. In some countries advice on cereal fertilising in spring is based on measurements of the

Table 9
The use of nitrogen fertilisers in England and Wales in 1983 compared with the range of recommendations

	Amounts applied		*Range of recommendations*
	Average	*Range*	
	(kg ha^{-1} of N)		*(kg ha^{-1} of N)*
Spring wheat	140	0–250	0–150
Winter wheat	181	0–400	0–225
Spring barley	107	0–250	0–125
Winter barley	150	25–250	0–160
Oilseed rape	272	0–400	200–250
Potatoes	203	0–400	50–220
Sugar beet	155	0–400	0–140
Temporary grass	181	0–400+	300–450[a]
Permanent grass	98	0–400+	300–450[a]

[a] For swards that are predominantly ryegrass.

amounts of mineral nitrogen (nitrate + ammonium) in the soil profile at the end of winter; Becker and Aufhammer [23] described the N-min. method used in Germany. The result is that recommendations vary greatly; as it is not possible to define average conditions over large areas of differing nitrogen status in a country, comparisons of average application rates with an average of recommended rates is not relevant. Only in England and Wales, so far as I am aware, are farmers' fertiliser practices regularly recorded in surveys; the latest results of surveys made annually for the last 40 years in England and Wales are published by Church [24]; his results are used in this section. Table 9 compares the range of recommendations made by ADAS [14] with the average use and the ranges of amounts applied for major crops in 1983.

For most arable crops the average use of N lies within the range of recommendations, exceptions are for oilseed rape and sugar beet. Average use on grass is less than the lowest recommendation. For all arable crops the upper limits of amounts actually applied, and recorded in the Survey, exceed the highest amount in the range of recommendations; this raises the question as to whether some farmers are applying more N than is justified by the needs of their crops and whether this extra N is wasted by leaching.

7.2. Phosphorus and potassium: recommendations and use compared

The use of fertilisers has a long history in most of western Europe and P and

Table 10

	$NaHCO_3$-soluble P		*Exchangeable K*	
	Arable land	*Grassland*	*Arable land*	*Grassland*
	(ppm)		*(ppm)*	
Rothamsted (clay soil)	11	15	170	200
Woburn (sandy soil)	20	20	120	120

K fertilisers have been regularly applied for many years. The result, as was stated earlier (Section 4), is that many soils have been enriched in these two nutrients and for much of the arable land a policy of using P and K fertilisers to maintain satisfactory levels of these two nutrients in the soils is sufficient. This implies that gains of P and K, and losses in crops removed will be balanced. As advice will be based on soil analysis, it is important to establish the amounts of soluble P and K in the soils at which this balance is achieved and fertilisation should be planned to maintain these levels. The Reference Experiments described earlier (Section 1.2) provide such information and Williams [25] stated the amounts of P soluble in 0·5 M $NaHCO_3$ solution and of exchangeable K which maintained a balance; they were as shown in Table 10.

Recommendations for the use of P and K made by ADAS [14] for cereals take account of the yield level expected, soil analysis, and the use made of straw. (If straw is removed more K-fertiliser must be used as was stated in Section 1, consequently twice as much K is recommended where straw is removed as is recommended where the straw is ploughed in or burned.) In Table 11 the rates of application to certain crops in England and Wales recorded by Church [24] are compared with ADAS recommendations [14].

The yields of crops on the fields included in the Survey are not known. If we assume that cereal yields were close to national averages (Table 2) and that most straw was ploughed in or burned (which returns P and K to the soil) the amounts of P and K fertilisers used were not greatly different from the recommendations. The amounts used for oilseed rape and potatoes were close to the recommended amounts. For sugar beet the P used was as recommended; about half of the crop area received salt so the average amount of K used was rather more than was necessary.

7.3. Survey of fertiliser practice in Scotland

Surveys along the lines of those made in England and Wales were made in Scotland some years ago; the work was resumed in 1983 with a survey of

Table 11

Comparisons of average amounts of P and K fertilisers used on arable crops with amounts recommended to maintain satisfactory soil P and K levels

	Actual use in 1983		*Recommendations*[a]			
	P	*K*	*P*	*K*	*Yield level*	*Straw use*
	(kg ha^{-1})		*(kg ha^{-1})*		*(t ha^{-1})*	*etc....*
Spring wheat	16	30	18	25	5·0	returned
			18	50	5·0	removed
Winter wheat	22	38	26	38	7·5	returned
			26	76	7·5	removed
Spring barley	17	37	35	50	10·0	returned
			35	100	10·0	removed
Winter barley	23	45	for all cereals[a]			
Oilseed rape	27	48	22	33		
Potatoes	90	222	110	208		
Sugar beet	32	133	33	83	with salt[b]	
			33	166	without salt	

[a] The P and K recommendations were the same for all cereals, being adjusted as shown to allow for expected yield and the fate of the straw.
[b] With salt (sodium chloride) to give 150 kg ha^{-1} of Na for sugar beet.

Table 12

	Fertilisers and farmyard manure used in 1983			
	N	*P* *(kg ha^{-1})*	*K*	*% of area receiving FYM*
Spring barley				
S[a]	92	21	39	23
E and W[a]	107	17	37	17
Turnips				
S	77	68	81	53
E and W	76	29	51	38
All tillage				
S	113	27	51	23
E and W	154	24	50	15
All grass				
S	131	16	28	34
E and W	125	12	33	40
All crops and grass				
S	124	20	38	30
E and W	139	17	37	28

[a] S = Scotland, E and W = England and Wales.

250 farms; the authors emphasise that this was a small sample. Data are given in Table 12, quoted from the Report [42] which indicates that practice in Scotland is, in general, similar to that in England and Wales. Data are given for spring barley as this occupies 75 % of the cereal area and 60 % of the total tillage area, and for turnips which are a very important root crop in Scotland; they receive much more P and K than the same crop does in England and Wales.

8. FERTILISERS APPLIED TO GRASSLAND IN RELATION TO AMOUNTS REQUIRED

8.1. Grassland in UK

Large amounts of nutrients are involved in the production of good yields from grassland; the quantities of N and K are particularly large. Table 13 shows data published by Morrison and Russell [26]; the total amounts are removed from the field when grass is cut, but when grazed two-thirds of the plant nutrients are returned in excreta.

Nitrogen is the key to growth of grass and the results of a series of national grassland manuring experiments made in recent years are reflected in current recommendations on the use of N; these take account of the water-holding capacity of the soil and the estimated potential yield on the site. Recommendations range up to 300 kg ha^{-1} of N annually in intensive grazing systems, and to 300–500 kg ha^{-1} for grass that is cut.

It is not easy to assess average fertiliser use on grassland since the amounts needed vary greatly with type of sward, climate and soil. The use to be made of the crop has a large influence on the nutrient cycle as Table 13 shows. When grass is cut and removed much more P and K must be applied

Table 13
Amounts of nutrients in a typical grass crop yielding 7·5 t ha^{-1} of dry matter annually

	Total in herbage (kg ha^{-1})	*Returned by grazing animals (kg ha^{-1})*
N	187	131
P	26	18
K	150	105
Ca	45	31
Mg	15	10

to maintain reserves of these nutrients. Leech and Hughes [27] give the following data from the 1983 Survey of Fertilizer Practice which indicate that a sufficient allowance is not made by farmers in England and Wales for differences in the use of grass when they plan the use of P and K fertilisers:

	Average use ($kg\ ha^{-1}$)			*Recommendation* ($kg\ ha^{-1}$)		
	N	*P*	*K*	*N*	*P*	*K*
All grazed grass	124	11	23	300	13	25
All cut grass	156	14	36	500	88	250

The recommendations quoted are for average conditions for the maintenance of grazed grass and for mown grass they are appropriate to the removal of 10 t ha^{-1} of grass dry matter in the season.

Milk production. The role of fertilisers in this system was described by Thompson and Blair [28]. They concluded that there are good opportunities for using more fertiliser to produce more conserved grass fodder and thereby save on the cost of purchased concentrated feeds. They state that more attention should be given to using fertiliser to cheapen the cost of cattle feed than to increasing the yield per cow. Experiments show that dairy farmers could profitably use twice as much N as is used at present on average. In their review of experimental work they indicate the large range in yields and in amounts of N-fertilisers justified (a full account of this work is given by Morrison *et al.* [29]). For example for swards cut six times a year maximum yields ranged from 6·5 to 15·0 t ha^{-1} of dry matter at different sites. Yields at an optimum point where the response was at the rate of 10 kg of dry matter/kg of N (designated Y_{10}) ranged from 5·2 to 14·4 t ha^{-1}. The N required for maximum yield ranged from 540 to 678 kg ha^{-1} of N, but the amount needed to produce Y_{10} was from 260 to 530 kg ha^{-1} of N and the response at this optimum ranged from 15 to 26 kg dry matter/kg of N. While much of this variation was identified as being associated with differences in the quantities of water available to the crops, there were other causes. In all branches of farming, including production from grassland, it is essential to identify the reasons for site-to-site variations in experimental results; the knowledge thus acquired should be applied to practical conditions and until this is accomplished on grassland farms we cannot expect the fertilisers purchased to be used with full efficiency. Much of the variation is associated with soil type; Thompson and Blair [28] quote results for contrasted soil types where response to N ranged from 18 to 28 kg dry matter/kg of N. Soil maps have been published for all of Britain [30]; therefore the way is open to aid farmers by developing ways of overcoming local difficulties in securing high fertiliser efficiency due to the properties of particular soils. To achieve this experiments are needed to relate crop production and response to fertilisers to the physical and chemical properties of soils which are related to differences between soil types.

Meat production. Baker [31] described the role of fertilisers in meat production. He identified a close relationship between the use of N-fertiliser and stocking rate and profitability in the production of meat from grass. When less than 30 kg ha^{-1} of N was used annually he found that there were 405 Livestock Unit Grazing Days (LUGDs) in the year and the Stocking Rate was 1·2 Livestock Units (LUs) per hectare; when 240–270 kg ha^{-1} of N was used in the year LUGD rose to 764 days and LU^{-1} ha to 3·3. In 1979 the Stocking Rate in UK was only 1·56 LU ha^{-1}.

8.2. Grassland in other European countries

Baker [31] stated that information on the use of fertilisers on grass in Europe was sparse; he quoted the figures on which Table 14 is based.

Baker concluded that response to N in 'lowland Europe' is similar to that measured in UK. Experiments in European countries had confirmed that the use of more N would lead to the production of more grass, higher stocking rates, and greater meat output per hectare. Baker concluded 'Both France and Ireland, with large areas of grassland currently receiving low

Table 14

The use of fertilisers on grass in some European Countries

	% *Fertilised*	*Amounts applied* (*kg* ha^{-1})		
		N	*P*	*K*
Belgium	100	120	35	58
France	20	20–25	18–22	21–25
Germany (meadows, leys)			26	
Ireland (pasture)	75	40	5	17
(silage)		116	10	56
Netherlands	99–100	245	7	17

levels of N are countries where the potential for increases in output could be considerable'.

Van Burg *et al.* [32] reviewed the use of nitrogen and the intensification of livestock farming in EEC countries. They pointed out that intensification is an important tool to attain a sufficient level of employment for the available labour force and consequently to increase productivity. The average yield of a cow in EEC countries increased from 2943 kg annually in 1960 to 4055 kg in 1979. The summary of fertiliser recommendations which they prepared is in Table 15 together with some data which they collected on the amounts actually used in the 1978–80 period. They wrote that these latter rates were generally less than the recommendations.

Table 15
Fertiliser recommendations for grassland and estimates of the amounts actually used

	Recommendations ($kg\ ha^{-1}$)		
	N	*P*	*K*
Belgium	200	35	42
France	80–100	31	58
Germany	250–300	24–57	n.a.
Netherlands	250–400	20	100
Ireland	0–250	15–43	30–183
Denmark	350	n.a.	n.a.
UK	350	13	25
(Many recommendations depend on soil type and stocking rates)			

	Amounts actually used ($kg\ ha^{-1}$)		
	N	*P*	*K*
Belgium	120	35	58
France, for leys	70	18–22	21–25
for permanent grass	32		
Germany, for hay	60	26	n.a.
for grazing	130		
Netherlands, for permanent grass	265	7	17
Denmark, for clover/grass leys	250		
for permanent grass	150		
Ireland	40	4–9	17–58
UK	88	11	17

n.a. = not available.

These authors [32] showed how the efficiency of N used on grass is improved by the choice of the correct amount of N, by irrigation where the supply of water is insufficient for crop needs, by timing spring applications in relation to accumulated temperature, by choosing the best form of N to fit local conditions (for example ammonia is liable to be lost from top-dressings of urea unless adequate rain falls immediately after the application), and by careful management of stocking rate, grazing systems and supplementary feeds. Correct decisions on these factors lead to good quality grass for grazing or conservation which provides higher profit and labour income. They also discussed the effect of intensification of grassland farming on environmental problems. A serious difficulty is that slurry should be applied to land to return the nutrients to the soil, but it is difficult to incorporate slurry applications in intensive management systems. If the slurry is applied in winter to avoid these problems pollution of surface and underground waters is liable to occur; therefore injection techniques for use during the growing season are being investigated. Direct losses of fertiliser-N are negligible if amounts and timings are correctly decided; the greatest risks of loss are from sandy soils and when large rates (above 400 kg ha^{-1} of N annually) are applied.

Van Burg *et al.* [32] concluded that, in the European Community as a whole, forage crops are the most important basic resource for animal production, amounting to 55 million ha (41 million ha are in permanent pasture, but they say this is 'grossly under-utilised'). There is a high potential for using more N in terms of increased animal production; in the range from 200 to 400 kg ha^{-1} of N annually increases of 12 kg of milk and 1 kg of liveweight gain will result from 1 kg of N. Conserved grass feeds are important so that expenditure on other feeds can be lessened; the trend from hay to silage making should be continued to provide better quality feed. They state however that in general in the Community, farm structure is the bottleneck as farmers cannot afford large investments and the buildings that are available limit the use that can be made of the land.

9. NUTRIENT BALANCE SHEETS

In assessing the needs for fertilisers, and in judging whether current practices supply too little or too much nutrient, attempt must be made to construct national nutrient balance sheets. Examples of balance sheets constructed from the results of field experiments have already been given in Table 4. Such long-term experiments provide an ideal basis for constructing nutrient cycles, and for monitoring the changes in nutrient reserves in the soils; the results from each experiment apply exactly and directly only to the site and the experimental system, but when conducted on major soil types and specific cropping systems, a series of such experiments does provide the guidance the agricultural industry requires.

9.1. An example of a national nutrient balance sheet

A simple example of balance sheets for three crops grown in England and Wales is given in Table 16. The amounts of fertilisers applied are those given by Church [24], published [1] average crop yields are used and the crop compositions given in Table 1 (contents of crops are as in Table 2).

These data show the very great differences at present in fertilising cereals and potatoes. Potatoes receive five times as much P as they remove and about 40% more K than is removed, also about 40% more N. Cereals receive 20–30% more N than the harvested crops contain, but little more P than is contained in the grain and straw at harvest. By contrast the cereals receive only about a third to a half as much K as in the total harvest. Data given for the amounts of K removed in cereal grain only emphasise the importance of considering the fate of cereal straw. If all the straw is returned to the land, current applications of fertilisers are sufficient to replace the K which is removed in grain, but if the straw is removed and is not returned in farmyard manure, the soil reserves of K will be speedily depleted by cereal growing. Lack of attention to nutrient balances can result in serious problems with potassium nutrition. The problem does require the results from long-term experiments to provide data for nutrient balance sheets; national assessments of the fertilisers applied, and of the

Table 16

Average yields of wheat, barley and potatoes in UK in 1982, average fertiliser dressings applied to these crops in England and Wales, and estimates of N, P and K in the harvested crops

	National average yield (*t ha*$^{-1}$)	*Fertilisers applied on average*			*Nutrients in harvested crops*		
		N	*P*	*K*	*N*	*P*	*K*
		(*kg ha*$^{-1}$)					
Wheat	6·2 (grain)	166	22	37	118 (87)[a]	21 (17)[a]	105 (25)[a]
Barley	4·9 (grain)	120	19	40	98 (74)[a]	15 (13)[a]	88 (23)[a]
Potatoes	35·8 (tubers)	199	87	222	115	18	161

[a] Quantities in parentheses are those removed in grain only.

proportions of cereal crops where straw is removed or returned, are also required. On a wider scale for farming systems as a whole other information will be required on crop and urban and industrial wastes applied to land, together with estimates of the plant nutrients which enter a country in imported feeding stuffs. National balance sheets are essential for planning fertiliser use, manufacturing capacity, and imports of raw materials. An extreme example is a calculation I have made that Thailand has exported to Western Europe, in cassava purchased for animal feed, more K than is used as fertiliser in the whole of Thailand.

9.2. Nutrient balance sheets in European agriculture

Hébert [5] reviewed the levels of fertiliser inputs and soil nutrient status in European agriculture by calculating nutrient balances. It was assumed that 'offtakes' were made up of all the grain (cereals and oilseeds), sugar beet roots, potato tubers, and meat and milk. Table 17 contains a summary that is derived from Hébert's paper, the data refer to 1980.

In all countries less N leaves the farms in produce than is supplied by fertilisers, the difference is notable in the Netherlands. By contrast the produce removes more P than is applied in Denmark and the Netherlands. In eight countries more K is applied than is removed in produce (there are no data for offtakes of K in Greece where fertilisers supply only 3 kg ha^{-1} of K). For the 10 countries of the EEC Hébert calculated that removals account for only 51 %, 68 % and 32 % of the N, P and K respectively that is applied in fertilisers. The balance would be even greater if allowances were

Table 17

Amounts of N, P and K applied as fertilisers in EEC countries and offtakes of these nutrients in produce sold from farms in 1980

	N		*P*		*K*	
	Applied	*Offtake*	*Applied*	*Offtake*	*Applied*	*Offtake*
	(all data are in kg^{-1} ha^{-1})					
Belgium	122	75	29	23	74	25
Denmark	129	62	17	18	41	17
France	68	42	25	10	45	11
Germany	127	65	30	15	77	18
Greece	36	15	7	4	3	na
Ireland	47	33	11	5	26	3
Italy	57	31	18	9	17	7
Netherlands	239	90	18	29	46	14
UK	67	32	10	8	18	9

made for the nutrients supplied in purchased feeds and in any urban and industrial wastes applied to agricultural land. In some areas with positive balances, and where soil analyses indicate satisfactory levels of available P, some economies in the P applied could be made without ill-effects as surplus P is well retained in soil. On the other hand reduction in the use of K could result in soil-K levels falling to amounts that were less than the status classed as 'satisfactory' since no allowances are made for losses of K by fixation by soil and by leaching. On grazed land the returns of nutrients are uneven and losses from the small areas receiving excreta may be large (as was stated earlier in Section 3). Hébert emphasises the importance of returns in animal excreta, he quoted these figures for the Netherlands:

	N	*P*	*K*
	(Millions of tonnes)		
Offtakes from farms in animal products	125	47	20
Returns in animal manures	277	75	214

He states that although the theoretical balance sheet shows a deficit in P (Table 17) in fact, because of the input of P in purchased feeds into the very heavily-stocked country, the P balance is positive and in 1977 only 11 % of soils were low in available P while 50 % tested 'rich'; Ansiaux [16] found a similar proportion of Netherlands soils to be rich in P.

Hébert [5] considers that improvements in recommendations which take account of nutrient cycles require improved methods of soil analysis and a better understanding of the fate of nutrients in soil and subsoil. Simple experimental work is required to 'buttress the recommendations' and 'in view of the cost of such work international cooperation could be very fruitful'. (The Rothamsted Reference Experiments noted previously in this paper (Section 1) are good examples of the type of experiment that is very useful in this context.) Research is also required to provide a better evaluation of the availability of the N in organic manures and wastes and of the pathways of nitrogen in the soil/crop/animal system as a whole. The report of the Royal Society's Study Group on the Nitrogen Cycle of the United Kingdom [33] emphasised the importance of such studies. They stated that large quantities of N are lost from animal excreta by ammonia volatilisation, by denitrification and possibly by leaching. They recommended: 'Because nitrogen losses from livestock wastes constitute a major loss of nitrogen from UK agriculture we recommend that work aimed at quantifying the main pathways of loss, devising practical ways of minimising them, and improving the conversion of nitrogen in livestock feed to protein for human food be accorded high priority'.

10. MAXIMISING THE RETURNS FROM FERTILISERS: PRESENT POSITION, FUTURE PROSPECTS

10.1. The mathematical basis for economic advice

A sound basis for assembling the results of field experiments, and for making economic assessments of the responses of crops to fertilisers as measured in the experiments, was established by Crowther and Yates [34] at Rothamsted in 1941. They showed how a standard curve relating crop responses to fertiliser dressings could be used to assemble, for any one crop, the results of uncoordinated experiments testing fertilisers at several rates, and then to calculate responses to standard quantities. The outcome was a series of national recommendations which ensured the rational fertilising of British crops in the 1940s. The exponential equation they used showed how response to unit fertiliser diminished with increasing fertiliser rates until maximum yield was reached. This expression of the 'Law of Diminishing Returns' suited economists and at a time when the fertiliser rates tested were only moderate, and farmers used even less in practice, it sufficed to provide practical recommendations. In later work in the 1950s as larger fertiliser rates were tested, and crop yields increased, it became clear that interactions between nutrients, and between fertilisers and other factors affecting crop growth, could not be ignored. The exponential equation also became inadequate as excess fertiliser depressed crop yields; the alternatives developed (which I have discussed [18]) were parabolic, quadratic, and split straight line, relationships. Normally a curvilinear relationship indicates that other constraints to growth which have not been corrected are interfering with the response to the nutrient tested through their interaction effects. A straight line response model indicates that the factor tested is dominating growth and it facilitates identification of the economic optimum rate. Wimble [35] discussed the theoretical basis of recommendations based on response relationships. His advice was 'Expect to use a four-parameter family, choosing between two straight lines and one of the others in the light of the data'.

Advances in the use of soil analysis to measure reserves of P, K and other nutrients, and allowances for the effect of farming system, soil type, and previous cropping, on the N available from soil, refined the general advice derived from series of experiments so that it could be applied to individual farms. Later developments [14] have introduced variations in recommendations to allow for the level of yield expected and for the return, if any, of crop by-products such as cereal straw or sugar beet tops.

10.2. Crop yield increases as measures of fertiliser efficiency

It is essential that the increased production that can be attributed to the use of fertilisers should more than cover the cost of the inputs. For mobile nutrients, of which N is the most important, the return must be assessed from the performance of the crop to which the fertiliser is applied. Thus efficiency is often expressed as kg of produce/kg of N applied. Tinker [36] has stated that for cereals 70–75 % of the N taken up remains in the grain (the remainder being in straw, chaff and roots). If 75 % of the N applied goes into the grain, and this contains 2 % of N, then the efficiency is 37·5 kg of grain/kg of N and this can be regarded as a maximum. Efficiency calculated in this way depends on the rate of N applied, small rates giving higher returns than larger amounts do. Tinker quotes experiments where the return was 20–30 kg of wheat grain/kg of N and considers the national average return in England and Wales has been 24 kg of grain/kg of N over long periods. The immobile and the less mobile nutrients (P and K for example) are sorbed and precipitated by soil and, in the well-established systems common in Europe, we must plan fertiliser additions to maintain suitable levels of available nutrients as measured by soil analysis. If this level is allowed to fall too far by omitting fertiliser, deficiency will result and crop yields will be reduced. In these conditions the efficiency of immobile nutrients, and the consequent economic returns, can only be assessed on a long-term basis—and this requires data from long-term field experiments on appropriate crop systems.

10.3. Nutrient uptakes as measures of fertiliser efficiency

The best possible cost/benefit ratio is achieved when the whole of a fertiliser dressing serves the purpose for which it was intended—that the whole amount enters a crop, or a series of crops, and stimulates a proportional increase in yield, and that none is wasted by loss from the system or by reaction with soil to form compounds that remain unavailable to crops for long.

Nitrogen. The present efficiency of N-fertilisers as a whole in UK, and probably in most of northwestern Europe too, in terms of the percentage of the applied nutrient appearing in the harvested produce, is probably no greater than the apparent efficiencies shown in Table 6. Estimates have been made that on the world scale the present efficiency of N is no more than 40 % for the first crop grown and that most of the remaining 60 % is lost to the environment. This is a great challenge to agricultural science. The way forward must be by modelling of soil and crop uptake processes—which requires information on soil properties which is not at present generally available. These models will be associated with accurate estimates of the amounts of N required by a particular crop yield and on measurements of reserves in the soils (by the N-min. method for cereals) and supplemented where possible with measurements of the nitrate concentrations in the plants. Together with careful assessment of the past and the expected future weather (rainfall and temperature) these methods will lead to recommendations for amounts and times of application that will lead to maximum efficiency in uptake by the crops, and therefore to maximum profit on the expenditure on the input.

Phosphorus and potassium. Here we will be concerned with management of the reserves inevitably accumulated in soil—as assessed by soil analysis. Recommendations will be based on expected crop yields—high yields of some crops will require higher applications of P and K to achieve these yields. Account must also be taken of soil properties that influence nutrient solubility and plant uptake (such as P absorption capacities), and of the release of reserves that are not indicated by conventional soil analyses for available nutrients (such as the release of fixed K). Again models of soil/nutrient reactions and of plant-uptake processes will be required to achieve maximum efficiency which, for immobile nutrients, can only be assessed on a long-term basis. Appropriate long-term field experiments to monitor changes in soil properties and crop performance will be an essential feature of the systems which will be devised for improved nutrient management.

11. IMPROVING THE RETURNS FROM INPUTS TO FARMING SYSTEMS

11.1. Costs of production

Fixed costs

These costs (which account for rent and rates, labour, machinery and power, and general overheads) are now at high levels on most farms in European countries; these costs are difficult to reduce without impairing the ability to farm well according to the principles of good husbandry. The

University of Reading [37] made a survey in which fixed costs ranged from £309 ha^{-1} on cereal farms to £490 ha^{-1} on dairy farms. The Recorded Farms Scheme of ICI PLC [38] showed a similar range from £329 ha^{-1} where cereal growing was the main enterprise to £530 ha^{-1} in dairy farms. These fixed costs must be more than covered by the gross margin for the farmer to have a profit to reward his management.

Variable costs

These are made up of the costs of fertilisers, agrochemicals, and seed. These inputs must be used efficiently by adopting all correct practices to secure a good return on the expenditure. A comprehensive approach is essential because, if the money spent on fertiliser is to be repaid the crop must be protected from weeds, pests and diseases to secure the maximum benefit from each input.

11.2. Interactions between inputs

These interactions must be recognised as a vital factor in increasing yields (and therefore the efficiency of fertilisers and all other inputs). This is well illustrated by the data in Table 18 which were presented by Dilz *et al.* [39]; in this experiment made in Netherlands in 1971 the effects of the fungicide Benomyl (sprayed at heading) on the yields and nitrogen contents of winter wheat were measured.

These results show that the maximum benefit from N-fertiliser in terms of both yield and uptake of N could not be obtained unless fungicide was also applied. Dilz *et al.* [39] also quote the results of Belgian work where the return from fungicide applications depended on the rate of N applied: when only 50 kg ha^{-1} of N was applied the return from fungicide was an increase in grain yield of 500 kg ha^{-1}, with 150 kg ha^{-1} of N the grain

Table 18

The effects of fungicide on yields and nitrogen contents of winter wheat receiving increasing amounts of nitrogen fertiliser

N applied (kg ha^{-1} fungicide)	*Yield of grain (t ha^{-1})*		*Amount of N in grain (kg ha^{-1})*	
	Without	*With*	*Without*	*With*
35	4·7	4·8	75	79
70	5·5	5·7	89	99
105	5·8	5·9	100	103
140	5·3	6·2	103	118

doubled to an increase of 1000 kg ha^{-1} of grain. They point out that the use of fungicide extends the duration of leaf area thus prolonging photosynthesis and uptake and relocation of N in the grain. As crop protection extends the period of grain filling it favours the effect of a late top-dressing of N on both grain yield and recovery of N. Figures quoted by Dilz *et al.* are 20–35 % recovery from a late dressing of N without crop protection, but 35–55 % recovered when protection was provided.

Other interactions occur with management practices such as choice of tillage practices or cropping systems. For example cereal yields are improved when a 'break' of a non-cereal crop diminishes infection by pests or diseases of roots; they are also increased when a preceding legume or heavily-fertilised root crop leaves large residues of N in the soil which then benefit the cereal. These effects are well illustrated by the results of an experiment reported by Darby *et al.* [40] which are summarised in Table 19. There were large differences between the yields of winter wheat grown after beans and of wheat grown after wheat. Growing beans before the wheat controlled the severe take-all disease which seriously damaged the second wheat crop and the beans also supplied much N for the wheat. There were consistent gains from the spray to control aphids and fungal diseases of the leaves. The gains of the pesticide were generally greater on the wheat grown after beans, indicating that there are interactions between the effects of controlling damage to the leaves and measures taken to avoid root disease. Maximum yield was achieved when both leaf and root diseases were controlled and the correct amount of N was applied.

Much evidence from experiments made in Europe confirms the findings

Table 19

Average yields of winter wheat from two varieties grown in a multifactorial experiment at Saxmundham (Suffolk) in 1982

Previous crop	*Winter wheat*		*Winter beans*	
Aphicide/fungicide spray	*Without*	*With*	*Without*	*With*
N-fertiliser applied (kg ha^{-1} of N)	*Yields of wheat grain (t ha^{-1})*			
0	2·3	2·4	7·5	7·9
70	—	—	9·2	9·6
100	—	—	9·4	10·2
130	6·1	6·4	9·9	10·6
160	6·4	6·9	9·6	10·8
190	6·8	7·4	—	—
220	7·5	7·6	—	—

Table 20

Results of the survey of fields of winter barley grown in England and Wales in 1981 made by ICI PLC

Group of farms	*Yield* ($t\ ha^{-1}$)	*Gross margin* (£ ha^{-1})	*Variable cost* (£ ha^{-1})	*Variable cost as £/tonne of grain*
Top 25%	6·9	462	196	28
Average	5·5	332	189	35
Bottom 25%	4·1	220	177	43

outlined above and shows that the returns from each input are increased by the presence of other necessary inputs when these are correctly applied. Therefore it is false economy to restrict inputs that are justified by the available scientific evidence from modern production systems. These inputs, which include improved varieties, improved soil management and crop establishment practices, fertilisers, and agrochemicals to control weeds, pests and diseases, and growth processes, all interact to raise yields. Conversely if one input that is justified scientifically is omitted the returns from the other inputs that are used will be diminished and there will be less profit on the outlay that is made, and the cost per tonne of produce will be greater.

Data collected by ICI PLC [41] in a survey of 610 fields of winter barley in 1981 illustrate these points on the returns from inputs. Average fertiliser applications were 161 kg of N, 25 kg of P, and 46 kg of K, per hectare; these fertilisers cost £75 ha^{-1} out of total variable costs of £189 ha^{-1}. The whole group of farms was divided into the top 25% with the largest yields and the bottom 25% with the lowest yields; gross margins and variable costs are shown in Table 20.

Table 20 shows that the bottom 25% of farmers would have had difficulty in meeting their fixed costs and that they produced grain which cost 50% more per tonne for variable costs than that produced by the top 25% of farms. The 10 farms with the highest yields were also identified; their average yield was 8·2 t ha^{-1} of grain, variable costs were £196 ha^{-1} (no more than the average costs of the top 25%), and their average gross margin was £582 ha^{-1}. The success of these 10 farmers was due to their efficient use of inputs, no doubt aided by good soil conditions.

11.3. The way ahead

Surveys of the type shown in Table 20 have shown clearly that a reduction in variable costs by cutting inputs that are essential to crop growth and yield is a certain way to lower margins and profits, and to food that is more expensive per tonne. If we are concerned by surplus production the remedy is not to grow lower yields but to sow less hectares. On the land that is cropped or stocked effort must be concentrated on using inputs efficiently at optimum levels to maximise yields and so minimise costs of production per tonne. Research and development must be planned to develop methods of application that achieve high efficiency in the use of each input and to aid farmers to apply these methods effectively so that there becomes little difference between 'top' and 'bottom' groupings in statements such as that in Table 20.

Advances in the potential yielding ability of crops, in the abilities of farmers to use inputs to achieve these higher yields, and in the availabilities of these inputs—which interact to raise yields as constraints are overcome—mean that economic assessment of fertiliser response curves is no longer a sufficient basis to secure the higher efficiency that we require, both for economic success, and to move nearer to the potential for production of our crops. In aiming for greater profitability (or cheaper produce) at required total production levels we should not consider reducing inputs, but we should be prepared with the information to justify their increased use to meet the higher potentials that have been established in practice on farms.

To make progress the emphasis must be on models of the production systems which will be based on the hierarchy of limiting factors and which require full use of measurements of the properties of the inputs to the system, such as soil and fertiliser, and on the system itself. In this way progress will be made by the pathways established in modern science and engineering where objectives (such as flying to the moon!) are achieved by having a full knowledge of the properties of the materials that will be used, and of the system in which we expect the materials to operate. To reconcile the needs of agriculture for efficient intensification to secure economic production with the need to avoid all environmental contamination requires increased research on the efficient use of inputs. A benefit of the models we will develop and use will be that they will reveal the nature of the new research required.

12. CONCLUDING REMARKS AND RECOMMENDATIONS

The aims of those research workers concerned with the efficient use of fertilisers in agricultural systems, and of those concerned with abating environmental problems, are identical. The whole quantity of plant nutrients applied should serve the intended purpose by being taken up by the crop, or a series of crops, for which the fertilisers were applied, so that plant growth is increased; none of the nutrients applied should escape directly into the environment. Losses of nitrogen from agricultural systems are inevitable; some losses may occur in arable systems when organic materials are microbially decomposed to release inorganic nitrogen when no crops are growing to take it up. In systems where livestock graze herbage crops much larger losses will occur from the excreta they deposit, and from animals wastes applied to grassland. These losses of N are not direct losses from fertilisers, but they are associated with fertiliser use since fertilisers increase the amounts of nutrients in the crop/animal/soil cycle and research on methods of diminishing the losses is required. Losses of phosphorus from agricultural systems will be minimised by managing land so that all soil erosion is prevented and any surface run-off avoided. Research on all these topics by agricultural and environmental organisations should be closely coordinated.

12.1. Some agricultural research that is required

The immediate purpose of agricultural research must be to develop scientific management systems in which fertilisers and other inputs are used to overcome all constraints to crop growth and achieve the required yield levels which maximise returns to the farmer and minimise costs to the consumer. The essential outcome must be that all inputs are used with maximum efficiency and none escapes directly into the environment. The emphasis must then be on improved technological advice to farmers on the forms, amounts, times and methods of application of fertilisers which achieve this objective. Some research topics and other activities which will be required are noted below.

Reference experiments on the main farming systems should be established. These will provide information on crop responses to fertilisers, on the interactions of fertilisers with other inputs to the system, and on variability in yield due to climatic factors. They will also provide soil and plant materials for the associated multidisciplinary research by soil scientists, plant physiologists, and crop protection workers. In particular they will afford opportunities for the testing and improvement of methods of analysing soils for available nutrients and for correlating soil analyses with crop responses to fresh fertilisers at varying levels of yield potential. They will also provide material for correlating analyses of plant tissues with the responses of the crops to fertilisers. Both analytical methods (on soil and on plant material) will be used in developing advice to farmers.

Experimental and investigational work on the pathways of nutrients in animal farming systems, and particularly in animal excreta and organic wastes, is required. Very large quantities of nutrients are involved in these systems which, as was noted above, now suffer considerable losses, particularly of nitrogen.

Models of nutrients in soil/crop systems, and particularly those involving soil/root relationships, will result from this work on arable farming systems. In animal farming systems models of nutrient pathways in the soil to crop to animal system will be extended to the return of wastes. These models will form the basis of future advice to farmers.

12.2. Application of the research

The improved technologies that will be developed from the research noted above will require other scientific, social and economic studies for their application.

Advice to farmers on crop production will be based on soil survey maps and related measurements of soil properties, on knowledge of local cropping systems, and on climatic information—in particular we need better short-term and long-term weather forecasts than are at present available.

National fertiliser policies must be developed to provide sufficient nutrients for the levels of production required. These policies will require knowledge of farmers' present practices and where surveys of current fertiliser use are not made they should be initiated. Nutrient balance sheets must also be calculated; these are required at both local level (to improve advice to farmers), and at national levels for over-all planning. Account must be taken in these balance studies of all sources of nutrients—mainly but also purchased feeds and urban and industrial wastes which may be applied to land, and of offtakes in all produce that leaves the farms.

Economic assessment of fertiliser use for advisory purposes must recognise the vital role of these inputs in interacting with other inputs in building yields. Withdrawing one essential input, or using less than the optimum amount, lowers the efficiency of other inputs, diminishes gross margins, and raises the unit cost of produce. The economics of fertiliser use cannot be considered in isolation, they must be based on models of the crop production system as affected by inputs. When all inputs are fully efficient the maximum economic returns will be achieved.

12.3. The future

The outcome of the activity which is now required, and which is noted above, will be a stable social and economic background against which a scientific policy for the management of soil fertility will be developed. This policy will give us the power to solve present problems in using fertilisers; they will then be completely efficient and will result in no direct pollution of environment, while performing their vital role of producing the amounts of food which people require at costs they can afford. The problems we now face are common to countries of the European Community; the research required will benefit greatly by a measure of international cooperation to extend the work by interaction between those involved, while avoiding unnecessary overlap in attacks on particular problems. In this context this Symposium has an important part to play by initiating the necessary discussions.

REFERENCES

1. ANNUAL REVIEW OF AGRICULTURE (1984). Command 9137, London, HMSO.
2. Widdowson, F. V., Penny, A. and Bird, E. (1980). Results from the Rothamsted Reference Experiment. II. Yields of the crops and recoveries of N, P and K from manures and soil, 1975–1979. Rothamsted Experimental Station Report for 1979, Part 2, pp. 63–75.
3. Widdowson, F. V., Penny, A. and Hewitt, M. V. (1982). Results from the Woburn Reference Experiment. III. Yields of the crops and recoveries of N, P, K and Mg from manures and soil, 1975–1979. Report of Rothamsted Experimental Station for 1981, Part 2, pp. 5–21.
4. Holmes, M. R. J. (1980). *Nutrition of the Oilseed Rape Crop*, London, Applied Science Publishers.
5. Hébert, J. (1984). Levels of fertilizer input and soil nutrient status in European agriculture. Preprints for 18th Colloquium of the International Potash Institute: 'Nutrient balances and fertilizer needs in temperate agriculture'. Bern, Switzerland, International Potash Institute, pp. 201–24.
6. Wetselaar, R. and Farquhar, G. D. (1980). Nitrogen losses from tops of plants. *Advances in Agronomy*, **33**, 263–302.
7. Prew, R. D., Church, B. M., Dewar, A. M., Lacey, J., Penny, A., Plumb, R. T., Thorne, G. N., Todd, A. D. and Williams, T. D. (1983). Effects of eight factors on the growth and nutrient uptake of winter wheat and on the incidence of pests and diseases. *Journal of Agricultural Science*, Cambridge, **100**, 363–82.
8. Lidgate, H. J. (1984). Benefits of fertiliser input to arable cropping. *Chemistry and Industry*, no. 18, 649–52.
9. Ryden, J. C. (1984). Fertilisers for grassland. *Chemistry and Industry*, no. 18, 652–7.
10. AGRICULTURAL RESEARCH COUNCIL (1983). The fate of fertiliser nitrogen applied to cereals on a chalk soil, especially loss by leaching. In: *Annual Report of the Agricultural Research Council for 1982–83*, pp. 14–15.
11. Powlson, D. S., Jenkinson, D. S., Pruden, G. and Johnston, A. E. (1983).

Studies with ^{15}N labelled fertilizers. Report of Rothamsted Experimental Station for 1982, Part 1, p. 163.

12. SLANGEN, J. H. G. and KERKHOFF, P. (1984). Nitrification inhibitors in agriculture and horticulture. A literature review. *Fertilizer Research*, **5**, 1–76.
13. RYDEN, J. C., BALL, P. R. and GARWOOD, E. A. (1984). Nitrate leaching from grassland. *Nature, London*, **311**, 50–3.
14. AGRICULTURAL DEVELOPMENT AND ADVISORY SERVICE (ADAS) (1983). 1983–84 Fertiliser recommendations. Reference Book 209, London, HMSO.
15. ENGELSTAD, O. P. and TERMAN, G. L. (1980). Agronomic effectiveness of phosphate fertilizers. In: *The Role of Phosphorus in Agriculture* (Eds Khasawneh, F. E., Sample, E. C. and Kamprath, E. J.).
16. ANSIAUX, J. R. (1977). The level of soil phosphorus status. *Phosphorus in Agriculture*, **70**, 1–10.
17. JOHNSTON, A. E. and POULTON, P. R. (1977). Yields on the exhaustion of land and changes in the NPK contents of the soils due to cropping and manuring, 1852–1975. Rothamsted Experimental Station Report for 1976, Part 2, pp. 53–85.
18. COOKE, G. W. (1982). *Fertilizing for Maximum Yield*, 3rd edn, London, Granada Publishing Ltd, pp. 34–7.
19. VETTER, H. (1977). The importance for soil fertility of adequate phosphate contents in the soil. *Phosphorus in Agriculture*, **70**, 11–24.
20. GACHON, L. (1977). The usefulness of a good level of soil phosphate reserves. *Phosphorus in Agriculture*, **70**, 25–30.
21. MATTINGLY, G. E. G. (1980). The reliability of soil phosphorus analysis in relation to fertilizer recommendations. *Chemistry and Industry*, 6 September 1980, 690–3.
22. FAO (1983). *Fertilizer Yearbook for 1982*, Vol. 32, Rome, Italy, Food and Agriculture Organisation of the United Nations.
23. BECKER, F. A. and AUFHAMMER, W. (1982). Nitrogen fertilisation and methods of predicting the N requirements of winter wheat in the Federal Republic of Germany. *Proceedings of the Fertiliser Society*, no. 211, pp. 33–66.
24. CHURCH, B. M. (1983). Use of fertilizers in England and Wales, 1982. Rothamsted Experimental Station Report for 1982, Part 2, pp. 161–8; see also Report for 1983, part 2, pp. 295–300.
25. WILLIAMS, R. J. B. (1973). Changes in the nutrient reserves of the Rothamsted and Woburn Reference Experiment soils. Rothamsted Experimental Report for 1972, Part 2, pp. 86–101.
26. MORRISON, J. and RUSSELL, R. D. (1980). Fertiliser recommendations for grassland. *Chemistry and Industry*, 6 September 1980, 686–8.
27. LEECH, P. K. and HUGHES, A. D. (1984). Survey of fertiliser practice. Fertiliser use on farm crops in England and Wales 1983 (SS/CH/15). Agricultural Development and Advisory Service, Rothamsted Experimental Station, Fertiliser Manufacturers Association.
28. THOMPSON, W. and BLAIR, T. (1981). The role of fertilizers in milk production. *Proceedings of the Fertiliser Society*, no. 197, pp. 1–30.
29. MORRISON, J., JACKSON, M. V. and SPARROW, P. E. (1980). The response of perennial ryegrass to fertilizer nitrogen in relation to climate and soil. Report of

the joint ADAS/GRI Grassland Manuring Trial—GM 20. Technical Report no. 27, Hurley, Berkshire, Grassland Research Institute.

30. SOIL SURVEY OF ENGLAND AND WALES (1983). 1: 250,000 Soil Map of England and Wales (Sheets 1–6). Rothamsted Experimental Station: Soil Survey of England and Wales.
SOIL SURVEY OF SCOTLAND (1983). 1: 250,000 Soil Map of Scotland (sheets 1–7). Macaulay Institute for Soil Research, Aberdeen, Soil Survey of Scotland.
31. BAKER, H. K. (1981). The role of fertilizers in meat production. *Proceedings of the Fertiliser Society*, no. 198, pp. 1–29.
32. VAN BURG, P. F. J., PRINS, W. H., DEN BOER, D. J. and SLUIMAN, W. J. (1981). Nitrogen and intensification of livestock farming in EEC Countries. *Proceedings of the Fertiliser Society*, no. 199, pp. 1–78.
33. THE ROYAL SOCIETY (1983). *The Nitrogen Cycle of the United Kingdom*, London, The Royal Society.
34. CROWTHER, E. M. and YATES, F. (1941). Fertilizer policy in war-time. *Empire Journal of Experimental Agriculture*, **9**, 77–97.
35. WIMBLE, R. (1980). Theoretical basis of fertiliser recommendations. *Chemistry and Industry*, 6 September 1980, 680–3.
36. TINKER, P. B. (1983). Nutrient and micronutrient requirements of the cereal crop. In: *The Yield of Cereals* (Ed. D. W. Wright), London, Royal Agricultural Society of England, pp. 59–67.
37. UNIVERSITY OF READING, DEPARTMENT OF AGRICULTURAL ECONOMICS AND MANAGEMENT (1983). Farm business data. Reading, The University.
38. ICI PLC (1981). Recorded farms, 1981 Crop year. Technical Services Group Report 15, Billingham, Cleveland, Imperial Chemical Industries PLC.
39. DILZ, K., DARWINKEL, A., BOON, R. and VERSTRAETEN, L. M. J. (1982). Intensive wheat production as related to nitrogen fertilisation, crop protection and soil nitrogen: experience in the Benelux. *Proceedings of the Fertiliser Society*, no. 211, pp. 93–124.
40. DARBY, R. J., WIDDOWSON, F. V., PENNY, A. and HEWITT, M. V. (1983). Studies on yield variation; field experimentation on outside sites. Rothamsted Experimental Station Report for 1982, Part 1, pp. 257–8.
41. ICI PLC (1981). *Pointers to Profitable Winter Barley*, Billingham, Cleveland, Imperial Chemical Industries PLC.
42. CHURCH, B. M. and LEECH, P. K. (1984). *Survey of Fertiliser Practice: Fertiliser Use on Farm Crops in Scotland 1983*, Scottish Colleges of Agriculture, Rothamsted Experimental Station, and Fertiliser Manufacturers Association.

10

The Industrial Significance of Fertilisers in Terms of Economic Activity and Employment

J. W. MARSHALL

ICI Agricultural Division, Cleveland, UK

SUMMARY

The paper examines a wide range of issues concerned with fertiliser production and use to put the economic significance of the industry into perspective. While the industry may appear small in terms of direct employment its contribution to the European economy is significant.

Fertiliser production adds considerable value to Europe's scarce energy resources. The use of fertiliser enables European agriculture efficiently to harness solar energy to produce food energy with an energy input/output ratio of 2 or 3 to 1.

Fertilisers reduce food prices by as much as 50% by reducing the unit cost of production. The judicious use of fertilisers now means that European agriculture can produce grain profitably at world market prices. This has considerable implications on the future of the Common Agricultural Policy. Europe can now export grain onto the world market without subsidies while maintaining cereal farmers' incomes through production gains. This benefits not only Europe's farmers but also Europe's taxpayers.

Fertilisers are the key to modern agriculture. They lead to efficient and economic food production which gives the fertiliser industry an economic significance far beyond that of a straightforward manufacturing industry.

1. INTRODUCTION

To fulfil the requirements of the title it is necessary to take a broader approach to the subject than the wording would at first suggest. The economic impact of the fertiliser industry in terms of its contribution as a creator of wealth and employer of labour while significant is only one part of the story. The major economic importance of the industry comes as a result of the processes used and the effect of the products manufactured. This is due to the industry's influence on Europe's energy balances, on its agriculture and on food prices, and it is these areas on which this paper will concentrate.

The paper looks first at the areas of direct influence on economic activity and employment. This section will consider the effect of the industry on GDP, on taxes, on employment and on wages. It then investigates the more important area of influence on energy and agriculture and examines the way in which the industry helps to expand and add value to Europe's energy resources and reduces the real cost of food production in Europe by enhancing the efficiency of European agriculture.

1.1. Employment

The EEC fertiliser industry directly employs some 110 000 people. The industry is highly capital intensive and one of the most automated in the world, which accounts for the low level of employment.

The 110 000 directly employed in fertiliser manufacture within the EEC are, however, not the only workers dependent on a successful indigenous industry. Many of the 1·6 million other EEC chemical industry employees will be involved in the supply of raw materials to the fertiliser industry or, as is more likely, in the processing of by-products from that industry. Carbon dioxide is a by-product with many important uses, in the food industry as a coolant and for the carbonation of drinks and in the electricity supply industry as the vital coolant for the cores of nuclear reactors. Many of the production processes used in the oil industry owe much to fertiliser technology. The production of methanol and its derivatives is one example.

Fertilisers are heavy bulk chemicals and as such require a complex distribution system. Many of those employed in European haulage, storage and shipping thus rely on the fertiliser industry for their livelihoods.

European agriculture and the agricultural supply industry employ over eight million people, all of whom rely indirectly on an indigenous fertiliser industry for the security of fertiliser supply which guarantees stability and their long term prosperity.

At the end of the food production chain, the food processing industry's employees are highly dependent on fertiliser which directly accounts for half the food produced within the EEC and thus for half the food industry's throughput.

It is impossible accurately to assess the exact number of people whose employment depends directly on the European fertiliser industry. What is clear, however, is that the initial statistics, which showed 110 000 people directly employed in the industry, seriously understate the case.

1.2. Economic activity

The EEC fertiliser industry has a turnover of \$8·4 billion which makes it a significant contributor to the Community GDP.

To put the turnover into some perspective the \$8·4 billion is equivalent to 15% of the total GDP of Denmark and to 50% of that of Ireland.

The fertiliser industry is also an important taxpayer within Europe contributing taxes directly to government via corporation and employee taxation as well as indirectly by adding value to the raw materials used.

2. ENERGY AND THE EUROPEAN FERTILISER INDUSTRY

The basic process involved in the manufacture of fertilisers is shown in Fig. 1. The production of nitrogen, the most important of the three major fertiliser nutrients, requires a source of hydrocarbon energy. Most modern ammonia plants fix atmospheric nitrogen by using natural gas, though some use naphtha from oil and could indirectly use coal. The important feature is that the energy is supplied from a non-renewable source.

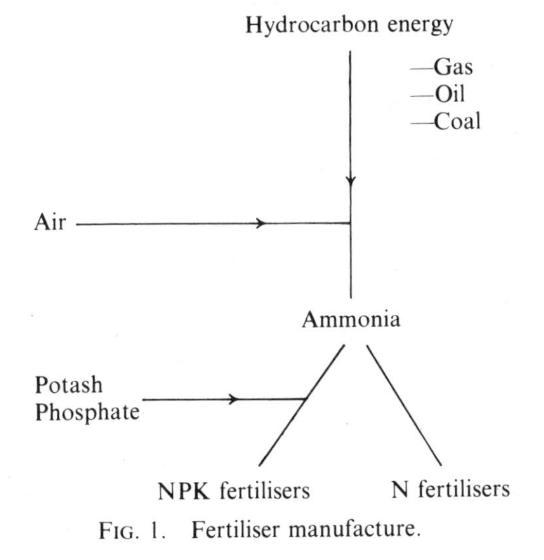

FIG. 1. Fertiliser manufacture.

The industry is often accused of burning scarce non-renewable energy to produce food which is already in surplus within Europe. The truth is that the industry in fact consumes a very small proportion of total energy and uses it in the most efficient manner possible, namely to produce more energy. The industry is perceived as manufacturing fertilisers, it would be more accurate to think of it as adding value to energy.

2.1. Energy balances

Fertilisers are the key to the energy efficiency of modern agriculture. They allow a plant to achieve its fullest yield potential by expanding leaf growth such that the plant's ability to capture solar radiation is enhanced. By encouraging optimum use of solar energy they are responsible for producing considerably more energy than was required for their manufacture.

Some of the most thorough work on the energy balances of fertiliser and food production was conducted in the late 1970s in the UK by Lewis and Tatchell [1]. They looked at the increment in energy produced as a result of fertiliser use on various crops. In the case of cereals 4·5 GJ of fertiliser energy produces one tonne of extra output, containing 15 GJ of food energy, a response of 3·3 to 1. In the UK agriculture consumes some 4 % of total energy and fertilisers some 25 % of that. In return for that energy fertilisers are responsible for 50 % of food produced. Far more energy (10 % of total UK fossil fuel usage) is consumed in processing, packaging and distributing food than in actual food production.

2.2. Fertiliser technology

In the 1960s when fertiliser production was on a smaller scale than today, the energy requirement to produce ammonia was 85M Btus/ton. At the time of the Lewis and Tatchell work in the 1970s this requirement had fallen to 40M Btus/ton.

The latest technology has reduced this requirement to less than 30 Btus/tonne. The theoretical thermodynamic requirement is 20M Btus/tonne and the progress towards this minimum illustrates the tremendous strides the industry has taken in its endeavours to achieve a highly energy efficient process. This is shown graphically in Fig. 2.

This technology-led progress does not stop at ammonia production. Finished nitrogen fertiliser reflects the dramatic downward trend in energy requirement, partly because of the contribution from ammonia efficiency and partly due to improvements in the production technology used to convert ammonia to nitrogen fertiliser. Figure 3 illustrates the progress made on urea and ammonium nitrate.

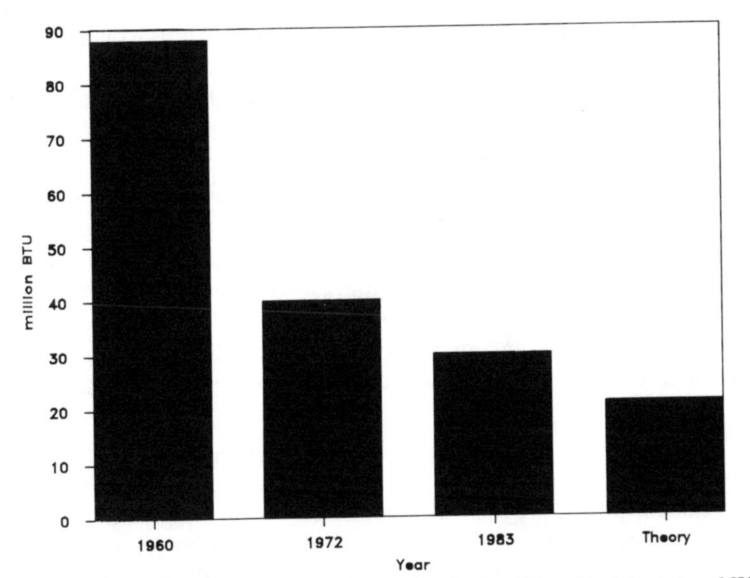

FIG. 2. Energy required for ammonia production, million Btu/short ton NH_3. (1 short ton = 0·907 metric tons (tonne); 1 British thermal unit (Btu) = $1{\cdot}054 \times 10^3$ Joules (J).)

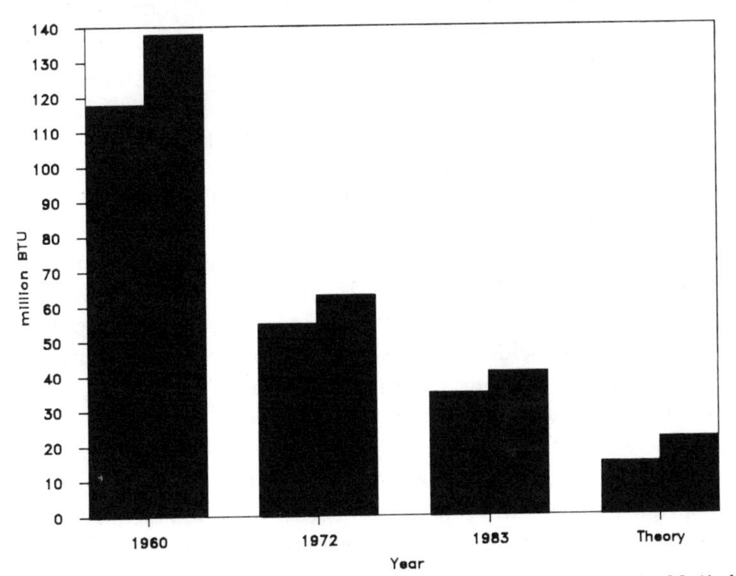

FIG. 3. Energy for urea and ammonium nitrate, million Btu/short ton N. (1 short ton = 0·907 metric tons (tonne); 1 British thermal unit (Btu) = $1{\cdot}054 \times 10^3$ Joules (J).) (Left-hand columns, ammonium nitrate; right hand, urea.)

Ammonium nitrate is a more energy efficient process than the urea process which is why the European industry is largely ammonium nitrate based. It also suggests that new capacity should be ammonium nitrate based as this has the lower theoretical energy requirement.

Modern fertiliser production started in Western Europe with the development at the start of this century of the Haber Bosch process for N-fixation in Germany. Since that time Europe has led the world in fertiliser technology and today over 60% of new projects, either planned or under construction, use European technology. This technology is particularly important in the Third World, helping developing countries to feed growing populations and in the Eastern Bloc where reliance on West European technology has considerable strategic significance.

3. FERTILISER AND FOOD PRICES

For years the principal benefit attached to the use of fertilisers has been seen as the increase in crop yields resulting from their use. In many developing countries this remains the case. However, in view of the level of EEC food surpluses this is seen as a somewhat dubious benefit in the circumstances facing European agriculture today. There is, however, another major benefit arising from fertiliser use which is just as valid today as it has ever been—the reduction in the cost of producing foodstuffs which fertilisers have brought about. This point needs emphasising since an input seen only a few years ago as a key component of an efficient European agricultural system is now being questioned in some quarters regarding its future role.

A powerful illustration of the impact of fertilisers in reducing food production costs is provided by translating the results of the UK's unique Broadbalk Experiment into financial terms.

The Broadbalk Field is situated at the Lawes Agricultural Trust Experimental Station at Rothamsted, Hertfordshire. It has been subject to unchanged fertiliser practice for 140 years in order to investigate the long-term effects on the yields of wheat and potatoes of fertiliser use. Yields obtained with and without the three major plant nutrients, nitrogen, phosphorus and potassium, are shown in Table 1.

Yield increases of 391% and 455% for wheat and potatoes respectively are associated with application of N, P and K. It is important to note the interdependence of the three nutrients in order to obtain the highest yields. Furthermore, this trial does not investigate the full range of yield response to fertiliser application. Higher yields would almost certainly be obtained at higher rates of nitrogen on wheat and potash on potatoes.

Table 1
Broadbalk yield data

	Long-term treatment[a] (*kg ha*$^{-1}$)			*Yield* (*t ha*$^{-1}$)	
	N	*P205*	*K20*	*Wheat*	*Potatoes*
None	0	0	0	1·69	8·47
N	96	0	0	3·68	8·30
PK	0	77	107	2·04	16·63
NPK	96	77	107	6·60	38·57

[a] Annually since 1844 at these rates of application.

A recent UK study [2] examined the contribution of fertilisers to the country's total agricultural earnings. The results of this study for cereals are shown in Table 2. From 1980 to 1983 approximately 9·3–10·5 million tonnes of cereal grain produced each year in the UK can be attributed to the use of fertilisers. The annual net benefit to the UK of the use of fertilisers costing £181M to £301M from 1980 to 1983 respectively is of the order of £538M to £685M, an economic response in excess of 3 to 1.

Returning to the Broadbalk results, what are the financial consequences of the yield increases? Let us assume that the non-fertiliser costs of growing the two crops do not change as a result of using fertilisers. This is not too big an assumption to make: some costs (e.g. harvesting) would be marginally lower but others (e.g. agrochemicals because of greater weed competition)

Table 2
Value of fertilisers to UK cereal production (£ million)

	1980	*1981*	*1982*	*1983*
Gross output				
Wheat	786	858	1 076	1 124[a]
Barley	651	798	849	849[a]
Total	1 437	1 656	1 925	1 973
Supported by fertiliser[b]	719	828	962	986
Cost of fertiliser used	181	230	262	301
Net benefit	538	598	700	685

[a] Estimates based on value of national production at L104/te for wheat and gross output for barley equal to 1982.
[b] Based on fertiliser supports 50% of production.

Table 3
Broadbalk costs of production

Long-term treatment (kg ha⁻¹)			*Costs*[a] *(£ ha⁻¹)*	*Yield (t ha⁻¹)*	*Cost (£ t⁻¹)*
N	*P205*	*K20*			
Wheat					
0	0	0	599	1·69	354
96	77	107	680	6·60	103
Potatoes					
0	0	0	1 330	8·47	157
96	77	107	1 411	38·57	37

[a] Non-fertiliser costs taken from *Farming in the Eastern Counties of England 1982/83*, Cambridge University. Fertiliser costs based on 1984 market prices.

could increase. Typical growing costs, with and without fertiliser, are shown in Table 3.

This analysis demonstrates the major benefit of using fertilisers in today's circumstances—the dramatic effect they have on production costs per unit of output. The cost of producing wheat is reduced by 71 %, potatoes by 74 %. Put another way, the implication of withdrawing fertiliser use long term would be a three or four-fold increase in the cost of producing basic foodstuffs. This is the real social benefit arising from fertiliser use—lower prices to the consumer and, of course, profit to the farmer.

European agriculture has made considerable progress in recent years in reducing the cost of producing many important crops. The best example is provided by winter wheat. In the UK the application of technology has reduced the real cost of wheat production by 25 % over the last 5 years as shown in Table 4.

The reduction in EEC costs of production is of considerable significance to the finances of the Community. European grain, produced under

Table 4
The real cost of UK wheat production (at constant 1983 prices)

	1979	*1980*	*1981*	*1982*	*1983*	*1984*
Yield (t ha⁻¹)	5·46	6·42	6·29	6·45	6·87	7·56
Real cost (£ tonne⁻¹)	107·20	92·60	96·60	92·50	85·80	78·20

Source: Cambridge Agricultural Economics Unit: Report(s) on Farming.

Table 5
Comparative USA[a] and UK[b] wheat production costs

	US exchange rate *$1·3*	*US* exchange rate *$1·6*	*UK*
Yield (tonnes ha^{-1})	2·19		6·45
Var. costs (£ ha^{-1})	55·7	45·8	199·20
Fixed costs (£ ha^{-1})	209·2	172·4	358·40
Total costs (£ ha^{-1})	264·9	218·2	557·60
Total costs (£ $tonne^{-1}$)	121·0	99·6	86·50

Sources:
[a] *US Costs of Production*, 1982, USDA.
[b] *Farming in the Eastern Counties of England 1982/83*, Cambridge University.

intensive high input/high output systems can now compete with US produce from their low input/low output units. A comparison of wheat production costs in the UK and USA shows that, although US costs per hectare are much lower, when yields are taken into account UK production costs per tonne are actually lower. Table 5 illustrates this comparison and also highlights the impact of exchange rates. Figure 4 would suggest that

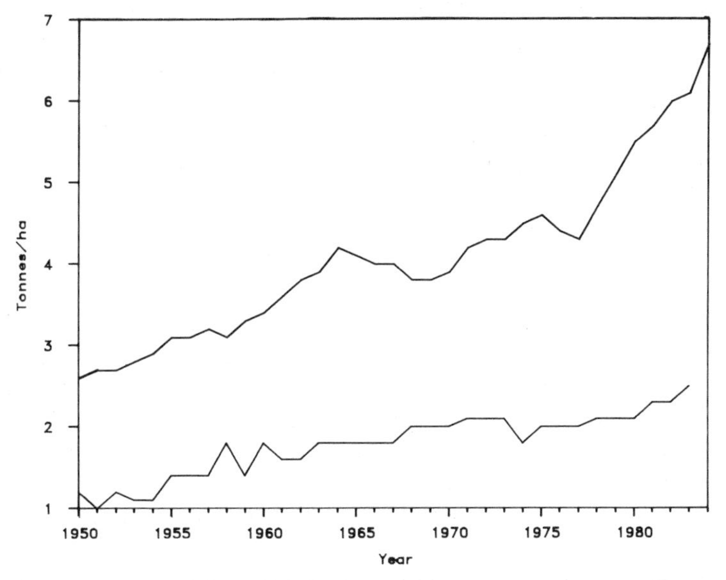

FIG. 4. Comparative UK and USA wheat yields, tonnes per hectare. (Top curve, UK; bottom curve, US.)

this situation is set to continue showing as it does the way in which yield improvement in the UK has far outstripped that achieved in the USA in recent years.

4. WORLD POPULATION GROWTH

A final area worthy of mention when considering the economic significance of the fertiliser industry is the more global one of growing world population who require an increasing supply of food.

The latest UN estimates suggest that the surge in population growth has now slowed due to improved population control in China and the Third World. Current world population is 4·7 billion. By the year 2000 it will be 6·1 billion and it is forecast to stabilise at around 10 billion around 2100. Although it is reassuring to read predictions of a stable population, the expansion in numbers of people to that point is still at the rate of 45 million people per year equivalent to over 5000 extra mouths every hour. Those extra mouths all require extra food and that can only be obtained in one of two ways. It can be obtained either by increasing the area of land farmed or by increasing the yield of food produced by each hectare. The arable land available per head of population is actually declining and the world is therefore forced to look to yield increases to achieve its extra food. This yield increase will depend entirely on the judicious use of fertilisers both in the developed nations to provide food for export and in the developing countries to improve their levels of self sufficiency. The economic significance of fertiliser to world development is thus immeasurable.

5. CONCLUSIONS

To clarify the economic significance of fertilisers it has been necessary to look at a wide range of issues. While the industry itself may appear small in terms of its employment in relation to, say, steel or mining, its contribution to the economy of Europe, and indeed the world, is considerable.

Fertilisers are perhaps the most useful end product of Europe's fossil energy raw material. The use of this non-renewable energy source in fertiliser allows European agriculture to multiply that energy by a factor of 2 or 3.

Fertiliser plants are among the most refined users of energy, modern plants having been developed such that they are now operating at close to their theoretical energy optima. This refinement in technology has come from Europe such that the European industry now provides plant and expertise to the bulk of the world's fertiliser industry.

Fertilisers reduce food prices. Without their use the European consumer could face food prices at twice their current level. Fertilisers achieve this by reducing the unit cost of production. This also enhances the international competitiveness of European grain on world export markets and will lead to a bright future for Europe as a significant and profitable world cereal exporter.

Fertilisers are, in short, the key to modern agriculture. They lead to efficient and economic food production, which in turn means:

cheaper food for the Western consumer and survival for the Third World consumer;
a better living for Europe's farmers.
a more interesting and responsible job for Europe's farm workers through less arduous manual labour.

Fertiliser technology is also totally compatible with the European Commission's objective of reducing the cost of agricultural support, an objective of considerable economic significance to all of Europe's taxpayers.

REFERENCES

1. Lewis, D. A. and Tatchell, J. A. (1979). Energy in UK Agriculture. *Journal of Science, Food and Agriculture*, **30**, 449–57.
2. Lidgate, H. J. (1984). Presentation to Society of Chemical Industry, July.

11

L'Industrie Phytosanitaire: Réalités Présentes, Perspectives à Moyen Terme

F. Calmejane

Union des Industries de la Protection des Plantes,
Paris, France

1. INTRODUCTION

Si le rôle de l'industrie phytosanitaire–encore trop souvent méconnu–est facile à mettre en évidence, il est plus mal aisé de situer cette industrie au sein de l'industrie chimique et de l'agro-fourniture.

Ses activités directes prolifèrent aussi bien en amont—depuis la chimie lourde jusqu'à la chimie fine et la parachimie—qu'en aval, par ses apports substantiels à la distribution, au transport et à l'application des produits.

Les rapports entre les effectifs directement employés et les chiffres de production par pays ne donnent qu'une indication qu'il faut compléter par comparaison avec les mêmes rapports dans les autres secteurs de la chimie.

Malheureusement, les données disponibles ne permettent pas une étude exhaustive englobant tous les pays industrialisés dotés d'une industrie phytosanitaire importante. Les perspectives à moyen terme sont la résultante de deux tendances un peu contradictoires:

une dynamique propre encore très porteuse avec des possibilités d'innovation certaines à explorer;

des contraintes financières et administratives de plus en plus lourdes qu'il convient de contenir dans les limites compatibles avec les impératifs de la recherche et du développement.

A la suite de la défection de M. Woodburn, il m'a été demandé de parler de l'industrie phytosanitaire en termes d'activité économique et d'emploi.

Autant le premier aspect se prête à une multitude de développements, autant le second se caractérise par une carence de données homogènes, ce qui devrait limiter mon propos. Pour que mon intervention présente néammoins quelque intérêt il m'est apparu nécessaire d'en élargir un peu le thème.

Comme l'a dit M. King en ouvrant la première séance de travail, il faut 'réenvisager l'approche du problème'. C'est dans ce but que nous aiderons à une meilleure compréhension par une information plus complète. C'est pourquoi nous avons pour projet de vous présenter quelques aspects de l'industrie phytosanitaire.

Des éléments publiés, utilisés récemment lors d'une étude servant de base à une projection à court et moyen terme de l'évolution de l'industrie qui nous intéresse vont étayer notre discussion.

L'utilisation des produits phytosanitaires n'est pas limitée aux seules cultures alimentaires. Le coton qui figure au second rang des cultures par ordre commercial d'importance ainsi que l'hévéa en sont deux exemples.

Toutefois le but premier des produits antiparasitaires est bien de contribuer à réduire et, espérons-le, à vaincre la faim dans le monde. Le nombre d'habitants sous-alimentés, estimé à 1·5 milliard sur les 6 milliards attendus au début du XXIème siècle dépend directement du développement de l'agriculture, maintenant que les épidémies qui décimaient les populations sont efficacement combattues.

Parmi les dépenses de la production agricole, les achats représentent 54 % des sommes engagées selon la répartition suivante:

Energie	23 %
Services	20 %
Engrais; Semences	5 %
Produits de protection	2·1 %
Divers	3·7 %

Le marché mondial des produits phytosanitaires peut être estimé cette dernière saison entre 5 et 6 milliards de $. Selon la nature et l'usage d'une spécialité phytosanitaire, on peut estimer son coût entre 25 et 30 millions de $. Dans ce total, si la synthèse, la recherche et le développement représentent environ 50 %, la part de la 'toxicologie' au sens large représente presque 20 % et ne cesse de croître.

Il est en effet nécessaire de connaître aussi vite et aussi complètement que possible la nature des risques encourrus pour mieux diriger l'utilisation des produits et conseiller les personnes qui sont en contact avec eux à tout moment. Ceux qui sont au fait des recherches réalisées par les sociétés qui créent de nouveaux produits anti-parasitaires savent combien de grands produits ont été arrêtés en cours de développement, ceci en raison de données d'ordre toxicologique qui sont apparues au fur et à mesure du déroulement des études à long terme.

Chaque jour, quelque six mille substances chimiques sont identifiées. Parmi elles, mille en moyenne sont nouvelles. A la fin 1982, le C.A.S. (Chemical Abstracts Service) avait enregistré quelque six millions de substances chimiques mentionnées dans les publications depuis 1965. En fait, le nombre de substances d'usage courant n'excède guère soixante à soixante-dix mille, trois mille d'entre elles représentent 90% de l'ensemble en poids.

2. LES PHYTOSANITAIRES: UN FACTEUR DE PRODUCTION

Nul ne conteste plus aujourd'hui les progrès spectaculaires réalisés au cours des quarante dernières années grâce à l'optimisation des facteurs de production en général et à la protection des plantes en particulier.

Des chiffres très éloquents ont été montrés ce matin et je n'ai rien qui puisse être ajouté à la démonstration qui a été faite, si ce n'est reproduire l'évolution de la production moyenne de blé à l'hectare en France de 1945/46 à 1984/85 (Tab. 1).

Ces progrès ne sont pas le seul fait de la phyto-pharmacie mais il faut cependant rendre à celle-ci la part qui lui est due. La protection chimique des cultures est une assurance de sécurité pour les agriculteurs qui sont en droit d'attendre un profit de leurs investissements, ceci pour la plupart des pays de la Communauté Européenne. N'oublions pas qu'ailleurs la protection des cultures est vitale pour les cultures vivrières et les cultures industrielles.

3. AMÉLIORATION DE LA QUALITÉ

Quant à la qualité, il est dans cette salle des personnes plus qualifiées que moi pour en parler, je pense notamment à MM. les Responsables des Services Officiels de la Protection des Végétaux. Je ne crains pas cependant d'affirmer sous leur contrôle, que l'utilisation des moyens modernes de production agricole et notamment de protection chimique des cultures, lorsqu'ils sont mis en oeuvre au stade voulu et dans les conditions légales d'application, ne portent aucune atteinte à la qualité des productions.

Il y aurait également beaucoup à dire sur l'apport des produits de protection des récoltes à la qualité des productions. Quant aux risques pour l'environnement, ils sont largement pris en compte, comme en témoignent les obligations draconiennes imposées à l'industrie pour la fabrication des

Table 1
Collecte de Blé en France

Années	*Surfaces ensemencées en hectares*	*Rendement moyen à l'hectare en quintaux*	*Quantités commercialisées en quintaux collecte*
45/46	3 783 003	11·13	25 780 197
46/47	4 131 119	16·36	43 052 200
47/48	3 392 948	9·62	17 836 126
48/49	4 231 194	18·05	47 444 174
49/50	4 222 694	19·14	59 282 802
50/51	4 318 784	17·83	53 768 135
51/52	4 250 400	16·70	49 174 115
52/53	4 296 500	19·60	55 366 879
53/54	4 219 400	21·30	63 332 146
54/55	4 491 320	23·50	77 786 710
55/56	4 553 630	22·76	70 884 058
56/57	2 745 080	20·70	43 344 273
57/58	4 667 620	23·74	81 646 821
58/59	4 615 000	20·80	64 916 232
59/60	4 425 750	26·10	
60/61	4 311 840	25·38	77 661 779
61/62	3 894 100	24·00	67 762 000
62/63	4 522 500	30·88	100 738 000
63/64	3 812 300	26·70	70 742 720
64/65	4 353 500	31·60	97 945 709
65/66	4 480 900	32·70	110 068 690
66/67	3 931 800	28·40	78 784 023
67/68	3 847 500	37	105 350 000
68/69	3 983 400	36·80	114 910 577
69/70	3 908 200	41·6	110 392 317
70/71	3 597 000	34·6	96 348 016
71/72	3 800 200	39	121 862 153
72/73	3 788 900	46	149 001 166
73/74	3 807 100	46	146 382 751
74/75	3 961 500	46	160 594 002
75/76	3 585 700	40	117 969 002
76/77	4 080 800	38	130 252 664
77/78	4 008 000	43	141 134 000
78/79	4 072 000	51	171 720 000
79/80	3 987 000	47	165 398 000
80/81	4 466 000	52	202 747 000
81/82	4 629 000	49	190 245 000
82/83	4 734 000	53	215 860 000
83/84	4 749 000	51	213 000 000
84/85	4 929 000	63	271 000 000

produits phytosanitaires, et aux utilisateurs à travers les prescriptions d'homologation. Ces quelques rappels me paraissent nécessaires pour aborder les réalités de notre industrie. Nous y reviendrons plus loin.

4. LA PHYTO-PHARMACIE ET L'EMPLOI

Les pays producteurs indiquent en moyenne un nombre de 10 000 personnes concernées; c'est le cas du Japon et de la France. L'Allemagne en compte un peu plus et le Royaume-Uni un peu moins. Les Etats-Unis, avec un peu plus du tiers de la production et de l'utilisation mondiales, doit compter pour le double soit au total environ 70 000 à 80 000 emplois. Mais ceci peut être au moins doublé car tous les pays se sont dotés au niveau agronomique, toxicologique, physicochimique et autres de moyens d'évaluation et de contrôle en vue de l'homologation et de la préconisation, qui nécessitent une main d'oeuvre qualifiée et nombreuse.

Ceci a permis une meilleure compréhension et une collaboration accrue entre l'industrie et les Pouvoirs Publics.

En raison des aspects nouveaux qui apparaissent continuellement, relatifs à l'emploi et aux conséquences directes ou indirectes des anti-parasitaires, il faut élargir en conséquence le domain d'activité.

Les travaux qui se développent ici depuis hier le démontrent ainsi que votre présence, vous tous spécialistes avertis, et vos interventions.

Parmi ces champs d'action qui se développent, citons notamment:

L'application des traitements. Ceci concerne les appareils, les méthodes de traitement, les associations de produits, la variation des volumes, les sites,...

La récupération des emballages vides, l'enlèvement des produits non utilisés ou au moins les recommandations pour leur destruction ou leur neutralisation, autant de nouvelles mesures prises récemment.

L'information à tous les niveaux:
- dossiers officiels;
- utilisateurs;
- préconisateurs;

par des conférences ou des documents écrits est aussi en voie de nette amélioration et de développement.

Il faut signaler sur ce point les travaux réalisés non seulement par les associations nationales, mais aussi ceux coordonnés au niveau du GIFAP et notamment ceux visant à l'information des populations des pays en voie de développement.

Je voudrais profiter de l'occasion qui est donnée à l'industrie de mieux se présenter en rappelant à ceux qui n'ont parfois qu'une vue parcellaire de notre activité le schéma, même simplifié de la vie d'un produit.

4.1. Synthèse

On peut estimer qu'une sur les 9000 matières actives découvertes peut atteindre le niveau commercial

La découverte se continue pour les échelons suivants:

Détermination de l'activité biologique.

Etude quantitative de cette valeur biologique avec étude préliminaire:
- comparative;
- spécificité ou polyvalence;
- sélectivité, inocuité

Essais de plein champs:
- étude exhaustive des facteurs précédents;
- répétition sur deux à six ans.

Parallèlement:
- étude des actions secondaires (abeilles, prédateurs);
- étude des risques de résistance.

4.2. Etudes spécifiques sur la toxicologie

(a) Directe—évaluation des risques

- aiguë;
- chronique;
- action cancérigène;
- action mutagène;
- action tératogène.

(b) Induite—évaluation

- résidus sur parties traitées;
- métabolites systémiques;
- eau;
- sol;
- déchets sur parties non utilisées;
- déchets dans produits finis: conserves, vin, bière.

4.3. Partie chimique proprement dite

Procédé de fabrication.

Coût du produit:
- matières premières;

main d'oeuvre;
fabrication et matériel;
rendement;
résidus de fabrication et coût des éliminations vis à vis de l'air, de l'eau et du sol;
station d'épuration, de traitement des eaux perdues;
incinération des phases solides après coagulation.

Médecine du travail: exposition des employés à tout moment de la fabrication.
Conditionnement.
Transport et stockage.

4.4. Formulations—évolution

Nous venons d'une période, il y a trente ans à peine, où l'on mettait en oeuvre, avec 1000 litre/ha de bouillie Bordelaise à 2 kg de Sulfate de cuivre/100 litre plus la chaux de neutralisation, 25 à 30 kg de 'produit' par hectare.

Après une période de transition où l'on mettait en oeuvre 2 à 4 kg/ha, nous arrivons, sous la pression toxicologique s'appliquant aux résidus après traitement, à des produits qui s'utilisent à 150 g/ha, et déjà arrivent les spécialités pouvant être utilisées à 15 et 30 g/ha.

4.5. Protections, evolutions, limitations

(a) Economiques: prix de revient du produit et des études de la fabrication, de l'évaluation des propriétés biologiques, des études toxicologiques...
(b) Administratives—homologation qui étudie tous les facteurs déjà mentionnés:
    propriétés physico-chimiques des spécialités phytosanitaires offertes à l'utilisateur:
        étude du dossier toxicologique incluant la formulation, c'est à dire éventuellement les adjuvants:
            dispersants
            mouillants
            émulsifiants
            colorants
            stabilisants
            adhésifs
            etc...
        et les métabolites des matières chimiques utilisées dans les matières comestibles, dans les animaux, l'eau et les sols.

(c) En conséquence, des législations ont été mises en place pour l'évaluation de tous les risques. Deux facteurs importants ont influé favorablement sur une évolution positive:

la prise de conscience par toutes les parties concernées des risques à minimiser;

la progression du perfectionnement des méthodes d'évaluation de ces risques; par exemple les analyses de micro-résidus de l'ordre du ppb.

Pour mémoire je rappellerai la course contre le temps qu'il faut mener pour que l'industrie puisse satisfaire à toutes les études et à la durée d'un brevet. Sur ce point, nul doute que nos demandes relatives à certaines protections seront entendues tant elles sont fondées.

5. CONCLUSIONS

En conclusion, je dirai que l'utilisation des produits antiparasitaires doit être assimilée dans un programme de lutte raisonné. C'est à dire qu'elle est la conclusion de l'évaluation des trois bilans principaux:

bilan économique: profit pour l'agriculture;

bilan toxicologique: balance risque/profit;

bilan de l'action sur l'environnement, grâce notamment à la connaissance des problèmes et des solutions trouvées, domaine sans cesse en voie d'amélioration.

On oppose parfois à la lutte chimique même raisonnée des méthodes qui doivent évidemments être examinées selon les trois critères exposés ci-dessus. Il ne faut cependant pas perdre de vue le but final que l'agriculteur vise lorsqu'il commence une culture. Trop souvent malheureusement, les représentants des agriculteurs sont absents de discussions qui les concernent au premier chef.

L'industrie a montré qu'elle restait un partenaire majeur de l'agriculture et montre journellement qu'elle entend maintenir son rôle.

SESSION III

Ways for Economical and Ecological Optimisation of the Use of Fertilisers and Pesticides

Chairman: E. P. CUNNINGHAM
(*The Agricultural Institute, Dublin, Ireland*)

12

Évaluation et Procédés de Mise sur le Marché des Pesticides Agricoles

M. HASCOET

Institut National de la Recherche Agronomique, Versailles, France

RÉSUMÉ

Si le bénéfice de l'emploi des pesticides est considéré comme un acquis, des craintes persistent quant à leurs effets sur l'environnement. Le caractère subjectif des données exigibles en matière d'environnement est souligné de même que l'apport important que constituent les données de base dans la prévision du risque pour le milieu naturel. Quelques exemples sont présentés concernant la position des autorités nationales sur les données complémentaires exigibles dans chaque pays. La nécessité des études de post homologation est ensuite évoquée et des exemples de réalisation en Europe donnés. L'opposition parfois entretenue entre les essais de laboratoire et de plein champ est discutée ainsi que les problèmes posés et les voies possibles d'une actualisation de critères d'homologation.

1. INTRODUCTION

Les pesticides constituent un 'mal nécessaire' dans la mesure où les autres approches se révèlent insuffisantes pour résoudre les problèmes phytosanitaires qui se présentent à nous. Ces problèmes se posent avec une acuité particulière dans l'agriculture intensive qui prédomine dans le marché commun.

Or si le bénéfice de l'emploi de ces produits sur la production agricole est considéré comme un acquis, des craintes persistent dans le grand public et même dans certains milieux scientifiques quant à l'impact de leur usage généralisé sur notre environnement.

Des informations et divers essais sont demandés à ce sujet pour l'homologation d'un nouveau produit par les autorités nationales.

Il faut cependant reconnaître que le choix des données exigibles en matière d'environnement est plus subjectif que dans d'autres domaines. Ceci est dû au caractère éminemment complexe du milieu naturel, à la sensibilité particulière de certaines gens (pêcheurs, chasseurs, apiculteurs) vis-à-vis de l'emploi des pesticides ou enfin à l'inégale connaissance que nous avons de l'impact de ces produits sur les différents écosystèmes.

Ce caractère subjectif du choix pourrait entraîner des différences importantes entre les exigences nationales ce qui créerait une contrainte évidente pour les industriels qui homologuent un produit dans plusieurs pays.

En ce qui concerne la Communauté, les exigencies nationales tentent dans leurs grandes lignes, de se conformer aux recommandations du Conseil de l'Europe (CE). Ces recommandations concernent tous les aspects de l'homologation: identité du produit et de la formulation, propriétés physicochimiques, toxicologie expérimentale, impact sur l'environnement et efficacité.

Ces aspects ne sont d'ailleurs pas indépendants et une partie importante des données de base doit être prise en compte lorsqu'il s'agit de prévoir l'effet d'un pesticide sur l'environnement.

2. INTERET DES DONNEES DE BASE DANS L'EVALUATION DE L'ECOTOXICITE

La connaissance des usages demandés, de l'époque et des conditions d'application permet de prévoir quelles sortes de populations naturelles seront exposées au risque.

La nature de la formulation elle-même n'est pas indifférente. On connaît le risque accru pour les oiseaux granivores de l'emploi des granulés au lieu des poudres ou liquides. La nature du support de ces granulés est elle-même importante. Ainsi on a constaté en France que le remplacement de la rafle de maïs par un support minéral pour des formulations à base d'aldicarbe réduisait sensiblement la fréquence des accidents. De même les formulations microencapsulées d'insecticides, qui accroissent la sécurité de l'applicateur, pourraient constituer un danger pour les abeilles.

Des relations existent aussi entre les propriétés physicochimiques d'une molécule et son aptitude à la dissémination. Une tension de vapeur élevée, en favorisant la volatilisation d'un pesticide réduit sa persistance sur les cultures ou les sols traités. La solubilité dans l'eau, en favorisant le lessivage comme l'entraînement superficiel est un facteur notable de dispersion. Les essais de comportement physicochimique (hydrolyse, photodégradation) renseignent sur la persistance et sur les voies de dégradation tandis que le coefficient de partage octanol-eau est considéré comme indicatif de la faculté d'une molécule à s'accumuler dans les organismes vivants. Les valeurs de résidus et surtout les cinétiques d'évolution sont immédiatement utilisables pour les espèces fréquentant le biotope traité (arthropodes) et par extrapolation renseignent sur la rémanence dans l'environnement en général.

Concernant la toxicité expérimentale, la somme considérable de données obtenues sur rongeurs est extrapolable aux mammifères sauvages. En particulier les données de mètabolisme et les effets observés sur la reproduction constituent une indication de valeur.

3. DONNEES COMPLEMENTAIRES CONCERNANT LE RISQUE POUR L'ENVIRONNEMENT—POSITION DES AUTORITES NATIONALES

Les informations précédentes aident à définir le profil écotoxicologique d'une substance mais ne permettent pas de juger des risques particuliers que l'on rencontrera dans un biotope donné. En fonction du type d'utilisation et de l'interprétation des données de base un complément d'informations est donc demandé par les autorités nationales concernant:

- le sol et les organismes vivants dans le sol;
- le milieu aquatique;
- le milieu terrestre (oiseaux, mammifères, arthropodes).

Dans la CEE il s'inspire, avons-nous dit des recommandations du Conseil de l'Europe, mais les exigences varient néanmoins selon les états.

3.1. Sol et microorganismes vivants dans le sol

Le sol constitue une cible 'privilégiée' lors de l'application des produits antiparasitaires. Tout ou partie du traitement appliqué aux cultures parvient au sol où il se retrouve fixé, partiellement redistribué et progressivement dégradé.

Deux problèmes se posent:

- celui du comportement de la molécule (mobilité, persistance, métabolisme);
- celui de son effet sur les organismes vivants dans le sol, surtout s'il s'agit d'un produit stable ou fréquemment employé.

Tableau 1
Sol et microorganismes du sol—données complémentaires: Position de la RFA, de la GB, des PB et du CE

		CE	*RFA*	*PB*	*GB*
Comportement					
mobilité		+	colonne 3 sols standards allemands	4 méthodes proposées 2 ou 3 sols types	pas de protocoles définis mais données exigibles sur les 3 points
persistance		+	2 sols standards	2 sols types 2 températures	
dégradation		+	2 sols standards	2 sols + compléments	
Effets sur faune et microflore					
macrofaune	Indicateurs	carabidés vers de terre limace	—	—	Indicateurs carabidés staphylinidés vers de terre limace
microflore		activité microbienne globale inhibition de la nitrification	—	activité microbienne globale ($O^2\downarrow$, $CO^2\uparrow$) inhibition de la nitrification	décomposition de la M.O. litter-bag test

Dans le Tableau 1, sont schématisées les positions des autorités nationales concernant ce deux aspects [1, 2, 3].

Ainsi concernant les études de comportement dans les sols les états représentés dans le Tableau 1 (les seuls qui aient une législation formelle) se conforment (ou se réservent le droit de le faire) aux recommandations du Conseil de l'Europe. On peut simplement regretter que le choix de sols expérimentaux allemands soit imposé. Sur ce point précis la législation hollandaise est plus souple, mais par ailleurs elle manifeste des exigences plus grandes pour les informations de base et surtout exige des compléments 'modulables' dans certaines situations (dégradation aux basses températures, en climat sec, en anaérobie, comportement au delà de la zone racinaire . . .). Cependant ces positions (et aussi celles d'autres états: France, Belgique) n'étant pas très éloignées, il serait souhaitable de parvenir à un consensus communautaire. Ceci supposerait réglé un préalable: établir les caractéristiques pédologiques des sols considérés comme représentatifs des sols européens et donc utilisables dans tous les pays.

Par contre en ce qui concerne les effets sur la macrofaune et la microflore du sol, les experts sont certainement conscients des risques potentiels que peuvent présenter certaines molécules pesticides mais les avis sont partagés sur le choix des indicateurs ou la valeur prévisionnelle des méthodes actuellement proposées. Ceci explique l'absence de normes d'évaluation à ce point de vue dans plusieurs pays (Allemagne mais aussi France).

3.2. Milieu aquatique

Hormis le cas d'application directe (destruction des plantes aquatiques, accidents . . .) la pollution de l'eau par les produits phytosanitaires lors d'un traitement agricole est sinon improbable, du moins fort peu importante qu'il s'agisse de lessivage ou de ruissellement superficiel. En effet on peut considérer que dans nos régions le lessivage profond n'est pas quantitativement important ce qui limite le risque de pollution des nappes phréatiques. Fryer *et al.* [4] ont montré que le pichlorame, considéré comme mobile à cause de sa solubilité et de ses caractéristiques d'absorption n'a pas été retrouvé en dessous de 75 cm de profondeur 2 ans après une application de 1·68 kg/ha. De même lors de plusieurs séries d'essais réalisés en France sur vignes, l'application de 20 kg/ha d'aldicarbe n'entraîne jamais l'apparition de résidus au delà de 70 cm, bien que l'aldicarbe et ses métabolites d'oxydation soient aussi considérés comme mobiles. Les faibles quantités de triazines ($<2\%$ de la dose d'emploi et inférieures au seuil de signification biologique) qui ont été observées dans

Tableau 2
Milieu aquatique—essais de base et données complémentaires: position du CE, des PB et de la GB

	CE	*PB*	*GB*
Comportement: (eau + sédiments)			
essais de base:	—	—	—
	(données physicochimiques + adsorption, métabolisme/sol, suffisent)		
situations à risque	—cinétique de dégradation: eaux naturelles		résidus: (courbe de disparition) eau
	—adsorption: sédiments, végétaux aquatiques	adsorption, métabolisme: sédiments . . .	
Toxicité: milieu aquatique			
essais de base:	—	CE 50 algue CL 50 daphnie CL 50 poisson	CL 50, 96 heures poisson daphnie?
situations à risque	CL 50, 96 heures poissons plusieurs espèces CL 50 daphnie, cyclopes, larves: tubifex, chironomes . . .	—N_1 2 espèces par niveau trophique —N_2 seuil de toxicité: (sur plusieurs critères) algue, daphnie, poisson	toxicité cumulative poisson

N_1, N_2, niveaux de risques.

les tuyaux de drainage situés à des profondeurs variant de 0·9 à 1·60 m [5, 6] confirment l'existence d'une mobilité modérée pour ce type d'herbicide.

En ce qui concerne le ruissellement superficiel, des essais réalisés en Belgique sur des terrains en pente, ont montré que les quantités de néburon et de nitrofène entraînées hors de la parcelle traitée ne représentaient respectivement que 0·4–1·5 % et 0·04–0·15 % de la quantité appliquée. Cette étude a été complétée par la surveillance des eaux de la Méhaigne qui traverse un bassin à vocation agricole de 2000 ha. Au cours des années 1976–77, du chlortoluron a été mis en évidence en octobre et en décembre 1976 et du méthabenzthiazuron dans 1 seul prélèvement (5 ppb). Les autres herbicides utilisés dans la région (urées substituées, nitrofène) n'ont pu être décelés au niveau du ppb. Des essais réalisés en France (Champagne 1984) ont conduit à des résultats analogues. L'entraînement superficiel reste donc dans nos régions peu important. Il peut le devenir occasionnellement dans les régions peu important. Il peut le devenir occasionnellement dans les régions méditerranéennes si des pluies violentes surviennent peu de temps après le traitement lorsque le terrain est en forte pente.

Cette distinction entre le risque très limité que présente l'application en cultures et celui évident dû à l'emploi quasi direct sur les eaux de surface, est bien marquée dans les recommandations du CE comme dans les législations nationales ainsi qu'il apparaît dans le Tableau 2 qui présente les positions respectives du CE, des PB et de la GB [1, 2, 3].

A propos des études de comportement, on constate que seules les situations à risque font l'objet d'un complément d'information. Pour les usages agricoles courants, la plupart des pays considèrent comme suffisantes les données physicochimiques de base et les études réalisées sur sol (adsorption et métabolisme).

En matière de toxicité pour le milieu aquatique l'essai de base se limite généralement à une toxicité aiguë sur poisson, mais la panoplie des essais à mettre en oeuvre s'élargit considérablement dès que le risque augmente. Les exigences sont particulièrement sévères aux PB, peut-être parce que ce pays possède des surfaces importantes de terres drainées, quadrillées de fossés qui peuvent être facilement pollués lors des traitements de culture.

3.3. Mammifères sauvages et oiseaux

Nous avons vu que la toxicité espérimentale apportait des informations très complètes extrapolables dans une certaine mesure aux mammifères sauvages bien que ceux-ci par exemple le gros gibier puissent avoir un comportement alimentaire différent des rongeurs. Compte tenu de leur milieu naturel, le risque de contamination du gros gibier est d'ailleurs peu important comme le démontre une étude allemande concernant leur contamination par les OC.

La situation est moins favorable, a priori, pour les oiseaux. Bien que l'on puisse évoquer la possibilité d'une contamination directe par exposition aux embruns de traitement ou l'intoxication indirecte par ingestion d'insectes contaminés, la cause d'intoxication la plus fréquente (au moins chez les granivores) réside dans la consommation des granulés ou des grains traités.

Aujourd'hui il s'agit surtout d'intoxications aiguës car les pesticides récents sont généralement biodégradables et les risques d'accumulation, à travers une chaîne alimentaire, pratiquement inexistants.

Les recommandations du CE tiennent cependant toujours compte de ce risque et prévoit la recherche des résidus dans les tissus et dans les oeufs. La sensibilité des oiseaux se révélant souvent différente d'une expèce à l'autre, la toxicité aiguë est généralement établie sur 2 espèces (par ex. caille Japonaise et canard Mallard). Lorsque les conditions d'utilisation (appâts, traitements de graines) ou de formulation (granulés) introduisent un risque supplémentaire, les PB demandent une étude de toxicité subaiguë sur 2 espèces pendant 14 jours, tandis que les autres pays (GB, France, Belgique) se satisfont d'une toxicité cumulative pendant 5 jours sur 1 seule espèce.

Enfin pour les produits à haut risque (?) chez certaines espèces les PB prévoient une étude de toxicité semi-chronique sur 10 % de la durée de vie de l'oiseau.

L'action sur le comportement est surtout pris en compte en GB qui souligne l'intérêt des observations dans les zones traitées.

Une bonne harmonie existe donc entre les législations nationales concernant le risque pour les oiseaux, risque qui est maintenant bien cerné au moins en ce qui concerne les applications agricoles.

3.4. Arthropodes

3.4.1. Abeille

L'abeille est l'espèce représentative des insectes dans les études de base et d'autre part c'est une pollinisatrice et une productrice de miel. C'est pourquoi une DL 50 abeille, orale et/ou de contact est demandée dans la plupart des pays européens. Lorsque cette toxicité aiguë se révèle élevée et si les produits sont appliqués en situation de risque pour les abeilles, un complément d'information est demandé qui est maintenant à peu près normalisé dans la CEE. Il comprend 2 étapes (la 1ère facultative):

essais sous-abri où sont en présence les plantes traitées en fleurs et les abeilles;

essais en plein champ, réalisés sur des parcelles isolées de grande dimension, près desquelles sont disposées des ruches. L'essai permet de juger de la mortalité des abeilles et de la contamination éventuelle du rucher.

Ce type d'essai s'est révélé particulièrement riche d'enseignements dans le cas des pyréthroïdes. Il a permis de mettre en évidence l'effet répulsif de ce type de molécule et a démontré que la mortalité des abeilles visitant les parcelles traitées, même en pleine floraison, était pratiquement négligeable.

3.4.2. Auxiliaires entomophages et espèces indifférentes

Il est bien connu que la destruction des auxiliaires entomophages peut être à l'origine de la multiplication de certains ravageurs. De même qu'un effet toxique vis-à-vis de l'entomofaune dite 'indifférente' se révèle préjudiciable; en appauvrissant la composition de l'agrocénose, elle intervient négativement sur l'équilibre dynamique de l'agrosystème.

Nous avons rappelé précédemment que lors des essais d'efficacité, des informations sont obtenues sur la toxicité d'un produit pour différentes espèces d'arthropodes. On peut donc, par observation directe ou extrapolation estimer l'impact du pesticide sur les insectes du biotope traité. Il ne s'agit là que de données qualitatives mais il ne nous semble pas que les différents pays européens se soient penchés plus avant sur le problème, sans doute par suite des difficultés que présente ce type d'investigation.

Un effort important est réalisé dans ce domaine par l'OILB et en particulier par le groupe de travail 'Pesticides et arthropodes utiles' qui a pu proposer des protocoles expérimentaux standardisés pour estimer au laboratoire la sélectivité des pesticides [10].

D'autres méthodes de laboratoire permettent d'apprécier l'effet à plus ou moins long terme d'un pesticide sur un couple proie-prédateur ou parasite-hôte [11, 12].

Enfin malgré les contraintes expérimentales particulières à ce genre d'essais, le groupe de travail 'Protection intégrée en verger' tente d'élaborer une méthode d'essai de plein champ qui soit 'faisable et fiable' [13, 14, 15].

4. ETUDES POST-HOMOLOGATION

La complexité de l'environnement rend illusoire la possibilité de réaliser un nombre d'essais suffisants au titre de l'homologation pour rendre compte de toutes les situations que l'on peut trouver dans la réalité. D'autre part il est souhaitable de disposer de réseaux de surveillance qui puissent contrôler la validité des essais d'évaluation. Des travaux complémentaires ou de surveillance sont donc réalisés, indépendamment du contexte de l'homologation.

Ainsi en GB le 'Joint Cereal Ecosystem Project' [16] surveille l'impact des techniques agricoles modernes (y compris les traitements) sur certaines espèces d'invertébrés, en particulier les parasites et leurs prédateurs dans les cultures de céréales.

En France des travaux sont réalisés depuis 1979, par l'INRA sur l'incidence des interventions insecticides sur les composants de l'entomofaune dans les biocénoses céréalières [17].

L'ACTA développe également dans le cadre du programme 'Conséquences des techniques de protection des cultures sur l'environnement' des travaux sur l'impact des insecticides et des acaricides sur la faune auxiliaire en verger et celui des pyréthroïdes sur céréales [18].

En ce qui concerne plus particulièrement la surveillance, l'INRA et l'ONC suivent depuis 1957 la mortalité du gibier due aux pesticides par analyse des cadavres d'animaux [19].

De même, à la suite de l'APV donnée à l'emploi des pyréthrinoïdes pour le traitement du colza en début de floraison, la PV en relation avec les apiculteurs, a créé un réseau d'observation destiné à renseigner sur les mortalités anormales d'abeilles consécutives au traitement.

5. LES ESSAIS DE LABORATOIRE ET DE PLEIN CHAMP

Une opposition est souvent entretenue entre essais de laboratoire et de plein champ; les premiers sont considérés comme peu représentatifs de la réalité et les seconds (longs et coûteux) d'interprétation parfois difficile. Que doit-on penser en particulier de la recommandation des experts de la FAO qui conseillent de pondérer les essais de laboratoire donnant à penser que le risque pour l'environnement est élevé et de les doubler par des essais sur le terrain.

Il est évident qu'un essai de laboratoire est un compromis entre la faisabilité et la représentation fidèle du phénomène biologique évalué. En ce sens l'interprétation des résultats par intercomparaison avec d'autres données est toujours considérée comme très importante. De plus les conclusions pessimistes sont souvent liées au fait que l'essai de laboratoire n'informe que sur un aspect de la toxicité. Ainsi les pyréthroïdes, sur la base des seules DL 50 doivent être classés comme dangereux pour les abeilles et les poissons alors que les essais 'de plein champ' permettent de nuancer parce qu'ils introduisent d'autres paramètres (adsorption sur les sédiments en suspension, demi-vie très courte dans l'eau, action répulsive vis-à-vis des 2 espèces). Les essais de terrain peuvent effectivement corriger une conclusion trop pessimiste mais en ce sens leur nécessité doit être mieux ressentie par le demandeur que par les autorités nationales! De toute manière le caractère séquentiel des essais doit être préservé et les essais de terrain envisagés seulement si le pronostic primaire est défavorable.

6. ACTUALISATION DES CRITERES D'HOMOLOGATION

Les critères d'évaluation sont choisis en fonction de l'acquis scientifique du moment et à ce titre ils peuvent à tout moment se voir remis en cause ou tout au moins voir leur valeur relativisée. Nous partageons tout à fait l'opinion de Greaves *et al.* [20] lorsqu'ils insistent sur la nécessité d'une législation très souple permettant la suppression des méthodes depassées et l'introduction de plus performantes. Mais la remise en question d'une technique expérimentale oblige perfois à repenser l'ensemble de l'évaluation pour un critère donné. Ainsi en est-il des essais in vitro de persistence dans le sol considérés comme non-valables par les microbiologistes si leur durée dépasse 2 mois. De nouveaux types d'essais extérieurs (par ex. microlysimètres) doivent donc être conçus et expérimentés qui permettent le développement normal des microorganismes et l'obtention d'un bilan aussi précis que celui réalisé au laboratoire.

Actuellement, se trouve aussi contesté le sacrifice, jugé non indispensable, de nombreux animaux de laboratoire et demandé leur remplacement par un matériel biologique 'insensible' [23]. Cette tendance conduit à développer les techniques in vitro dans la mesure où elles permettent d'obtenir des informations d'égale qualité. Des recherches sont effectuées dans ce sens par plusieurs laboratoires. Bunyan [21] a souligné l'intérêt des mesures d'activité enzymatique, praticables sur un petit volume de sang, qui permettraient l'observation non destructrice de l'effet des pesticides sur les animaux au cours des essais de plein champ. De même à Versailles Rivière [22] a pu constater, chez la Caille Japonaise que des modifications d'activité enzymatique (monooxygénases à cytochrome P-450) mesurées 'in vitro' vont de pair avec des modifications de sensibilité aux insecticides observées chez l'oiseau. D'autres études ont pour objectif la mesure des différences interspécifiques (spécificités de site d'action et métabolique). Ces données, en dehors de leur caractère cognitif pourraient nous fournir, par la suite, des outils biochimiques (cholinestérases, enzymes de métabolisation provenant de différentes espèces) permettant d'apprécier in vitro la sensibilité de ces espèces aux nouvelles molécules.

On doit noter aussi un déplacement des problèmes écotoxicologiques. Les phénomènes de bioaccumulation, au centre des préoccupations des écologistes comme des hygiénistes il y a une vingtaine d'année, ont beaucoup perdu de leur acuité. Sans doute parce que le caractère 'bioaccumulable' constitue un clignotant rouge pour les autorités et de ce fait peu de produits actuellement posent un problème sur ce point. Par contre, l'apparition des pyréthroïdes, insecticides extraordinairement efficaces dont le spectre est très large, ont sensibilisé les entomologistes à leurs effets potentiels sur l'entomofaune utile voire indifférente. Le groupe de travail de l'EPPO sur les pesticides utilisés pour la protection des plantes s'est d'ailleurs dit 'prêt à examiner des projets de directives pour déterminer les effets secondaires des pesticides sur les arthropodes utiles'.

7. CONCLUSION

L'effort considérable de réflexion et les connaissances acquises dans toutes les disciplines concernées par le milieu naturel nous ont permis de parvenir à une meilleure appréciation des risques que peut présenter l'introduction des pesticides dans ce millieu. A ce sujet, on ne soulignera jamais assez le rôle de l'ouvrage 'le printemps silencieux' qui malgré des inexactitudes notoires a permis de prendre conscience d'un problème auquel on n'accordait sans doute pas assez d'attention.

La réflexion menée en commun au sein des instances internationales FAO, OECD, OILB, EPPO... devrait se traduire à terme par une harmonisation accrue des législations nationales bien que comme le faisait remarquer Hudson des considérations à la fois politiques et scientifiques tendent à maintenir un certain particularisme au niveau des Etats.

RÉFÉRENCES

1. PSPS, Pesticides Branch, Ministry of Agriculture, Fisheries and Food, GB.
2. RFA, Communication du Dr W. Weinmann, BBA Braunschweig.
3. Pays-Bas—Commissie voor Fytofarmacie, Bureau Bestrijdingsmiddelen.
4. Fryer, J. D. *et al.* (1979). *Journal of Environmental Quality*, **8**, 83–6.
5. Schiavon, M. et Jacquin, F. (1973). Compte rendu de la 7ème conférence du COLUMA 35–43.

6. MUIR, D. C. G. et BAKER, B. E. (1978). *Weed Research*, **18**, 111–20.
7. COPIN, A. et DELEU, R. (1978). *Pédologie* XXVIII, 205–13 Gand.
8. BRUGGEMAN *et al.* (1974). *Z. Jagdwiss*, **20**, 70–4.
9. EISELE, W. (1972). Thèse Munich cité dans Etude de la contamination de la faune continentale par les OC et les PCB, CEE—EUR. 5888-EN.
10. FRANZ-HASSAN, cité dans le Bulletin SROP 1982/V/2.
11. COULON *et al.* (1979). *Phytiatr.-Phytopharm.*, **28**, 145–6.
12. DELORME, R. (1976). *Entomophaga* **21**(1), 19–29.
13. STEINER, cité dans le Bulletin SROP 1982/V/2.
14. SECHSER, B. et BATHE, P. A. (1978). *Rev. Zool. Agric. Pathol. Vég.*, **3**, 91–107.
15. REBOULET *et al.* (1981). *Défense des Végetaux*, **209**, 195–218.
16. POTTS, G. R. (1977). *The Origins of Pest, Parasite, Disease and Weed Problems*, Blackwell, Oxford pp. 183–202.
17. CHAMBON (1982). *Agronomy*, **2** No. 5, 405–16.
18. *ACTA Compte rendu d'expérimentation*, 1982–1983.
19. DE LAVAUR, E. et ARNOLD, A. (1981). Phytiatr.-Phytopharm., **30**, 89–95.
20. GREAVES, M. P. *et al.*, Weed Research Organization Techn, Report No. 59.
21. BUNYAN, P. J. (1980). Pesticide Residues Proceed. Confer., Ref. Book 347 ADAS.
22. RIVIERE, J. L. (1983). *Bull. Environ. Contam. Toxicol.*, **31**, 479–85.
23. HUDSON, G. M. (1984). Pesticides and human safety—Legislative aspects. Personal communication.

13

Integrierter Pflanzenschutz: ein Weg zur ökologisch und ökonomisch sinnvolleren Nutzung der Pestizide

R. DIERCKS

Chem. Bayerische Landesanstalt für Bodenkultur und Pflanzenbau, Freising-Müchen, Bundesrepublik Deutschland

ZUSAMMENFASSUNG

Trotz gesetzlicher Vorsorgeregelungen zur Minderung ökotoxischer Risiken verbleibt dem Anwender von Pestiziden noch ein Handlungsspielraum oft unnötiger Belastung der Agrarökosysteme. Die Folgegefahr ist ein sich selbstverstärkender Regelkreis immer häufigerer Krankheits- und Schädlingsprobleme und immer höherer chemischer Bekämpfungsaufwendungen. Integrierter Pflanzenschutz will dieser Gefahr begegnen durch eigenverantwortliche Beschränkung der 'Chimie' auf ein wirtschaftlich unvermeidbares Mindestmaß unter optimaler Schonung der natürlichen Regulationskräfte des Ökosystems. Dazu bedient er sich, neben voller Ausschöpfung acker- und pflanzenbaulicher Abwehrmöglichkeiten zur Herabsetzung der 'Schadenswahrscheinlichkeit', vor allem des Prinzips der 'wirtschaftlichen Schadensschwelle' bei Anwendung von Pestiziden, sodann ökologisch-selektiver Mittelwahl und Anwendungstechnik und schon ausgereifter biologischer Bekämpfungsverfahren. Zwischen Theorie und Praxis klafft noch eine breite Lücke. Daher stellen notwendige Fortschritte eine Herausforderung für Forschung, Beratung und Praxis dar, deren Aufgaben kurz umrissen werden. Fernziel ist die Entwicklung kompletter, gesamtintegrierter Produktionssysteme. Langfristig droht auch eine integrierte Pflanzenproduktion auf Grenzen zu stoßen, wenn der Landbau nicht vom wirtschaftlichen Druck permanenten Produktionswachstums befreit wird. Daher wäre eine Reform der EG-Agrarpolitik erforderlich, die auch der ökologischen Dimension im Landbau Rechnung tragen müßte.

1. EINFÜHRUNG

Über die Risiken einer Strapazierung der Agrarökosysteme durch Pestizide und über gesetzliche Regelungen, sie zu begrenzen, ist in vorausgegangenen Vorträgen schon berichtet worden. Meine Aufgabe wird es sein, in kurzen Zügen darzulegen, welche Möglichkeiten es gibt, um mit Hilfe der Strategie des integrierten Pflanzenschutzes diese Risiken noch weiter einzuengen und zugleich ein ökonomisches Optimum bei der Bekämpfung von Schadorganismen zu erzielen.

Kein Gesetz, keine noch so scharfen Zulassungsbestimmungen und keine Anwendungsvorschriften können vollständig unterbinden, daß dem Anwender von Pestiziden noch ein Handlungsspielraum oft unnötiger Belastung des Agrarökosystems verbleibt. Zwar unterliegen die Produktionsflächen dem Zwang wachsender Fremdregulation, ihrer natürlichen Regelmechanismen sind sie deshalb aber noch nicht völlig beraubt. Diese nach Möglichkeit zu nutzen, muß auch und besonders der moderne Landbau anstreben, wenn langfristig Rückschläge vermieden werden sollen [14, 19, 21].

Erschwert wird dieses Ziel durch ungewollte Nebeneffekte der Pestizide, nämlich durch Schädigung nützlicher oder auch indifferenter, aber stets ökologisch regulierend wirksamer Organismen. Die Folgegefahr ist ein sich selbstverstärkender Regelkreis immer häufigerer Schädlings- und Krankheitsprobleme und immer intensiverer Bemühungen, sie durch chemische Eingriffe zu bewältigen. Diese Gefahr droht besonders dann, wenn die Mittel routinemäßig und nur vorsorglich nach dem 'Versicherungsprinzip' und daher oft überintensiv zur Anwendung kommen, wie dies vor allem in pflanzenschutzintensiven Sonderkulturen seit langem üblich ist, und wenn überdies polytoxische Mittel mit stets gleichem Wirkstoff bevorzugt werden. Verstärkend wirken ferner flächendeckende und großräumige Applikationen sowie ausgeräumte, monotone Agrarlandschaften, weil dort häufiger als bei vielfältiger Anbaustruktur gleiche chemische Maßnahmen auf allen Flächen einer Feldflur zur gleichen Zeit getroffen werden und die Überlebenschancen für natürliche Antagonisten noch zusätzlich verringern.

Schadorganismen und chemische Abwehr drohen also in einen durch biokybernetische Rückkopplung immer rascheren 'Wettlauf' zu geraten, dessen endgültiger Ausgang ungewiß ist, ohzu daß restriktive Regelungen für Zulassung und Anwendung allein Abhilfe schaffen könnten. Problematisch ist auch die ökonomische Kehrseite, weil der Anteil der Kosten für Pestizide am gesamten Betriebsaufwand überproportional zu steigen, die Produktivität also zu sinken droht. Die Risiken—man spricht auch vom 'Pestizidsyndrom' [24]—mögen zwar generell heute noch nicht besorgniserregend sein; sie wachsen aber offensichtlich permanent und sind keineswegs mehr nur auf 'chemieintensive' Sonderkulturen beschränkt.

2. STRATEGIE UND METHODEN DES INTEGRIERTEN PFLANZENSCHUTZES

2.1. Zielsetzung

Weltweit ist man daher inzwischen bestrebt, diesem Pestizidsyndrom durch selbstverantwortliche Beschränkung der chemischen Pflanzenschutzmaßnahmen auf ein wirtschaftlich unvermeidliches Mindestmaß unter optimaler Schonung der Selbstregulationskräfte des Ökosystems wirksam vorzubeugen. Diese Zielsetzung allein verbirgt sich hinter dem arg strapazierten, oft auch falsch interpretierten Begriff 'Integrierter Pflanzenschutz', dessen Dimension gewissermaßen dort beginnt, wo die Anwendung der Pestizide 'nach Vorschrift' [14] endet und daher sehr viel mehr an Fachwissen und Einsicht erfordert als nur perfekte Spezialkenntnisse über Diagnostik, Mittelkunde und Gerätetechnik. Die international übliche Definition für diese ursprünglich nur im Obstbau [42], heute aber in fast allen Kulturen auf breite Resonanz stoßende Strategie des Pflanzenschutzes lautet: 'Die integrierte Bekämpfung ist ein Verfahren, bei dem alle wirtschaftlich, ökologisch und toxikologisch vertretbaren Methoden verwendet werden, um Schadorganismen unter der wirtschaftlichen Schadensschwelle zu halten, wobei die bewußte Ausnützung aller natürlichen Begrenzungsfaktoren im Vordergrund steht' [16, 44].

2.2. Das Prinzip der wirtschaftlichen Schadensschwelle

Schlüsselfunktion fällt dem in der Definition genannten Prinzip der wirtschaftlichen Schadensschwelle zu. Es beruht auf einem engen Bezugssystem zwischen Schädlingsdichte, Schadenshöhe und Bekämpfungskosten, um konkrete Anhaltspunkte für die örtliche Prognose zu gewinnen, ob und wann eine chemische Bekämpfung tatsächlich erforderlich ist, oder ob ein Befall geduldet werden kann, ohne zum chemischen Mittel greifen zu müssen, weil die Kosten dafür größer wären als die durch den Befall verursachten Erlösminderungen [11, 45].

Die Häufigkeit chemischer, das Ökosystem belastender Eingriffe wird geringer, das Potential natürlicher Begrenzungsfaktoren (z.B. Prädatoren und Parasiten unter den Nutzarthropoden) weniger stark beeinträchtigt, die Gefahr der Resistenzbildung zumindest verringert, und die Bekämpfungskosten wie auch der Fremdenergieaufwand sinken. An die Stelle des traditionellen, aber letzlich 'unbiologischen' Ziels einer radikalen Vernichtung von Schaderregern tritt ein dynamisches System nur der Regulierung von Krankheiten, Schädlingen und Unkräutern [43]. Die 'Chemie' wird also nicht, wie im 'Biologischen Landbau', prinzipiell in Frage gestellt, aber es wird angestrebt, sie intelligenter und systemkonformer, d.h. auch ökologisch verantwortungsbewußter, als traditionell üblich zu nutzen (='Pest Management').

Wie bedeutsam allein der betriebswirtschaftliche Aspekt dieses Prinzips ist, beweist eine Reihe jüngster Untersuchungen zur Frage der Kostendeckung bei praxisüblicher Anwendung von Pestiziden: So berichtet Barel [2] aus den Niederlanden, daß dort bei chemischen Bekämpfungsmaßnahmen gegen Blatt- und Ährenkrankheiten im Weizenbau in 50% aller untersuchten Fälle die Behandlung unrentabel gewesen war, weil die verhinderten Verluste die Bekämpfungskosten nicht abdeckten. Auf einen gleichhohen Prozentsatz wirtschaftlich unnötiger Anwendung von Herbiziden in Winterweizen und Wintergerste im norddeutschen Raum weist Heitefuss [23] hin. In seinen mehrjährigen Feldversuchen zur Ermittlung von Schadensschwellenwerten kam es teilweise sogar zu Mindererträgen durch Herbizide, nicht nur zu Mindererlösen als Folge unnötiger Kostenaufwendungen. Ganz ähnliche Befunde erbrachten umfangreiche Feldversuche in Baden-Württemberg [28] und im Raum Braunschweig [34]. In Baden-Württemberg z.B. war im Sommergetreide nur in 30–50%, im Wintergetreide in 45–75% der untersuchten Fälle eine Herbizidanwendung kostendeckend.

Es besteht also kein Zweifel über Notwendigkeit und Chancen des Prinzips der wirtschaftlichen Schadensschwelle. Es erlaubt, soweit schon praktizierbar, nicht unerhebliche Kosteneinsparungen gegenüber mehr oder weniger 'blinden' Routinespritzungen und verringert zugleich die toxikologische Belastung des Ökosystems. Der ökonomische Aspekt macht integrierten Pflanzenschutz für den Landwirt heute überhaupt erst attraktiv. Denn bei stagnierenden Erzeuger-, aber steigenden Betriebsmittelpreisen verbleibt ihm nur eine sehr scharfe Kostenkalkulation, um Einkommensverluste zu vermeiden oder wenigestens zu verringern. Schließlich trägt die Begrenzung chemischer Eingriffe in das Agrarökosystem auf ein wirtschaftlich notwendiges Mindestmaß auch ökoethischen Forderungen Rechnung, nämlich das Leben in seiner Gesamtheit, Kreaturen jeglicher Art, zu respektieren und nicht 'ohne Not' zu zerstören.

2.3. Problematik der Ermittlung und Anwendung von wirtschaftlichen Schadensschwellen

Die Höhe der wirtschaftlichen Schadensschwelle resultiert aus einer Fülle variabler ökologischer, ökonomischer und auch anbautechnischer Faktoren mit diversen Wechselbeziehungen, die sich positiv oder negativ auswirken können, also die Schadensschwelle anheben lassen oder sie tiefer anzusetzen erfordern. Unter den ökonomischen Faktoren des vorhergenannten Bezugssystems: Befallsdichte–Schadenshöhe–Bekämpfungskosten variieren vorwiegend die Erzeugerpreise und die Pestizidkosten. Das jedoch ist das geringste Problem, wenngleich es schon deutlich macht, daß anstelle fester, etwa absoluter Schwellenwerte eine Skala gleitender Daten erforderlich ist.

Sehr viel größer ist die Variabilität der anderen Faktoren und Wechselwirkungen. Darunter fallen zunächst Einflüsse, die neben dem Bekämpfungsobjekt unmittelbar die Höhe des Rohertrags einer Kultur bestimmen, nämlich der Standort im weitesten Sinne, Düngungsintensität, Sortenleistung, Saattechnik, Unkrautbesatz und Zweitschädlinge, um nur die wichtigsten Einflußgrößen zu nennen. Daraus ergeben sich Beziehungen zum tatsächlichen Ausmaß des von einem bestimmten Schaderreger verursachten prozentualen Ernteverlustes. Gleicher Befallsdichte, gleichen Erzeugerpreisen und gleichen Bekämpfungskosten können also, je nach Intensität der ertragssteigernden Produktionstechnik, wechselnde Geldeinbußen gegenüberstehen. Das ist eine weitere Schwierigkeit, die einzukalkulieren ist und wiederum die Notwendigkeit gleitender Schwellenwerte unter Berücksichtigung auch von Rohertragsdifferenzierungen erkennen läßt. Daß diese Schwierigkeit zu meistern ist, beweist eine große Zahl von Publikationen, vor allem auch die schon erfolgreiche Praktizierung in der Landwirtschaft [3, 4, 5, 17, 18, 22, 23, 34, 37].

Umwelt und Produktionstechnik können aber auch indirekt die Schadenshöhe dadurch beeinflussen, daß sie die Prädisposition der Wirtspflanze gegenüber Schadorganismen verändern (z.B. Erhöhung der Toleranz durch größeres Regenrationsvermögen, zeitliches Hinausschieben der Anfälligkeit, vor allem jedoch aktive Sortenresistenz). Wo solche ökologischen Momente eine positive Rolle spielen, könnte eine höhere Populationsdichte toleriert, die Schadensschwelle also heraufgesetzt werden.

Einen maßgebenden Einfluß übt die Umwelt sodann auf die Massenvermehrung und das Infektionspotential von Schadorganismen aus. Es sind zunächst meteorologische Faktoren, die fortlaufend in begrenzender, dezimierender oder fördernder Art in die Populationsdynamik eingreifen.

Witterungsbedingungen beispielsweise, von denen erwartet werden kann, daß sie zur Dezimierung eines Schädlings noch nach Beginn des Befalls führen, lassen die Schwelle kurzfristig heraufsetzen, wie auch umgekehrt, wenn optimale Witterung eine plötzliche Neuzuwanderung befürchten läßt. Steuerungsfunktionen nehmen auch die schon eingangs erwähnten biologischen Begrenzungsfaktoren wahr. Hohes örtliches Potential an 'dichteabhängigen' natürlichen Gegenspielern von Schädlingen würde daher ein Anheben der Schadensschwelle erlauben, wenn von ihnen eine sofortige 'Bremswirkung' erwartet werden kann.

Ein Problem besonderer Art stellt die nicht immer voraussehbare Witterung für die Beurteilung der Schadensschwelle bei pilzparasitären Krankheiten dar, insbesondere wenn die chemische Bekämpfung präventiv erfolgen muß, weil es an eradikativen Fungiziden, die auch postinfektionell, noch nach schon erkennbarem Krankheitsausbruch, wirksam eingesetzt werden können, vorerst vielfach mangelt. Dies ist der Hauptgrund, wenn bei der direkten Bekämpfung von Pflanzenkrankheiten die Versuche zur Schadensschwellenerarbeitung bisher am wenigsten erfolgreich waren [35, 38].

Auf dem Sektor der Herbizide spielt noch ein ganz anderer Aspekt eine Rolle: es gibt Unkrautarten, die schon bei relativ geringem Besatz die Erntetechnik, wenn sie vollmechanisiert ist, empfindlich stören können, so daß Ertragsverluste nicht immer das einzige Kriterium für die Bewertung der wirtschaftlichen Schadensschwelle sein müssen.

Es gibt noch weitere Einflußgrößen und Interaktionen, die ins Gewicht fallen können. Genannt wurden nur solche, die besonders typisch sind für die Gesamptproblematik der Ermittlung und Anwendung wirtschaftlicher Schadensschwellen. Ihre etwas ausführlichere Schilderung sollte vor allem verständlich machen, warum es auf diesem Sektor zwischen Theorie und Praxis, zwischen Anspruch und Wirklichkeit, noch eine große Lücke gibt, die langsam zu schließen, ein hohes Maß an gemeinsamer Anstrengung von Forschung, Beratung und Praxis erfordert.

2.4. Gesamtinstrumentarium des integrierten Pflanzenschutzes

Integrierter Pflanzenschutz erschöpft sich nicht in der Beachtung des Schadschwellenprinzips. Es gibt weitere essentielle Struktur- und Funktionselemente ökologischen Zuschnitts in diesem Konzept. Einige davon, soweit sie auch die Höhe der Schadensschwelle beeinflussen können, kamen schon kurz zur Sprache. Ungeachtet dessen seien auch sie nochmal genannt, wenn nachfolgend alle für ein integriertes Bekämpfungssystem geeignete und notwendige Einzelmaßnahmen, die angepaßt an die örtlichen Voraussetzungen auszuwählen und sinnvoll miteinander zu kombinieren sind, in ihrer Gesamtheit vorgestellt werden [8, 9, 10, 13, 15]:

Zentrales Bezugspaar bei allen anbautechnischen Überlegungen sind seit jeher Standort und Sorte. Der Einstieg in ein integriertes System erfolgt mit der richtigen Sortenwahl, wobei Ertragsleistung, Qualität und Resistenz gegenüber Schadorganismen als wichtigste Sortenmerkmale gegeneinander abzuwägen sind. Um die wirtschaftliche Schadensschwelle anzuheben, wären Sorten zu bevorzugen, bei denen hohes Leistungspotential, marktgerechte Qualität und geringe Anfälligkeit zusammentreffen. Wenn sich dies, wie noch vielfach der Fall, nicht voll verwirklichen läßt, so könnte auf die Resistenz umso eher verzichtet werden, je zahlreicher andere Elemente des integrierten Pflanzenschutzes verfügbar sind, um die Lebensgemeinschaften auf dem Acker nicht zu stark zu belasten.

Darunter fallen zunächst alle nichtchemischen Abwehrmöglichkeiten, soweit sie arbeitswirtschaftlich heute nicht auf enge Grenzen stoßen. Vorwiegend sind dies sorgfältige Bodenpflege und intensive Humuswirtschaft (zur Aktivierung des 'antiphytopathogenen Potentials'), vorbeugende Saattechnik und eine gezielte, dem sortenspezifischen Pflanzenbedarf angepaßte Düngung. Vor allem die Stickstoffdüngung erfordert das richtige Augenmaß, weil ein Zuviel nicht nur zum 'Luxuskonsum' durch die Pflanze führt und damit Geld- und Energieverschwendung bedeutet, sondern sehr oft auch das Risiko des Krankheits- und Schädlingsbefalls erhöht. Sinnvolle Nutzung solcher acker- und pflanzenbaulichen Möglichkeiten macht nur selten chemische Pflanzenschutzmaßnahmen völlig überflüssig, sie verringert aber die 'Schadenswahrscheinlichkeit' und kann auch—wie schon erwähnt—die wirtschaftliche Schadensschwelle für Pestizide heraufsetzen lassen, entweder auf Grund direkter antipathogener Effekte oder indirekt durch Erhöhung der Kondition oder auch Kompensationskraft der Pflanze. Eine Rückkehr auch zur vielseitigen, artenreichen und daher 'gesunden' Fruchtfolge wäre wünschenswert, widerspricht aber meistens dem ökonomischen Zwang zur Rationalisierung und Spezialisierung der Betriebe.

Für den Einsatz der Pestizide sodann sind die zwei wichtigsten Aspekte:

(a) Nutzung schon erprobter und bewährter Prognosemethoden auf der Basis sicherer Daten über die schon erörterten wirtschaftlichen Schadensschwellen. Solche zumindest vorläufigen Werte liegen vor allem für die Unkrautbekämpfung und die Bekämpfung tierischer Schädlinge bereits vor. Aber auch bei Krankheitserregern zeichnen sich von Jahr zu Jahr Fortschritte ab [51]. Ein jüngstes Beispiel ist die von uns in Bayern entwickelte Prognosemethode zur gezielten Bekämpfung der 'Hopfenperonospora' (Pseudoperonospora humuli). Sie beruht auf der örtlichen Ermittlung des Zoosporangiengehalts der Luft und der Regenbenetzungsdauer der Blätter. Die Zahl von früher im Hopfenbau üblichen stets mehr als 10 Fungizidapplikationen pro Jahr ließ sich mit Hilfe dieser Methode in den letzten Jahren um durchschnittlich 50 % senken [33].

(b) Der zweite Aspekt bei Anwendung von Pestiziden ist die Praktizierung einer 'ökologisch-selektiven' Mittelwahl und Anwendungstechnik. Dazu gehören vorwiegend die Wahl oligotoxischer Mittel mit schonender Wirkung gegen Nutzinsekten und partielle oder punktuelle Applikationen wie Randbehandlungen (gegen (flugträge Schadinsekten), Saatgutbehandlungen, Beidrillverfahren und Bandspritzungen, wodurch große Teile des Ökosystems unbelastet bleiben.

Biologische und biotechnische Bekämpfungsverfahren rangieren in einem integrierten Pflanzenschutzsystem selbstverständlich vor den Pestiziden [16, 26]. Die praktischen Anwendungsmöglichkeiten beschränken sich aber beim derzeitigen Stand der Entwicklung nur auf wenige Einzelfälle. Bekanntere Beipiele sind *Bacillus-thuringienis*—Präparate oder der Eiparasit Trichogramma evanescens gegen Lepidopteren und die Raubmilbe Phytoseiulus persimilis gegen 'Rote Spinne' in Unterglaskulturen.

Diese schematische Aufzählung könnte den Eindruck erwecken, daß integrierter Pflanzenschutz nicht anderes sei als nur Addition oder ein Aneinanderreihen von bestimmten Einzelmaßnahmen. Integration unter Einbeziehung der wirtschaftlichen Schadensschwelle für die Anwendung von Pestiziden ist aber sehr viel mehr, nämlich sinnvolle Verknüpfung und Verzahnung, und daher auch sehr viel schwerer zu verwirklichen. Sie setzt vor allem Abkehr von starrer 'Rezeptmentalität' voraus, bei der nur die eindimensionalen Beziehungen zwischen Schaderreger und Bekämpfungstechnik im Vordergrund stehen, um statt dessen, fußend auf einer Quantifizierung der vielschichtigen Wechselwirkungen zwischen kybernetischen Gesetzmäßigkeiten der Populationsdynamik von Schadorganismen einerseits und den Rück- und Nebenwirkungen produktionstechnischer Maßnahmen anderseits, den gesamten Anbau so sinnvoll zu integrieren und zugleich flexibel zu gestalten, daß es dem Betrieb gelingt, das wirtschaftliche Optimum mit einem Minimum an ökologischen Risiken zu erzielen. So gesehen ist integrierter Pflanzenschutz nahezu identisch mit einer gesamtintegrierten Pflanzenproduktion [10, 20, 25, 32], dessen Verwirklichung das Fernziel aller Landbauwissenschaften sein müßte. Gleichwohl sind viele der genannten Maßnahmen in Form von integrierten Teilsystemen schon heute praxisreif und finden auch im wachsenden Maße in Landwirtschaft und Gartenbau Anwendung.

3. VOLLE VERWIRKLICHUNG DES INTEGRIERTEN PFLANZENSCHUTZES—EINE HERAUSFODERUNG FÜR FORSCHUNG, BERATUNG UND PRAXIS

Um weitere Fortschritte in der praktischen Verwirklichung des integrierten Pflanzenschutzes zu erzielen und damit schrittweise die Risiken einer toxikologischen Überstrapazierung der Agrarökosysteme einzuschränken und schließlich ganz aufzuheben, sind Anstrengungen auf drei Ebenen erforderlich.

3.1. Forschung

Die Forschung müßte noch stärker als bisher zum 'Systemdenken' bereit sein und synökologische, biokybernetische Fragestellungen in den Vordergrund stellen. Anders sind die erforderlichen Entwicklungsarbeiten zur mathematischen, computergestützten Beschreibung wichtigster Input- und Outputvariablen, vor allem bislang vielfach nur qualitativ bekannter Interaktionen und ihre Programmierung zu dynamischen Prozeßmodellen nicht denkbar [30, 31, 40, 46, 49, 50, 51]. Das wiederum setzt zwangsläufig die Einsicht voraus, die unaufhaltsam auch in den Landbauwissenschaften fortschreitende Spezialisierung durch fächerübergreifende Kooperation und Integration ergänzen zu müssen [8]. Agrarische Spezial- und agrarische Komplexforschung wären daher künftig gleichrangig zu bewerten und zu fördern. Außerdem müßte bei den finanziellen Förderungsmodalitäten Beachtung finden, daß Ökosystemforschung einen 'langen Atem' benötigt und daher mit nur auf wenige Jahre begrenzter Mittelzuteilung unproduktiv bleiben muß.

3.2. Beratung

Innerhalb des pluralistischen Beratungsangebots nimmt aus der Sicht des integrierten Pflanzenschutzes die interessenunabhängige staatliche Beratung (amtlicher Pflanzenschutzdienst) eine zentrale Position ein. Ihr Einfluß wird aber bedrohlich gefährdet—dies trifft zumindest für die Bundesrepublik Deutschland zu—durch Überlastung mit immer zahlreicheren

Hoheitsaufgaben (Überwachung, Kontrolle, Gesetzesvollzug). Künftige legislative und exekutive Entscheidungen, für die in der EG zunehmend das Gemeinschaftsrecht maßgebend ist, erfordern daher angemessene Beschränkung, um eine Ausuferung von Reglementierungen für die Anwendung von Pestiziden zu verhüten und die staatliche Beratung vor einer Vernachlässigung ihrer zeitaufwendigen ökologisch orientierten Beratungstätigkeit zu bewahren. Weitere Verschärfungen restriktiver Regelungen sollten demnach schon im Vorfeld der Anwendung, d.h. bei der Zulassung einsetzen, um unter bestimmten Bedingungen nicht völlig unbedenkliche Pestizide von der Zulassung überhaupt auszuschließen, statt sie speziellen Anwendungsbeschränkungen zu unterwerfen, die dann—oft ohnehin mit nur fragwürdigem Erfolg—überwacht werden müssen [7].

3.3. Praxis

Ohne aktive, selbstverantwortliche Mitarbeit des Landwirts bleibt integrierter Pflanzenschutz rudimentär. Insbesondere die örtliche Entscheidung über das Ja oder Nein einer Pestizidanwendung erfordert Bereitschaft zur Selbsthilfe, um nach vorausgegangener Schulung und praktischer Anleitung auch eigene Beobachtungen, Zählungen und Messungen im Pflanzenbestand vorzunehmen und Entscheidungen zur 'Feinsteuerung' des chemischen Pflanzenschutzes selbst zu treffen. Das Risiko nicht immer ganz vermeidbarer Fehlprognosen beim Verzicht auf eine Pestizidanwendung ist weniger gravierend, wenn man den langfristigen Nutzen integrierter Verfahren in Rechnung stellt [7].

4. NOTWENDIGKEIT UND PERSPEKTIVEN EINER ÖKOLOGISCH ORIENTIERTEN EG-AGRARPOLITIK

Abschließend noch einige Anmerkungen über notwendige Korrekturen der EG-Agrarpolitik aus ökologischer Sicht: Die Schwierigkeiten einer Verwirklichung der Idealvorstellung vom integrierten Pflanzenschutz wachsen in dem Maße, wie die Produktionsleistung durch ertragssteigernde Maßnahmen hochgeschraubt wird. Diese Aussage trifft insbesondere für die Bemühungen um Verzicht auf unnötige Anwendung von Pestiziden zu, weil deren Effizienz, solange die Preise für die Präparate nicht überproportional steigen, mit wachsendem Ertragsniveau größer wird. Der Befall durch Schadorganismen überschreitet dann häufiger, wenn nicht ständig die wirtschaftliche Schadensschwelle. Auch integrierter Pflanzenschutz muß also mit Grenzen rechnen, wenn quantitatives Wachstum agrarpolitische Leitlinie bleibt! Völlig unberührt bleibt von einer generallen Anwendung des integrierten Pflanzenschutzes sodann die wachsende, immer schwerer finanzierbare Überproduktion in der EG. Es drängt sich somit die Frage auf, ob man den Landbau vom Zwang quantitativen Wachstums dadurch befreien kann, daß die ökonomischen Rahmenbedingungen für die Nahrungsmittelproduktion in der EG korrigiert werden.

Von den vielen Lösungsvorschlägen, die in der gegenwärtigen Diskussion über eine Reform der Agrarpolitik erörtert werden, verdient aus ökologischer Sicht das meiste Interesse eine künftige Trennung der Preis- und Einkommenspolitik in Verbindung mit direkten Ausgleichszahlungen an die Landwirtschaft [6, 10, 12, 29, 36, 39, 48]: das Senken der jetzt gestützten Erzeugerpreise auf Weltmarktpreisniveau würde zwangsläufig eine Intensitätsminderung, aber auch Produktions- und Einkommensverluste zur Folge haben [41]. Der dann notwendige Einkommenausgleich hätte den Charakter eines gesellschaftspolitisch durchaus legitimen 'Nebenhonorars' an den Landwirt für seine auch landschaftpflegerischen Leistungen, die ihm bislang nie vergütet worden sind, weil einzige Einnahmequelle die Nahrungsmittelproduktion ist. Die Betriebe stünden, wenn man dies ändert, nicht mehr unter dem wirtschaftlichen Druck, entweder ständig wachsen (und damit teilweise weichen) oder aber die Produktion, mit Hilfe auch der 'Chemie', steigern zu müssen, die Überproduktion würde gebremst und in der Landwirtschaft gäbe es wahrscheinlich keinen übergroßen Zielkonflikt zwischen Ökonomie und Ökologie. Flankierende Auflage- und Abgaberegelungen könnten den Effekt vergrößern [1].

Die Finanzierung einer solchen zweifellos radikalen Reform kann kein unlösbares Problem sein, wenn man in Rechnung stellt, daß jetzt viele Milliarden an Steuergeldern für Interventionen und Exporterstattungen zur 'Beseitigung' der Überschüsse verschwendet werden, die ohne betriebswirtschaftlichen Zwang zum Produktionswachstum entfallen würden. Voraussetzung wäre eine 'Renationalisierung' der Agrareinkommenspolitik, ohne daß deshalb Grundlagen und Fortdauer der gemeinsamen EG-Agrarpolitik in Frage gestellt sein müßten [47].

Die Landbauwissenschaften wären gut beraten, wenn sie sich mit den Forderungen von Ökologen insoweit identifizieren würden, daß 'den ökonomischen Leitbildern der Agrarpolitik künftig auch ökologische Leitlinien an die Seite zu stellen' sind [27].

LITERATUR

1. Anon. (1983). Abschlußbericht der Projektgruppe 'Aktionsprogramm Ökologie'. In: *Umweltbrief* 29, Der Bundesminister des Innern, Bonn.
2. Barel, C. J. A. (1981). Krankheitsprognosen sparen Zeit und Geld—Integrierter Pflanzenschutz in den Niederlanden. *DLG-Mitteilungen*, **96**, 20.
3. Beer, E. (1979). Ermittlung der Bekämpfungsschwellen und wirtschaftlichen Schadensschwellen von monokotylen und dikotylen Unkräutern in Winterweizen und Wintergerste anhand von Daten aus der amtlichen Mittelprüfung. Dissertation, Göttingen.
4. Beer, E. und Heitefuss, R. (1981). Ermittlung von Bekämpfungsschwellen und wirtschaftlichen Schadensschwellen für monokotyle und dikotyle Unkräuter in Winterweizen und -gerste. I. Zur Methodik der Bestimmung der Schwellenwerte unter Berücksichtigung wirtschaftlicher und biologisch-technischer Einflußgrößen am Beispiel des Winterweizens. *Zeitschrift für Pflanzenkrankheiten und Pflanzenschutz*, **87**, 65.
5. Beer, E. und Heitefuss, R. (1981). Ermittlung von Bekämpfungsschwellen und wirtschaftlichen Schadensschwellen für monokotyle und dikotyle Unkräuter Winterweizen und -gerste. II. Bekämpfungsschwellen und wirtschaftliche Schadensschwellen in Abhängigkeit von verschiedenen Bekämpfungskosten, Produktpreisen und Ertragsniveau. *Zeitschrift für Pflanzenkrankheiten und Pflanzenschutz*, **88**, 321.
6. Binswanger, H. C. und Müller, K. (1979). Vorschlag für die Einführung von Flächenbeiträgen. In: *Die europäische Agrarpolitik vor neuen Alternativen* (P. Haut), Bern, Stuttgart, s. 17.
7. Blaszyk, P. und Diercks, R. (1977). Pflanzenschutzberatung für den Landbau der Bundesrepublik Deutschland. DLG-Manuskript 030, Deutsche Landwirtschafts-Gesellschaft e.V., Frankfurt/M.
8. Diercks, R. (1980). Statusbericht Pflanzenschutz. Landwirtschaft-Angewandte Wissenschaft, H.244, Landwirtschaftsverlag GmbH, Münster-Hiltrup.
9. Diercks, R. (1982). Integrierter Pflanzenschutz. In: *Alternativen im Landbau*, Angewandte Wissenschaft, H.263, Landwirtschaftsverlag GmbH, Münster-Hiltrup, s. 158.
10. Diercks, R. (1983). *Alternativen im Landbau—Eine kritische Gesamtbilanz*, Verlag Eugen Ulmer, Stuttgart.
11. Diercks, R. und Heye, Ch. (1970). Notwendigkeit und Problematik der Ermittlung von Schadensschwellenwerten. *Zeitschrift für Pflanzenkrankheiten und Pflanzenschutz*, **77**, 610.
12. Durand, A. (1979). Die Trennung von Preispolitik und Ausgleichszahlungen im Ackerbau. In: *Die europäische Agrarpolitik vor neuen Alternativen*, Verlag Paul Haupt, Bern und Stuttgart, s. 37.
13. Franz, J. M. (1973). Gedanken zum integrierten Pflanzenschutz im Acker- und Gemüsebau. *Zeitschrift für Pflanzenkrankheiten und Pflanzenschutz*, **80**, 3.
14. Franz, J. M. (1975). Integration als Aufgabe—Bemühungen um eine zeitgemäße Schädlingsbekämpfung. *Zeitschrift für angewandte Entomologie*, **78**, 17.

15. FRANZ, J. M. (1978). Das Konzept des integrierten Pflanzenschutzes. *Gesunde Pflanzen*, **30**, 177.
16. FRANZ, J. M. und KRIEG, A. (1982). *Biologische Schädlingsbekämpfung*, Verlag Paul Parey, Berlin und Hamburg.
17. FUNCH, W. CH. (1974). Untersuchungen über ökonomische Schadensschwellen für Kraut- und Knollenfäule, Unkräuter und Viruskrankheiten im Kartoffelbau. Dissertation, Göttingen.
18. GARBURG, W. (1974). Untersuchungen zur Ermittlung der ökonomischen Schadensschwelle und der Bekämpfungsschwelle von Unkräutern im Getreide. Dissertation, Göttingen.
19. GIESE, R. J., PEART, R. M. and HUBER, R. T. (1975). Pest management. *Science*, **187**, 1045.
20. GRASS, K. (1984). Eine Alternative für alle—Integrierter Landbau in Praxis und Beratung. *DLG-Mitteilungen*, **99**, 901.
21. HASLER, A. D. (1971. *Man in the Living Environment*, The Institute of Ecology, University Wisconsin Press, p. 91, Madison/Wisconsin, p. 163.
22. HEITEFUSS, R. (1981). Schwierigkeiten der Erarbeitung von Schadensschwellen bei Unkräutern. In: *Wirtschaftliche Schadensschwellen in Pflanzenschutz—Schwierigkeiten und Notwendigkeit*, DLG-Manuskript 051, Deutsche Landwirtschafts-Gesellschaft e.V., Frankfurt/M., s. 47.
23. HEITEFUSS, R. (1982). Pflanzenschutz und Unkrautbekämpfung im konventionellen Landbau. In: *Alternativen im Landbau. Angewandte Wissenschaft*, H. 263, Landwirtschaftsverlag GmbH, Münster-Hiltrup, s. 121.
24. HUFFAKER, C. B. (1971). *Biological Control*, Plenum Press, New York and London.
25. HUFFAKER, C. B. (1982). Presidential Address: Some Current Concerns for the Future. *Bull. Ent. Soc. Amer.*, **28**, 13.
26. KLINGAUF, F. (1982). Möglichkeiten und Grenzen von Innovationen im Bereich der Pflanzenproduktion—Pflanzenschutz. *Innovationen im Agrarsektor, Agrarspektrum*, **5**, 335.
27. KNAUER, M. (1982). Landnutzung und Ökologie. In: *Alternativen im Landbau. Angewandte Wissenschaft*, H. 263, s. 240.
28. KOCH, W., HURLE, K., SANWALD, E. und WALTER, H. (1981). Konsequenzen für die Herbologie aus betriebswirtschaftlichen und pflanzenbaulichen Entwicklungen. *Zeitschrift für Pflanzenkrankheiten und Pflanzenschutz, Sonderb*, **IX**, 47.
29. KOESTLER, U. (1980). Nationale Eigeninteressen in der gemeinsamen Agrarpolitik und mögliche Reformen. In: *Alternativen zur EG-Preispolitik, Loccumer Protokolle*, **5**, 4.
30. KRANZ, J. (1977). Die Entwicklung von Pflanzenschutzsystemen. *Mitteilungen aus der Biologischen Bundesanstalt*, Berlin-Dahlem, Heft, **178**, 85.
31. KRANZ, J. (1979). Computermodelle als Hilfsmittel des integrierten Pflanzenschutzes. Proceedings—Internationales Symposium der IOBC/WPRS über integrierten Pflanzenschutz in der Land- und Forstwirtschaft, Wien, s. 165.
32. KRAUS, A. and DIERCKS, R. (1977). Integrierte Produktionssysteme—eine Aufgabe von Pflanzenbau und Pflanzenschutz. *Gesunde Pflanzen*, **29**, 1.
33. KREMHELLER, H. TH. und DIERCKS, R. (1983). Epidemiologie und Prognose des

Falschen Mehltaues (Pseudoperonospora humuli) an Hopfen. *Zeitschrift für Pflanzenkrankheiten und Pflanzenschutz*, **90**(6), 599.

34. Niemann, P. (1981). Schadschwellen bei der Unkrautbekämpfung. Angewandte Wissenschaft, Landwirtschaftsverlag GmbH, Münster-Hiltrup, H. 257.
35. Obst, A. (1981). Schwierigkeiten der Erarbeitung von Schadschwellen bei Pilzkrankheiten. In: Wirtschaftliche Schadensschwellen im Pflanzenschutz—Schwierigkeiten und Notwendigkeit. DLG-Manuskript 051. Deutsche Landwirtschafts-Gesellschaft e.V., Frankfurt/M., 21.
36. Priebe, H. (1979). Zur Lösung der Konflikte zwischen Einkommens- und Marktpolitik. In: *Die europäische Agrarpolitik vor neuen Alternativen*, Verlag P. Haupt, Bern und Stuttgart, s. 9.
37. Reschke, M. (1972). Untersuchungen zur Bestimmung von ökonomischen Schadensschwellen für Pflanzenschutzsysteme in Kartoffelbau. Dissertation, Göttingen.
38. Reschke, M. (1981). Schwierigkeiten der Anwendung von Schadensschwellen bei Pilzkrankheiten. In: Wirtschaftliche Schadensschwellen im Pflanzenschutz—Schwierigkeiten und Notwendigkeit. DLG-Manuskript 051. Deutsche Landwirtschafts-Gesellschaft e.V., Frankfurt/M., 29.
39. Riemsdijk, J. F. van (1979). Direkter Einkommenstransfer als zentrales Instrument der Agrarpolitik. In: *Die europäische Agrarpolitik vor neuen Alternativen*, Verlag P. Haupt, Bern und Stuttgart, s. 73.
40. Rijsdijk, F. H. (1980). Systems analysis at the crossroad of plant pathology and crop physiology. *Zeitschrift für Pflanzenkrankheiten und Pflanzenschutz*, **87**(7), 404.
41. Schulte, J. (1984). Begrenzter Einsatz von Handelsdüngern und Pflanzenschutzmitteln. In: *Angewandte Wissenschaft*, H. 294, Landwirtschaftsverlag GmbH, Münster-Hiltrup.
42. Steiner, H. (1965). Zwölf Jahre Arbeit am integrierten Pflanzenschutz im Obstbau. *Gesunde Pflanzen*, **17**, 5.
43. Steiner, H. (1968). Probleme der Bestimmung wirtschaftlicher Schadensschwellen bei der integrierten Bekämpfung. Twintigste Internationaal Symposium over Fytofarmacie en Eytiatrie 7.5.1968, Gent, Belgie, 577.
44. Steiner, H. (Hrsg.) (1975). *Aktuelle Probleme des Integrierten Pflanzenschutzes*, Dr. Dietrich Steinkopff Verlag, Darmstadt.
45. Stern, V. M., Smith, R. F., van den Bosch, R. and Hagen, K. S. (1959). The integrated control concept. *Hilgardia*, **29**, 81.
46. Teng, P. S. (1981). Validation of computer models of plant disease epidemics: A review of philosophy and methodology. *Zeitschrift für Pflanzenkrankheiten und Pflanzenschutz*, **88**(1), 49.
47. Thoroe, C. (1980). Renationalisierung der Einkommenspolitik—Ein integrationsfeindlicher Weg zur Reform der gemeinsamen Agrarpolitik? In: Alternativen zur EG-Preispolitik. Loccumer Protokolle Nr. 5, 32.
48. Weinschenck, G. (1980). Möglichkeiten und Grenzen einer Neuorientierung der Agrarpolitik. Agrarmarktsituation der 80er Jahre aus wissenschaftlicher Sicht. *Archiv. der DLG*, **66**, 49.
49. Wilbert, H. (1977). Fortschritte in der Entwicklung integrierter

Pflanzenschutzsysteme. Mitteilungen aus der Biologischen Bundesanstalt für Land- und Forstwirtschaft, *Berlin-Dahlem, H*, **178**, 66.

50. WÜSTEN, H., STEFFEN, G. und BERG, E. (1981). Stand und Entwicklung des Schadschwellenkonzepts als entscheidungsorientiertes System. *Zeitschrift für Pflanzenkrankheiten und Pflanzenschutz*, **88**(8/9), 465.
51. ZADOKS, J. C. (1981). EPIPRE: a disease and pest management system for winter wheat developed in the Netherlands. *EPPO Bull.*, **11**(3), 365.

14

Trends in Application and Formulation of Pesticides in Europe

D. SEAMAN

ICI Plant Protection Division, Bracknell, UK

SUMMARY

At present most pesticides are applied using high volumes of aqueous sprays having a broad droplet size distribution. To reduce volumes, drift and improve spray retention on targets, controlled droplet applications have been developed and electrically charged spray systems are at the research stage. Progress has been slow as the reduced volumes and altered distribution on the target have led to a changed biological response in many instances.

Much attention is now given to the best selection of formulation type to optimise biological effectiveness and improve handling characteristics for the farmer. Complex mixtures are now increasingly required and the technology to devise these products is being developed. Many formulations now contain built-in wetting agents to enhance biological activity. Improving the persistence of formulations by controlling release is now possible and can be applied for suitable pesticides.

1. PRESENT METHODS OF APPLICATION

Most pesticides are applied as dilute aqueous sprays from ground based sprayers. For arable crops hydraulic flat fan nozzles are most commonly used to generate the spray. For bush and tree crops mist blowers are used. Knapsack sprayers find their application for spot or directed treatment over small areas or for small bushes and trees.

The application of pesticide to seed is an effective method where the seed requires protection or where insects attack the young plant.

The application of granular formulations of pesticides in the furrow or broadcast finds limited application for control of soil borne insects and for volatile herbicides. Availability of pesticide is not immediate and relies on the volatility or water solubility of the chemical. Baits also find selective use for soil pests.

In Europe the aerial application of pesticides is limited mostly to late applications in cereals. Application is at higher concentrations of pesticide than for ground based application, using 20–50 litres of spray ha^{-1} compared with a normal 225 litres ha^{-1}. This is necessary to reduce the pay load on the aircraft to an economic level. Flat fan, cone or spinning disc or cage nozzles are used to generate the spray.

1.1. Application by hydraulic nozzles

This heading embraces ground and aerial spraying of arable crops and knapsack spraying. The process involved is

Atomisation → Transport to target → Impaction

The efficiency of this process is variable and depends on many factors as follows;

1. Condition of equipment, in particular nozzles.
2. Selection of nozzles.
3. Environmental conditions:
   wind speed;
   humidity;
   temperature.
4. Correct use of equipment.
5. Geometry and growth stage of crop.
6. For weed control, position within crop.
7. The formulation properties.
8. Where the chemical acts, on the foliage or by soil action.

A typical flat fan hydraulic nozzle generates spray with a volume median diameter (VMD) of about 250 μm. The droplet size range is broad, typically 15% by volume below 100 μm and 15% above 400 μm. The droplets emerge from the nozzle with a velocity of around 20 m s^{-1}. The small droplets rapidly slow down to velocities of around 1 m s^{-1} at the target. These small droplets can drift across the crop particularly at higher wind speeds. Drift is more likely with aerial spraying due to the aircraft speed and turbulence and the greater distance of the nozzle from the crop.

During transport to the target the water based droplets may lose much of their volume by evaporation. The smaller ones with their greater surface area to volume ratio will evaporate more readily to yet smaller droplets thereby increasing their propensity to drift.

On arrival at the crop, drops will impact with the target crop or weed if the target intercepts an individual droplet. For large drops a miss by a millimetre is as bad as a kilometre as the drop will pass by under gravity. Small droplets may find it difficult to impact on the upper parts of crops as the localised air currents present can carry them around the top surfaces and deeper into the crop into slower moving air where they can impact on lower surfaces. This is likely to be an advantage as medium size droplets are likely to attach themselves to the first surface they intercept. Large droplets are more likely to bounce or shatter into small droplets. The broad droplet distribution from this kind of nozzle can be quite effective in producing a penetrating spray within a crop canopy.

Typical arable sprayers are of simple construction, reliable and easy to use and generally effective. If they are to be superseded, any new sprayer should meet these considerations as well as providing technical and environmental improvements.

1.2. Application by mist blowers

The initial spray is generated by conventional nozzles, and carried by high velocity air to the target where it penetrates into the leaf canopy of trees and bushes. The average droplet size varies from machine to machine but is in the region of 100 μm. The same considerations with regard to drift, evaporation and spray retention apply as for arable spraying. The smaller droplets are more likely to drift but are also more likely to adhere to any surface they reach.

2. NEWER APPLICATION TECHNIQUES

There are two main ones:

(1) Spinning disc based systems.
(2) Electrically charged sprays.

The former became available in the last decade and are now on sale for the farmer to use. The latter are at the research stage.

2.1. Spinning disc systems

This method of application is commonly called Controlled Droplet Application (CDA). Spinning discs if properly used generate a droplet spectrum much closer to monosize than conventional systems. The droplet size depends on the disc speed, but typically for the application of herbicides to soil or arable crops, a droplet size of the same VMD as a flat fan nozzle of 250 μm is generated but 80 % lie within the range 200–300 μm. The relative absence of small drops and large drops enables lower spray volumes to be used whilst maintaining an adequate crop coverage. The absence of the very small drops effectively eliminates spray drift. The lower spray volumes help the farmer increase his work rate and reduce the weight of the spray tank thereby to some extent reducing soil compaction. Typically 20 litres ha^{-1} can be sprayed compared with about 200 litres ha^{-1} for hydraulic nozzles although volumes as low as 50 litres ha^{-1} can be sprayed in the latter case for suitable situations.

There are two types of spinning disc sprayers:

(1) Horizontal disc;
(2) Vertical disc.

Horizontal discs eject the spray horizontally over the crop and it then falls under gravity. Vertical discs are shielded to eject the spray directly downwards into the crop and the droplets are therefore travelling more quickly.

Adoption of spinning disc systems to date has been limited. Part of the reason for this is that each and every formulation needs to be biologically tested to demonstrate an equivalent or improved effect over the conventional sprayer. The biological effect can be different because:

(1) The spray is approximately ten times more concentrated.
(2) Spray coverage and distribution are different.

Formulations may in some cases require optimising for CDA. Experience to date has demonstrated the general trend that equivalent biological activity is not automatically achieved and the initial hope for improved activity has not been proved.

2.2. Electrically charged sprays

There are three methods under investigation:

(1) Post atomisation charging of aqueous sprays.
(2) Charging during rotary atomisation.
(3) Charging of oil based formulations to induce atomisation.

All are at the research stage. The first is closest to conventional application methods and uses conventional formulations as aqueous sprays. The hope is that the electrical charge persuades more pesticide to attach to the target, thereby reducing drift and the rate of pesticide necessary. The second is an extension of rotary atomisation methods. The last approach is a more revolutionary concept. Low conductivity organic liquids when highly charged will self atomise to generate essentially monosize droplets in the range 50–100 μm. It is therefore possible to obtain a satisfactory coverage of the target with as little as 1 litre of formulation per hectare. This therefore opens up the possibility of ultra low volume application of pesticides to a range of crops from the ground. In theory at least the recovery of the spray on the target should be much greater than with large quantities of aqueous sprays.

All the techniques are experimental at this stage. The water based systems, being less revolutionary, are more likely to make rapid progress if the equipment can be made economically. As with CDA many products will need biological testing to ensure that the likely changed distribution on the target is acceptable. Charged sprays are attracted to the closest object and this can lead to chemical being concentrated in the upper part of crops.

The oil based system requires completely new oil based formulations which have precise electrical and rheological characteristics. There is the huge task of designing and testing a whole range of products in order to establish this technique. The small charged droplets will tend to deposit on the closest object but have the advantage also of depositing on the underside of leaves to a much greater extent than conventional sprays. To penetrate the crop may require air assistance which to some extent defeats the simplicity of the original concept.

3. SPRAY TANK ADDITIVES

Returning to conventional aqueous sprays there are three general classes of additives:

(1) Polymers.
(2) Wetters.
(3) Oils.

Polymers can be added to the spray tank for two reasons. Firstly, by using, for example, polyacrylamide polymers to affect the break up of the fan spray and reduce the generation of fine drops, thereby reducing drift.

This works for ground based sprayers but for aerial application where the potential for drift is greater, their efficacy has been questioned. The greater shear generated by the speed of the aircraft is believed to largely overcome the effect of the polymers. Secondly, polymers can be added to the spray tank as 'stickers' to increase the adhesion of the pesticide deposit on the target in order to reduce losses due to rain or volatilisation. A wide range of adhesive polymers can be used including oils which polymerise in sunlight.

The use of polymers to control drift is carried out to a limited extent but 'stickers' have not been used in Europe to any extent. Any addition to the spray tank can modify the biological effect. In particular polymers can lock in the chemical. The effects are complicated and can differ for each chemical. Sufficient field testing is therefore necessary to prove the benefit of any additive/pesticide combination.

Wetters are much more common and proven additives to the spray tank. They are based normally on surface active agents of the non-ionic type. Common examples are octyl and nonyl-phenol ethoxylates with between 6 and 11 ethylene oxide units. When used on difficult-to-wet foliage, they increase the spray retention by reducing the contact angle on the target. Where improved spreading and coverage are desirable they will have this effect also. The degree of spreading depends on the 'wetting' properties of the substrate and the concentration of wetter in the spray. Thirdly they can enhance the uptake of the pesticide into the crop. Rates used are around 0·1 % in the spray. The mechanism is complex and not clearly understood but is probably a combination of solubilisation of the chemical and good wetting and perhaps disruption of the plant cuticle to allow entry. Wetters can have side effects. In particular they can encourage phytotoxicity by the chemical. So once again the benefit of using the additive needs to be proven by field testing. This is done by the pesticide manufacturer and becomes part of the label recommendations.

The use of oils is becoming more common and is necessary with a few herbicides to achieve a satisfactory result. The oils are mineral or vegetable oils and contain surface active agent to effect emulsion of the oil when added to an agitated spray tank. There are two kinds, those with about 1 % surfactant and those containing up to 25 % surfactant. The latter no doubt combine the effect of the wetter and oil as the use rate is about a quarter of the former kind which are generally added at around 1 % concentration to the spray. Their overall effect is to enhance the biological activity. The mechanism is not understood. The nature of the deposit on the foliage is likely to remain sticky and this may aid uptake of the chemical, possibly by the oil partitioning the chemical into the cuticle.

4. FACTORS DETERMINING THE CHOICE OF FORMULATION

The basic formulations for spray application are:

Wettable powers (WP).
Emulsifiable concentrates (EC).
Aqueous solutions (SL).
Suspension concentrates (SC).

The codes are GIFAP nomenclature.

The choice of formulation is determined by:

(1) Chemical and physical characteristics of the pesticide.
(2) Biological effect.
(3) Handling characteristics.
(4) Shelf stability.
(5) Cost.
(6) The feasibility of mixtures where required.

The chemical and physical characteristics of the pesticide determine the feasibility of a particular formulation. For instance a suspension concentrate cannot be made for a pesticide which is either too soluble in water or chemically unstable in that medium. If a chemical is too insoluble in water immiscible organic solvents, then an emulsifiable concentrate cannot be prepared.

Emulsifiable concentrates and aqueous solutions, which by definition are in molecular solution, are likely to be more active and potentially more phytotoxic than corresponding suspensions or wettable powders. The liquid formulations are preferred over the powder as they are easier to handle. For this reason there is the trend for suspension concentrates to replace wettable powders for pesticides with the right characteristics.

5. IMPROVED FORMULATIONS

New and improved formulations are being devised to give either improved handling and convenience, or improved biological effect.

5.1. Improved handling and convenience

The trend to suspension concentrates away from wettable powders has already been mentioned. In recent years manufacturers have introduced water dispersible granules (WG) as an improvement over both suspensions and powders. These are non-dusty, non-friable granules which disperse when added to the spray tank. They leave no residue in the package. They can be made by extrusion, agglomeration or fluid bed granulation processes.

To date there are only a few chemicals formulated in this way. They are more costly to make than wettable powders or suspension concentrates and require a considerable capital investment in plant. All the manufacturing processes are not straightforward and substantial development work is required for each product.

Another trend is towards pre-formulated mixtures. This saves the farmer separately measuring each component of his tank mix and ensures that the correct proportions of a fully tested mixture are used. With tank mixing each individual formulation has been optimised but not in combination. The requirements for mixtures can be challenging for the formulator. If all the pesticides are soluble in water or oils, then it can be quite simple to devise a formulation. If as an extreme case, the formulation chemist is asked to devise a mixture of a water soluble chemical, a water insoluble solid and a water immiscible liquid, then he has to create a mixed solution/suspension/emulsion formulation. Formulation chemists are grappling with the problem of devising products with good shelf-lives with mixed success.

In the past seeds have traditionally been dressed with powder seed dressings (DS) but now more and more liquids are being provided either as solutions (LS) or suspensions (FS). Powder seed dressings are very safe on seeds but have the tendency to fall off the seed during handling. Liquid formulations adhere very well and seed dressing equipment is now available which gives a good distribution on the seed. They can be more phytotoxic and careful selection and testing of formulations is necessary.

5.2. Improved biological effect

The efficacy of some chemicals is limited because uptake into the plant or weed is poor or they lack the necessary persistence. Uptake can be improved for systemic chemicals by including wetters as mentioned earlier. Persistence can be improved by producing controlled release formulations.

6. IMPROVING UPTAKE

Whether the wetter is included in the spray tank or built into the formulation, the end effect is equivalent. There are large numbers of wetters, surface active agents, of differing structures which can enhance activity. For a built in wetter, the formulation chemist and biologist are not limited to those sold commercially for tank mixing and can choose the one giving the best biological effect but only if it can be built into the formulation. For example the preferred wetters may not be sufficiently soluble in an aqueous solution and a more soluble one will have to be chosen. The general experience is that most wetters will enhance activity for suitable chemicals in a worthwhile way and that the difference between wetters is a second order effect. When considering use of wetters it must be remembered that phytotoxicity can be increased. A suitable balance between these desirable and undesirable effects will need to be sought. A number of formulations are now sold with built in wetters.

7. CONTROLLED RELEASE FORMULATIONS

Pesticides which are too volatile, soluble or unstable can very quickly lose their biological effect. This can be a benefit when persistence is not a requirement. However, there are cases when it is desirable to have a longer persistence. One example is with volatile soil applied herbicides. If sprayed directly onto the soil surface they can rapidly disappear. Incorporating the chemical into the soil increases persistence. If the chemical is incorporated onto a granular carrier this has the same effect as the granule holds back the release of the chemical. Release rate can be further reduced by coating the chemical within the granule with a coating of polymer through which it has to diffuse.

A very elegant way to control pesticide release is by microencapsulation. Tiny particles of pesticide of a few μm diameter are coated with a plastic film. The main technique adopted by the pesticide industry is the interfacial polymerisation process which is most suitable for chemicals which are water immiscible liquids or solids with good solubility in water immiscible solvents. The pesticide and one of the monomers is emulsified into an aqueous solution of a second monomer. The two monomers react with each other at the emulsion droplet interface to form a thin film of polymer around this emulsion droplet. Increasing the monomer concentrations increases the film thickness. Inclusion of a trifunctional monomer crosslinks the polymer chains. By controlling the polymer film thickness and the degree of crosslinking, the release rate of the chemical can be controlled to give a range of release rates. To become available, the pesticide diffuses across the film and is released at a steady rate until the core of the capsule is depleted. Pesticides which are only effective for a few hours have had their effectiveness extended to many days.

As well as increasing persistence, microencapsulation can reduce mammalian toxicity and plant phytotoxicity. Despite all these potential benefits there are very few products available. This is not necessarily disappointing as the principle only applies to a limited number of pesticides, in particular low persistence ones which for some applications require longer persistence. In the main, pesticides do not require persistence or, where sufficient persistence is required, chemicals have been selected which have sufficiently low vapour pressures, water solubilities and adequate chemical stability.

8. CONCLUSIONS

Conventional sprayers using aqueous sprays are very flexible and when used correctly are more efficient than is generally appreciated. Rotary atomisers can reduce spray volumes and drift but many chemicals are not performing as well as might be expected. The electrical charging of sprays to improve performance is in its infancy. The concept is attractive but it must be recognised that distribution on the target is different from conventional applications and can affect the overall biological effect.

Wetters added to the spray tank to improve spray retention and overall biological effect have been available for many years and have proved their benefit for a number of chemicals. Polymers to reduce spray drift require further testing to justify their use. Oil addition is beneficial for a small number of specific chemicals.

During the development of pesticide formulations, manufacturers examine the effect of formulation type on efficacy and any undesirable effects. For systemic chemicals the effect of surface active agents is regularly tested and where beneficial they are built into formulations. Improving persistence of pesticides is an area under study but it is difficult in the first place to define the need and secondly to devise and prove an improved formulation. However the technology of microencapsulation has developed considerably in recent years and a few chemicals are available in this form.

15

Quelques Éléments pour une Amélioration de la Gestion des Fertilisants

J. C. Remy

Institut National de la Recherche Agronomique, Laon, France

RÉSUMÉ

La gestion rationnelle des fertilisants peut permettre de concilier une agriculture intensive avec le respect de l'environnement. Cependant, l'intensification conduit immanquablement à des risques plus élevés, et à une nécessaire sophistication des techniques de maîtrise de la production.

Le cas de la fertilisation est étudié en confrontant les courbes de réponse à l'approche de type bilan.

L'analyse des difficultés rencontrées notamment en ce qui concerne l'appréciation de certains termes du bilan, montre la nécessité de progresser d'une part dans les techniques de prévision de rendements et des besoins des plantes, et d'autre part, dans une meilleure approche quantitative du fontionnement du cycle de l'azote dans le sol.

La démarche de type bilan pourrait évoluer dans un proche avenir en approche par simulation permettant de prendre en compte la situation actuelle et une évolution probable.

S'il est pratiquement impossible d'aboutir à un recul spectaculaire de l'emploi des engrais, il est vraisemblable que nous serons amenés à mettre en oeuvre manière plus courante, soit des méthodes d'adjustement rationnel de la fertilisation permettant de limiter la fréquence des erreurs, soit de recourir à des techniques de dépollution au champ.

1. INTRODUCTION

L'accroissement de la productivité végétale nécessite une adaptation constante aux besoins en éléments nutritifs et un recours plus important aux engrais minéraux. L'utilisation de doses croissantes d'engrais chimiques est soupçonnée de contribuer largement à la dégradation des écosystèmes sols-plantes-eau. S'il est certain que la pratique agricole est responsable par exemple de l'accroissement de la teneur en nitrates, on peut aussi montrer que les pratiques agricoles peuvent agir sensiblement sur les quantités mises en cause.

Nous tenterons dans notre approche d'analyser les enjeux économiques propres à l'activité agricole et les implications dans la dégradation de la qualité des eaux et des sols.

Nous terminerons, enfin, par une liste de recommandations dictée aux agriculteurs pour limiter l'accroissement des teneurs dans les nappes souterraines. Seuls les problèmes de l'azote et très accessoirement du phosphore seront abordés.

2. ACTIVITÉ AGRICOLE ET QUALITÉ DES EAUX

Une étude statistique effectuée par l'Agence de Bassin Seine Normandie (1978) à partir des teneurs en nitrates dans les nappes phréatiques a montré une nette liaison entre le mode d'occupation des sols et les teneurs en nitrates enregistrées, avec d'énormes variations (Tab. 1).

Dans le cas de nappes superficielles, on a pu mettre en évidence une évolution très nette en fonction du temps, dans des situations où l'agriculture est seule en cause. Nous citerons à titre d'exemple, celui de petits bassins versants expérimentaux de l'Agence Financière de Bassin Seine Normandie (Fig. 1).

Si la relation activité agricole et teneur en nitrates des eaux de surface est bien établie, il semble que le phénomène soit moins net avec les eaux profondes. Plusieurs hypothèses peuvent être retenues pour expliquer cette situation:

délai de transfert très long. Les experts s'accordent cependant sur des vitesses comprises entre 0·50 et 1 m par an;
dilution par des courants latéraux;
mécanisme de dénitrification en couches profondes.

La figure 2 illustre la variation de teneur en nitrates avec la profondeur, selon les sédiments traversés.

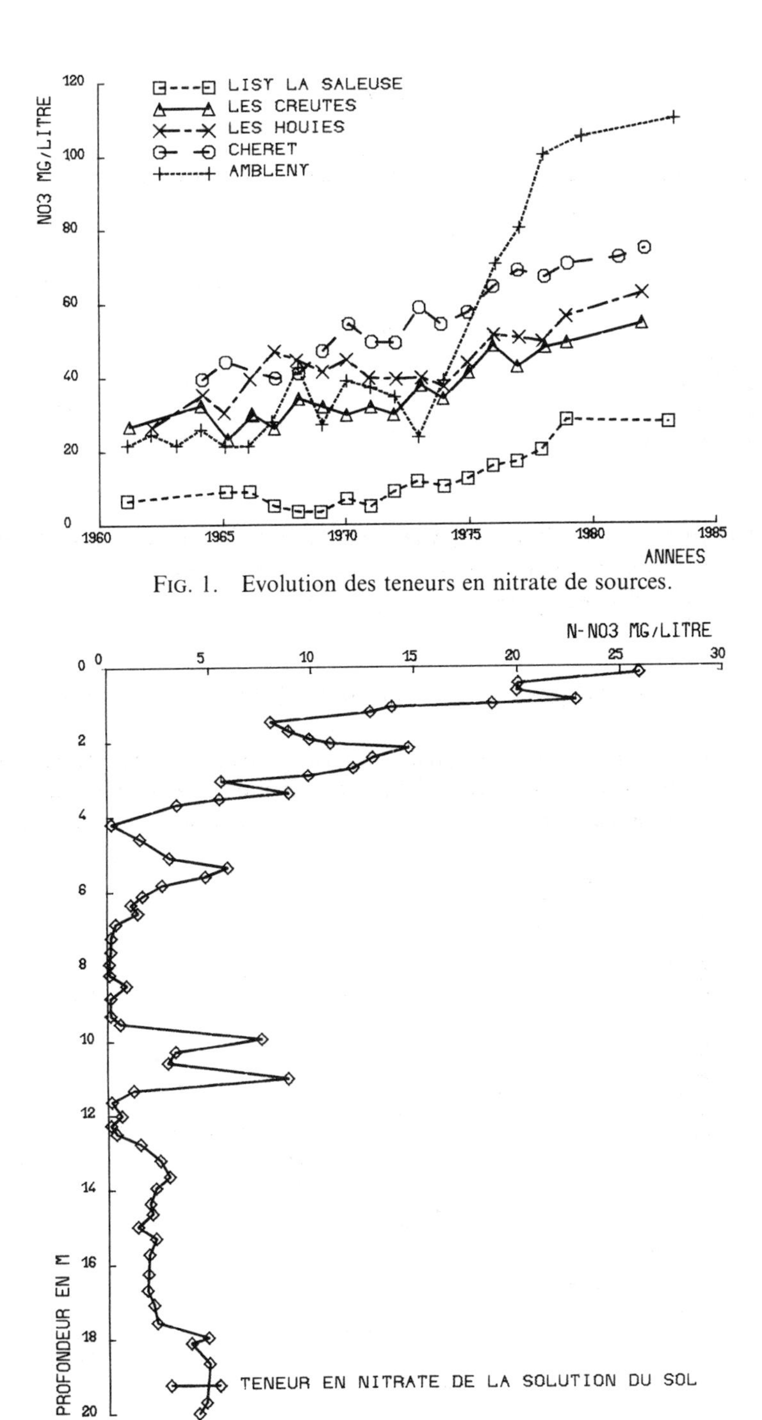

FIG. 1. Evolution des teneurs en nitrate de sources.

FIG. 2. Profil teneurs en nitrate en sol de limon.

Tableau 1
Teneur en nitrates des nappes phréatiques dans le bassin Seine Normandie 1978

	Nombre d'observations	NO_3 (*mg litre*$^{-1}$)	
		Minimum	*Maximum*
Forêts	30	0	8
Bocages	80	2	16
Cultures et pâtures	30	4	20
Cultures intensives	200	17	130
Zones urbaines et agr.	50	22	150
Zones urbaines et ind.	20	26	150

On constate sur la figure 2 que les teneurs de surface sont beaucoup plus élevées, jusque vers 4 m. Au-dessous, les teneurs demeurent basses, malgré des apports d'azote importants depuis 1960.

3. FERTILISATION ET PRODUCTIVITÉ AGRICOLE

La progression des rendements dans les systèmes d'agriculture intensive est due sensiblement pour moitié au progrès génètique, et pour moitié aux efforts pour améliorer la nutrition des plantes et leur protection sanitaire.

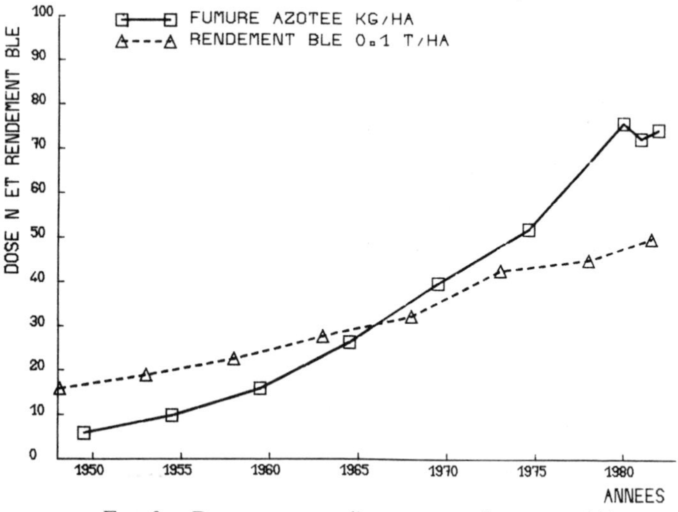

FIG. 3. Dose moyenne d'azote et rendement en blé.

Cette course à la productivité est rendue nécessaire pour amortir les charges fixes importantes de notre appareil productif.

Le rendement moyen en blé en France (Fig. 3) est passé de 1·5 t en 1950 à 6·3 t pour la récolte exceptionnelle de 1984. Pendant le même temps, la fertilisation azotée est passée de moins de 10 kg ha^{-1} à environ 80 kg ha^{-1}.

Dans la dernière décennie, l'accroissement de rendement (+1 t ha^{-1}) est à comparer à l'accroissement de fertilisation (+35 kg N ha^{-1}). Cette valeur est proche des normes habituellement utilisées pour caractériser les besoins de la culture (30 kg par t de grain).

4. ADÉQUATION ENTRE LES BESOINS DE LA PLANTE ET L'OFFRE DU SOL

La nutrition minérale appliquée à la production végétale requiert en général:

Une approche pragmatique au moyen des courbes de réponse à l'apport des fertilisants.
Une approche par bilan des éléments nutritifs au cours de la saison culturale.
Des moyens de contrôle et de diagnostic pendant le cycle végétatif.

Nous signalerons les conséquences de l'emploi de ces techniques sur le contrôle de l'environnement.

4.1. L'emploi des courbes de réponse

Très efficaces pour montrer l'intérêt et les limites d'un facteur de production, elles ont été exploitées:

sur le plan économique en recherchant l'optimum d'emploi des fertilisants;
sur le plan scientifique et technique en recherchant le déterminisme des anomalies rencontrées.

Les modèles les plus utilisés sont les suivants:

(*a*) *La parabole*

$$y = y_0 + bx - ax^2$$

Ce modèle a été utilisé surtout pour l'azote en France. Il présente l'inconvénient de la symétrie au niveau de la phase dépressive.

(*b*) *L'equation de Mitcherlich*

$$y = y_m(1 - \exp[-b\{x + s\}])$$

Ce modèle est très employe pour caracteriser la résponse au phosphore. Il rend très bien compte de l'hypothèse d'additivité entre la fumure minérale et la contribution du sol.

L'optimum économique d'une courbe de réponse à l'azote dépend de sa forme et du rapport de prix entre le fertilisant et le produit:

Exemple

$$y = 50 + 0{\cdot}444x - 0{\cdot}001\,25x^2$$

Si

$$P_y = 120F \text{ et } P_x = 4F$$

$$X_{\text{Opt}} = 164\,\text{kg ha}^{-1}$$

Si

$$P_y = 120F \text{ et } P_x = 8F$$

$$X_{\text{Opt}} = 151\,\text{kg ha}^{-1}$$

Si on doublait le prix de l'azote, l'optimum économique ne baisserait que de 13 kg. Il serait donc peu efficace de résoudre le problème de l'emploi des engrais azotés par une taxation.

Pour le phosphore, les courbes de réponse étant plus 'plates', le même calcul conduirait à une réduction beaucoup plus marquée des doses optimales. L'accroissement du prix des phosphates en 1974 a provoqué un recul durable de l'emploi des engrais phosphatés en France, malgré des disparités régionales importantes.

L'analyse des différents types de courbes de réponse (Fig. 4) a permis de progresser dans la rationalisation de l'emploi des engrais.

Dès à présent, on peut mesurer les risques moyens de surfertilisation par rapport à un pronostic initial:

existence d'un facteur limitant accidentel (maladies) (cas II);

mauvaise implantation de la culture avec récupération incomplète de l'azote disponible;

fourniture d'azote élevée. En général, ce risque est peu marqué, car lorsque les conditions climatiques sont favorables à la minéralisation, elles sont également favorables à la culture avec une amélioration du rendement.

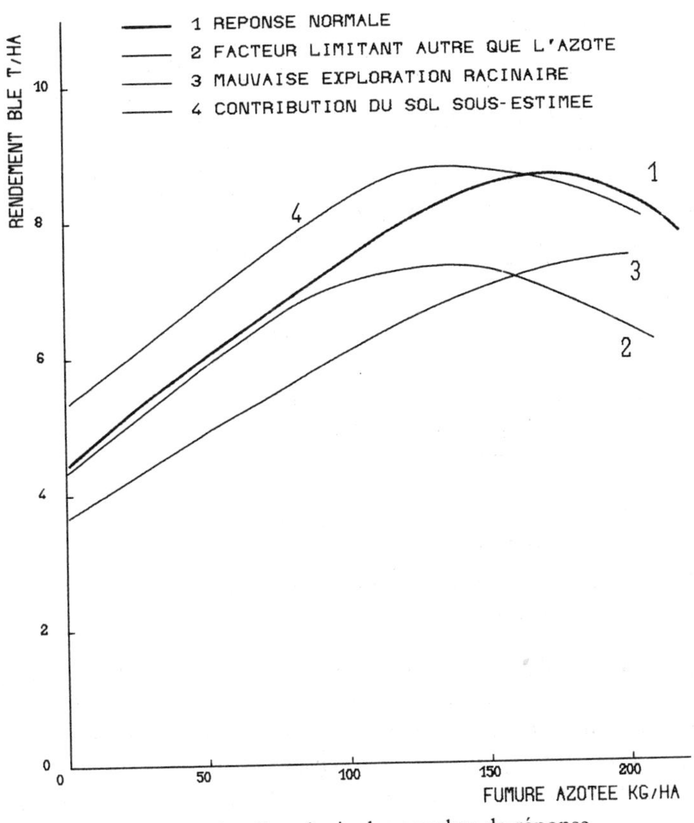

FIG. 4. Typologie des courbes de réponse.

Tableau 2
Besoins en azote pour les céréales

	kg t^{-1}
Blé dur	35
Blé tendre	30
Orge d'hiver	25
Maïs	23

4.2. La notion de bilan

Déjà bien connue pour les éléments phosphore et potassium où les bilans effectués en plain champ ou en cases lysimétriques ont montré l'importance des mecanismes d'insolubilisation et de lixiviation, cette notion a été mise à profit avec l'azote.

Le bilan a été réalisé sur l'azote minéral du sol au cours du cycle végétatif de la culture. Afin de s'affranchir des phénomènes hivernaux, le bilan de type budget a été réalisé entre la fin de l'hiver et l'époque du maximum de prélèvement.

$$(1/\rho)b.Y = R + M_s + M_r + M_a + X$$

ρ = coefficient d'interception racinaire
b = besoin en N par quantité produite
Y = rendement prévu
R = reliquat minéral sur 0·8 à 1·5 m
M_s = minéralisation de la matière organique du sol
M_r = minéralisation des résidus de récolte
M_a = minéralisation des amendments organiques
X = fumure minérale.

Des équations plus simples ont été proposées en fonction des milieux ou des objectifs. C'est le cas de la méthode dite 'N Min' où $X = \text{Cte} - R$. Le choix de la constante est fonction du site et du système de culture. Il est en général trouvé par expérimentation régionale.

L'estimation des différents termes a fait l'objet de nombreux travaux:

ρ — Lorsque le système racinaire est déficient, la récupération de l'azote minéral est incomplète. Ce coefficient peut varier de 0·5 à 0·9.

b — Les besoins par unité produite sont bien connus pour les céréales.

Y — Les prévisions de rendement sont difficiles à faire. Il existe des modèles climatiques simples, souvent insuffisants au niveau parcellaire où la technicité de l'agriculteur et son objectif de production devrait être pris en compte plus largement.

R — Le reliquat d'azote minéral de printemps mesuré en fin de période de lessivage et avant la phase d'absorption intense a fait l'objet de campagnes de mesures dans différents pays d'Europe de l'Ouest.

Dans la majorité des cas, la mesure est effectuée avec 3 à 6 échantillons sur une profondeur variant de 0·80 m à 1·50 m.

La tendance générale, dans tous les pays est de rechercher des équations de prédiction qui permettent de réduire le coût et l'importance des campagnes de prélèvement.

M_s, M_r, M_a La compréhension du cycle interne de l'azote, et notamment la quantification des processus de minéralisation immobilisation pourront apporter des progrès sensibles à la précision de l'estimation.

Par ailleurs, la contribution des résidus de récolte, essentielle dans le cadre des successions culturales commence à recevoir des réponses quantitatives en relation avec la composition des produits.

Les *amendements organiques* allant du lisier au fumier constituent un élément très important du bilan pour lequel les conseils pratiques sont encore insuffisants. Il est essentiel de pouvoir maîtriser pratiquement:

la composition et le type de l'amendement;
la régularité de l'épandage et l'incorporation au sol;
la cinétique d'évolution probable.

Il ne sert à rien de calculer à 10 kg près une fumure minérale si l'on n'est pas capable de maîtriser l'apport organique qui est majeur dans certains systèmes de production.

En ce qui concerne les lisiers, le problème des pertes est primordial. Des pertes dépassant 80 % des quantités d'azote apporté ont été signalées. A l'inverse, une bonne incorporation au sol permet d'aboutir à des pertes négligeables inférieures à 1 %.

4.3. Le devenir de l'azote excédentaire

L'utilisation directe de l'azote par la culture est loin d'être totale. L'emploi de l'azote 15 permet de suivre l'azote engrais apporté (Tab. 3).

Tableau 3
Devenir d'un engrais azoté marqué sous blé au champ

	% de l'engrais
Culture (partie aérienne)	52
Azote organique du sol (+ racines)	29
Azote minéral résiduel	3
Azote non retrouvé	16

Tableau 4
Coefficient d'utilisation (15 N) en fonction de la dose appliquée sur un blé

Dose (kg N ha^{-1})	*Coefficient d'utilisation*[a]
33	54·8
73	51·5
173	51·2
153	52·8

[a] Moyenne de 3 sites en France.

La valeur de 52 % est moyenne et l'on peut rencontrer des situations très diverses depuis moins de 20 % jusqu'à plus de 80 %.

Des travaux plus récents ont montré que l'influence de la dose était relativement peu marquée dans la gamme des apports usuels d'azote. Il existe donc un plateau assez important entre de très faibles doses et des doses très élevées (Tab. 4).

Par contre, on constate très régulièrement de grandes différences en fonction de la date d'apport. Ainsi, sur deux années consécutives, on a observé une variation continue depuis les apports précoces (20 à 30 %) jusqu'aux apports tardifs en mai (70 à 80 %) comme le montre la figure 5. Ce phénomène est dû à l'incorporation progressive de l'azote de l'engrais à la fraction organique. Plus le temps de séjour dans le sol est long avant la période d'absorption intense, plus le coefficient d'utilisation direct est faible. La probabilité de réutilisation ultérieure reste hypothétique.

Si l'on peut définir un coefficient d'utilisation directe, on peut aussi envisager de considérer un coefficient d'utilisation à long terme en cumulant les coefficients annuels successifs. Cependant, on dispose en

Tableau 5
Coefficient d'utilisation cumulé (1)

Années	*Case lysimatique maïs (1)*	*Champ blé–betterave*
1	49·2	51·9
2	9·5	3·8
3	1·8	1·7
4	1·6	1·3
5	1·4	
Cumul	63·5	58·7

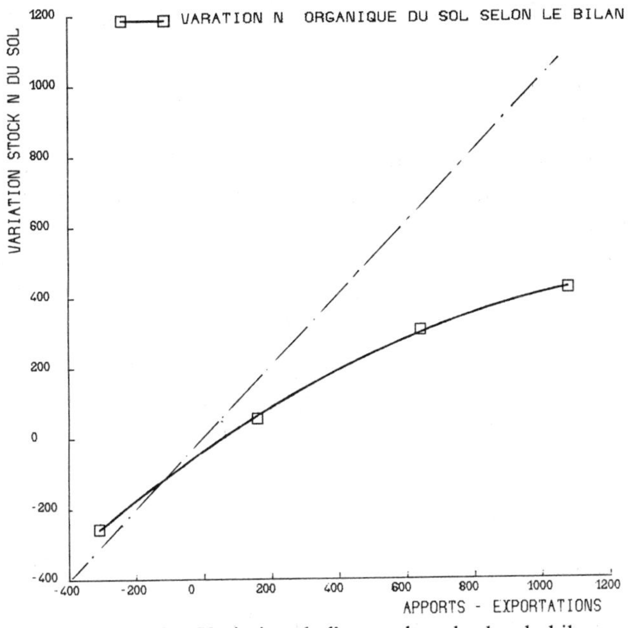

FIG. 5. Variation de l'azote du sol selon le bilan.

général de données assez limitées dans le temps. En outre, les dispositifs expérimentaux au champ présentent des erreurs de bordure de plus en plus importantes. Les résultats obtenus en lysimètres pourraient être plus représentatifs (Tab. 5).

Si le coefficient d'utilisation cumulé pouvait atteindre au mieux 85 à 90 %, il reste un défaut de bilan que l'on contrôle mal, la mesure directe étant difficile. On peut envisager:

(1) Une accumulation sous forme organique. Ce phénomène a été observé par Decau dans le Sud de la France après 11 ans d'expérimentation.

(2) Des pertes par dénitrification. En l'absence de conditions particulièrement favorables à la dénitrification, les mesures directes effectuées montrent des pertes assez limitées.

(3) Des pertes par volatilisation. En l'absence d'engrais ammoniacaux, elles sont négligeables. Elles demeurent très faubles si l'on prend soin d'incorporer les engrais aussitôt après leur épandage.

(4) Des pertes par lessivage. Elles sont difficilement quantifiables en l'absence de dispositif lysimétrique.

5. APPRÉCIATION DES PERTES PAR LESSIVAGE

Les deux principaux facteurs à prendre en compte pour caractériser les risques de perte par lessivage sont:

la présence effective de nitrates, avant la période hivernale provenant, soit d'un excès de bilan mal contrôlé, soit d'une minéralisation d'automne.

L'importance de la réserve en eau du sol qui mesure sa vulnérabilité par rapport à la protection des nappes. La réserve en eau comprend à la fois la texture et la profondeur du sol (Tab. 6).

Tableau 6
Pertes d'azote et profondeur du sol (moyennes sur 6 ans)

Profondeur du sol (cm)	*40*	*80*	*120*	*160*
Drainage (mm)	372	351	340	325
N lessivé mg ha^{-1}	57	47	47	32
Concentration mg $litre^{-1}$	66	57	62	44

Apport de 152 kg N ha^{-1} sur cultures fourragères en cases.

A titre expérimental, nous avons utilisé la formule de Burns (1975), pour définir les risques de lessivage d'azote en fonction des informations issues d'une carte des sols informatisée, selon un maillage régulier avec des éléments (Pixels) de 50 m × 50 m. Sur 13 000 hectares, nous avons pu répertorier différentes classes de sensibilité au lessivage (Tab. 7).

Il serait intéressant de vérifier la qualité des eaux, en fonction des dominantes des bassins versants pour vérifier l'efficacité de la méthode. Des conseils différenciés pourraient ensuite être donnés.

Tableau 7
Evaluation cartographique du risque de lessivage

	% de la surface	*Lessivage*
Zone bâtie	5·0	
Zone hydromorphe	6·1	
Perte 0 à 20%	0·1	Risque nul
Perte 20 à 40%	8·7	Risque faible
Perte 40 à 60%	46·3	Risque moyen
Perte 60 à 80%	29·6	Risque élevé
Perte 80 à 100%	4·4	Haut risque

6. LES TECHNIQUES DE CORRECTION

Lorsqu'une erreur de bilan a été commise, il faut rechercher par la pratique culturale des moyens de correction. Nous en citerons deux exemples:

les engrais verts;
les inhibiteurs de nitrification.

6.1. Les engrais verts

Ils permettent de consommer l'azote restant après une culture dont le rendement a été limité accidentellement ou pour utiliser l'azote minéralisé après une récolte précoce. Dans les deux cas, il y a soustraction d'azote minéral pour le stocker sous forme organique pendant la période automne-hiver.

Parmi les espèces les plus utilisées, nous citerons:

des crucifères (moutarde, radis chinois);
utilisés pour leur grande facilité de germination et d'implantation en condition sèche;
des légumineuses (Phaselia):
pour éviter l'apport d'azote;
des graminées (seigle):
pour profiter de l'effet bénéfique du système racinaire puissant (Tab. 8).

Le prélèvement d'azote représente environ 120 kg d'azote. La consommation d'eau se situant vers 100 mm venant ainsi en soustraction de la recharge des nappes. Il serait intéressant de comparer les 'pouvoirs épurateurs' des engrais verts.

$$PE\,(mg\ litre^{-1}) = \frac{N\ consommé}{Eau\ consommée} = 10^4\,\frac{N(\%)}{Q}$$

$N\%$ = teneur en azote de la matière sèche.
Q = quantité d'eau nécessaire pour produire 1 g de matière sèche.

La réutilisation de l'azote immobilisé par l'engrais vert pour la culture suivante a fait l'objet de nombreux travaux.

Tableau 8
Production de quelques engrais verts

Espèce	*M.S.* ($t\ ha^{-1}$)	*N* (%)	*Azote prélevé*
Moutarde rouge	3·4	3·1	105
Moutarde blanche	3·0	3·8	114
Radis slobolt	3·7	4·0	148
Phaselia	3·2	3·4	102

6.2. L'emploi des inhibiteurs de nitrification

L'intérêt des inhibiteurs de nitrification réside dans la possibilité de bloquer la première phase de la nitrification ($NH_4^+ \rightarrow NO_2^-$). La spécificité des inhibiteurs de nitrification n'est pas toujours assurée et certains ont un effet biocide prononcé pour l'ensemble de la microflore.

Les plus utilisés sont:

La Nitrapyrine (N Serve) qui a une action efficace, mais de durée assez brève (1 à 2 mois) en raison de la volatilité du produit qui est en outre insoluble dans l'eau. Une des formulations commerciales est cependant miscible et utilisée avec les solutions azotées. La dose appliquée se situe entre 2 et 4 kg ha^{-1}.

La Dicyandiamide (Didin) soluble dans l'eau, présente une assez bonne efficacité avec l'urée (Amberger, 1983) et une durée d'action de 2 à 4 mois selon la température. Les quantités à apporter sont cependant importantes (12% en N), l'inhibiteur étant lui-même un engrais azoté à long terme.

Le Sulfure de carbone est un biocide puissant permettant une inhibition efficace de la nitrification. Cependant, ses difficultés d'emploi empêcheront une utilisation sur une large échelle.

L'intérêt des inhibiteurs de nitrification est très limité dans les conditions normales de grande culture. Les essais ont montré que l'apport de fertilisants près de l'époque d'utilisation était presque toujours préférable.

Cependant, afin de freiner la nitrification d'amendements organiques tels que les effluents d'élevage ou les rejets de l'industrie agroalimentaire, les inhibiteurs pourraient permettre une meilleure valorisation de ceux-ci en cas d'apport d'automne. Il faut savoir que le maintien au stade ammonium est rarement complet et qu'il subsiste des risques de lessivage.

7. LES RECOMMANDATIONS AUX AGRICULTEURS

7.1. Eviter les excès de bilan

Estimer au plus juste les potentialités du site pédoclimatique, au besoin en améliorant la prévision de rendement au cours du cycle végétatif, lorsqu' il est encore possible d'intervenir.

Améliorer la connaissance et la pratique des amendements organiques.

Mieux caractériser la capacité de minéralisation des sols.

7.2. Améliorer l'interception de l'azote par les racines

Favoriser l'installation précoce du système racinaire des plantes.

Associer dans les rotations culturales des plantes à enracinement profond.

Effectuer les apports peu avant les périodes d'absorption importante.

Proscrire ou limiter au strict nécessaire les apports avant l'hiver.

Utiliser le fractionnement des apports pour éviter les doses excessives.

Eviter les irrigations excessives.

7.3. Mettre en oeuvre des techniques de correction

Incorporer au sol les substrats carbonés comme la paille.

Utiliser les engrais verts permettant d'assurer la couverture du sol pendant la période automnale.

8. CONCLUSION

Les risques d'accroissement des teneurs en nitrates sont liés à l'intensification de la production agricole. Toutefois, on peut sommairement définir 3 zones (Fig. 6):

Une zone où les pertes par lessivage décroissent en raison d'une amélioration du capteur racinaire pour les faibles niveaux d'azote.

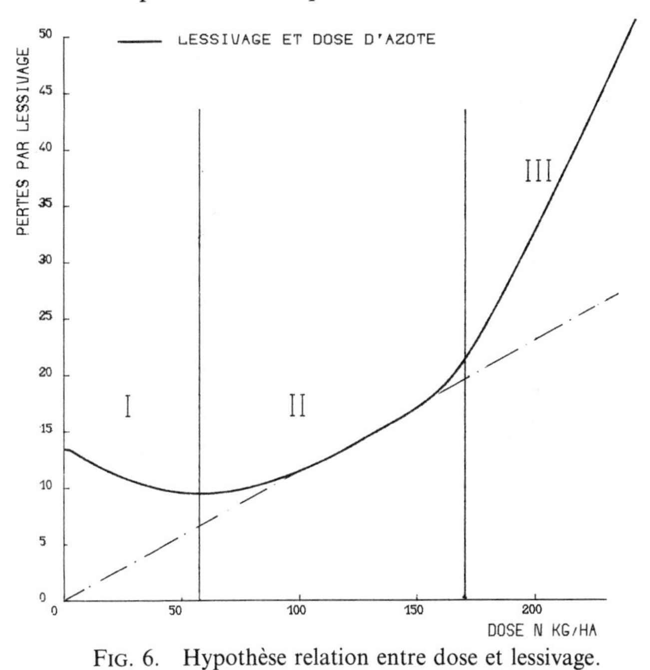

FIG. 6. Hypothèse relation entre dose et lessivage.

Une zone où les pertes sont sensiblement proportionnelles à la dose apportée et où le coefficient d'utilisation par les plantes est constant. C'est le domaine dans lequel l'Agronomie doit travailler.
Une zone où les pertes sont plus que proportionnelles et qui correspond aux excès de bilan. C'est le domaine que nous ne devons pas franchir et qu'il faut combattre.

En conclusion, nous devons travailler à préciser la zone optimum à la frontière des domaines II et III, en améliorant la technique des bilans prévisionnels et à promouvoir les techniques correctives les plus efficaces pour limiter les quantités d'azote pré-hivernales.

16

Can Agricultural Practice Reduce the Amount of Chemical Fertiliser?

HUBERT TUNNEY

Agricultural Institute, Johnstown Castle Research Centre, Wexford, Ireland

SUMMARY

Agricultural practice can undoubtedly reduce the amount of chemical fertiliser use. The reductions requested in the EC agricultural production could have an important bearing on the use of chemical fertiliser. There has been a dramatic and continuing increase in nitrogen fertiliser use, particularly in the latter part of this century. The use of phosphorus and potassium has also increased but to a lesser extent in recent years with the emphasis now on maintenance dressings. There is a need for improvements and standardisation of soil analyses as an aid to minimising fertiliser use. The total quantities of plant nutrients in animal manures and other organic manures are only a little less than applied in chemical fertilisers. Efficient recycling of these manures could reduce fertiliser needs. Nitrogen fixation by legumes is little used in the Community and it may offer the possibility of significant reductions in the use of fertiliser nitrogen. This possibility merits further investigation and research as, at present, it means lower yields and is less reliable than fertiliser nitrogen. There is evidence that high chemical inputs are not always related to increased returns. A number of other ideas which could possibly lead to lower fertiliser use are also introduced.

1. INTRODUCTION

There has been a dramatic and sustained increase in chemical fertiliser use in Europe since the end of World War II.

Most agricultural scientists have viewed this increase with pride and are happy to take credit that their Government Department or Institute has contributed. People in the fertiliser industry are no less proud of their achievement.

Many of us can remember the nutrient deficiency problems that severely limited crop and animal health and production up to a few decades ago. Indeed there are still significant areas of soils in the EC with phosphorus and other nutrient deficiency problems. If there is a demand for increased crop production at a profitable price in the future there is little doubt that chemical fertiliser use would continue to increase for many years. However, in the present situation of over production of most agricultural commodities in the EC and the absence of outside markets at profitable prices, it seems unlikely that fertiliser use will increase and may probably stabilise or even decrease.

Population growth in the EC is static with the result that demand for food is unlikely to increase in the near future. In addition dietary patterns may also be changing due to concern about health risks from too high a consumption of animal fats. If this should happen then demand for animal produce may decrease at a faster rate than would be suggested by population trends.

At the same time there is growing public awareness of the increasing use of chemicals in agricultural production and the possible adverse effects they may have on health and quality of the environment. It is in this framework that we should consider the possible influence of agricultural practice on reducing the amount of chemical fertiliser used.

It is important to keep in perspective that only about 25 % of the world population live in developed countries where chemical fertilisers are widely used and where there is concern with possible pollution and excessive use. About 50 % of increased crop production in developed countries in the past three decades has been attributed to fertiliser use [9].

However 75 % of the world population lives in developing countries where often little or no fertiliser is used and soil fertility status is low. Famine is widespread in some areas of the world and increased fertiliser use could contribute, perhaps, more than any other single factor to increased food production in the short term.

2. TRENDS IN FERTILISER USE

2.1. Nitrogen

There has been a dramatic increase in fertiliser use in most countries in Europe, particularly since the end of World War II . This increasing pattern

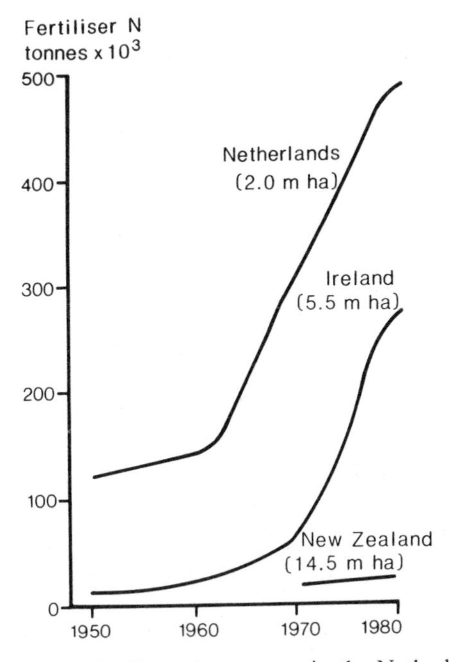

FIG. 1. Recent trends in fertiliser nitrogen use in the Netherlands, Ireland and New Zealand.

is illustrated in Fig. 1 which shows the trend in nitrogen fertiliser use for Ireland, the Netherlands and New Zealand between 1950 and 1980. This increase is still continuing though at a slower rate.

In the Netherlands the average rate of fertiliser nitrogen per hectare of agricultural land is over 240 kg per year and is perhaps one of the highest rates in the world. This is an average figure and some intensive grassland farms use over 500 kg fertiliser nitrogen per hectare per year. Ireland is at the other extreme with one of the lowest rates of chemical fertiliser use in the EC at 43 kg N ha^{-1} yr^{-1}. Figure 1 shows that there has been a four-fold increase in the Netherlands and a ten-fold increase in Ireland between 1960 and 1980. Most of this increase took place between 1970 and 1980. The other EC countries, with the exception of Greece which has the lowest, are intermediate in fertiliser use between the two extremes illustrated for Ireland and the Netherlands.

The relative rates of fertiliser use per hectare of agricultural land for the different EC countries are summarised in Table 1. Here we can see that the

Table 1
Consumption of commercial fertiliser per hectare of agricultural land in the European Community 1979/80 [10]

	WG	*F*	*I*	*N*	*B*	*L* (*kg*)	*Dk*	*Ir*	*UK*	*Gr*	*EC*
SF[a]	114	77	57	260	165	198	69	26	44	18	72
CF[a]	150	111	63	82	146	24	171	77	77	38	94
TF[a]	294	188	121	342	312	221	240	104	120	56	166
N	121	70	59	240	128	108	136	43	71	33	75
P_2O_5	75	62	40	41	70	51	46	27	24	20	46
K_2O	98	56	22	61	114	62	59	33	25	4	44

[a] SF = Straight fertiliser, CF = compound fertiliser, TF = Total fertilisers.

average nitrogen use for the EC is 75 kg N ha^{-1} yr^{-1} with the Netherlands, Denmark, Belgium, Germany and Luxembourg being above average.

It is important to emphasise that these are average figures with large areas in most countries receiving little or no fertiliser while other areas may receive many times the national average.

Figure 2 illustrates the regional distribution of nitrogen fertiliser on pasture for the Republic of Ireland. The rates are highest, at about the EC average, in the eastern and southern areas where there is a relatively higher proportion of intensive grassland. The rate in the west, with more extensive grassland and low tillage area, is only about a quarter the rate in the east of the country.

The use of nitrogen fertiliser in New Zealand, which also has a well developed agriculture, is almost negligible (1·5 kg N ha^{-1} yr^{-1}) by comparison with the EC countries. This is also illustrated in Fig. 1. The most generally accepted reason for this difference in that New Zealand agriculture is based on nitrogen fixation by clover whereas in Europe it is based on chemical fertiliser. The fact that New Zealand grassland farmers receive much lower prices for their produce than their EC counterparts may also play some part in this difference.

It is also possible to say that if all the farmland in the EC were to receive nitrogen fertiliser at the rate currently applied to our more intensive farms then the rate of growth in nitrogen fertiliser use could continue for many years. However, with the recent introduction of a milk super levy and adequate or over production in most commodities, large increase in fertiliser use is unlikely in the immediate future.

Intensive grassland receives the highest rates of nitrogen fertilisers.

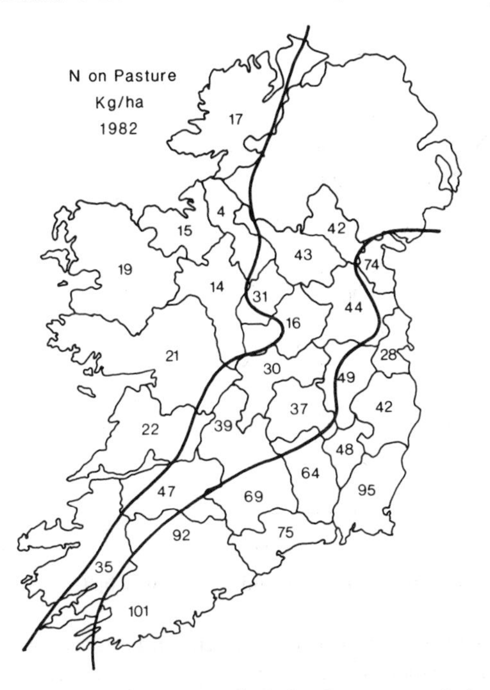

FIG. 2. Pattern of fertiliser nitrogen use in Ireland on pasture in kg ha^{-1} for 1982 [15].

Therefore economic factors that cause a change from other forms of agricultural production to intensive grassland, are likely to result in increased nitrogen use and vice versa. Figure 3 shows the increase in livestock numbers and the decrease in tillage area in Ireland between 1958 and 1980.

The rapid increase in livestock numbers between 1969 and 1974 was accompanied by a rapid increase in nitrogen use (Fig. 1) though tillage area decreased over the same period. It is also interesting to note that for the last four years, shown in Fig. 3, both cattle numbers and tillage area remained fairly static, however nitrogen use continued to increase sharply. A high proportion of increased nitrogen use on intensive grassland in recent years has been applied to land for dairying.

Experience in the advisory service in the Netherlands suggests that it is perhaps as difficult to reduce an established fertiliser application pattern on a farm even when the nutrient status is very high as it is to get fertiliser use increased on a farm where the nutrient status is too low.

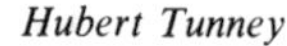

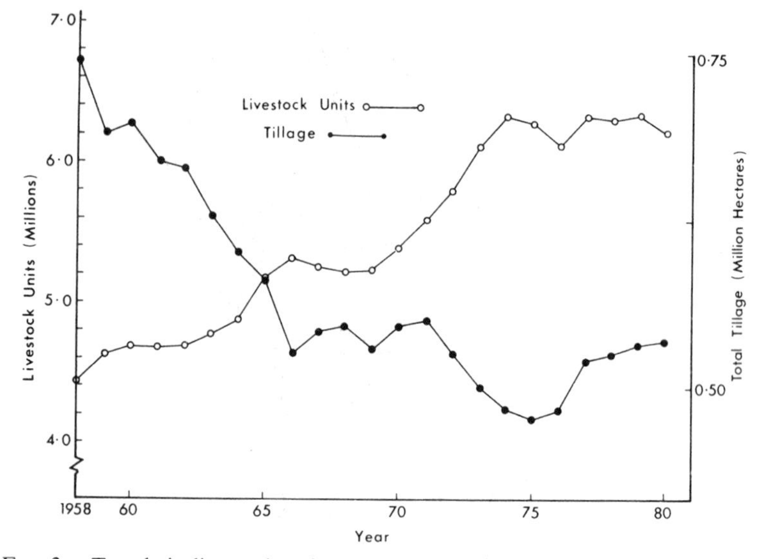

FIG. 3. Trends in livestock units and tillage area in Ireland 1958–1980 [15].

There has been similar experience in Ireland and other EC countries, for example, some farmers continue to apply excess nitrogen to sugar beet despite experimental results and advice to the contrary. This is probably due to the fact that the crop often looks better, with more leaves, but the total sugar yield is not increased. It is also true that, with high value crops, fertiliser is a relatively small part of total inputs and higher rates than necessary are used. Such excessive fertiliser and manure use occurs on only a very small percentage of the land area in the EC. It is difficult to quantify to what extent fertiliser use could be reduced in practice. Education and increased use of soil analyses could help avoid over use of chemical fertilisers on these farms.

2.2. Phosphorus and potassium

In contrast to the steep increase in fertiliser nitrogen, phosphorus and potassium use has remained relatively stable in recent years. This is illustrated in Fig. 4 with statistics for the United Kingdom [4].

For the fertiliser year 1982/83 the average UK consumption in kg per hectare of land (total 12·1 M ha) was 121, 37 and 41 for nitrogen, phosphate, and potash respectively [4]. This trend in phosphorus and potassium fertiliser use is the same for many countries in the EC. It appears that, on much of the intensively farmed agricultural land, P and K fertility

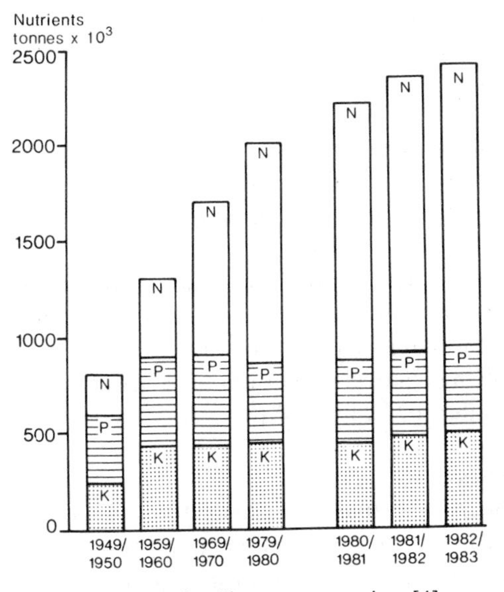

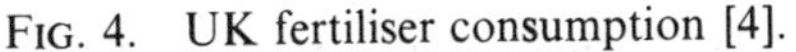

FIG. 4. UK fertiliser consumption [4].

is already at an acceptable level and that the annual application is aimed at maintaining soil fertility. In other words the application rate is related to the quantity of these nutrients removed by the crop and other sources. There is of course a difference between P and K in this regard as most good agricultural soils can supply up to 100 kg K ha^{-1} yr^{-1}, or more, from the soil reserves, almost indefinitely, due to weathering and release from soil minerals. Therefore K addition for maintenance can be significantly lower than removal by crop and water. In order to maintain soil P fertility it is generally necessary to replenish the quantity removed in the crop with P fertiliser or animal manures.

It is important to remember that there are still large areas within the EC where the P and K status is very low and deficiency symptoms would appear for most sensitive crops. Productivity, on these soils, could be improved greatly by increasing the soil fertility. It would be an interesting study to determine to what extent this improvement in P and K fertility could be used to compensate for the lower production that would occur with a reduction in nitrogen fertiliser within the community.

This approach would only be of interest if it was accepted that the increased P and K use would have economic and environmental advantages relative to the existing relatively high fertiliser nitrogen use.

Phosphorus is of particular importance to the EC as we are totally dependent on imports and it is also the most expensive of the major fertiliser nutrients.

2.3. Other fertiliser materials

In addition to N, P and K large quantities of ground limestone and other liming materials are applied to some soils in the EC each year to increase soil pH for optimum crop production. In addition other macro-nutrients including sulphur, magnesium and micronutrients such as copper, manganese and boron are often applied in special circumstances to correct deficiencies or meet the requirement of sensitive crops. These fertiliser materials, involving relatively small amounts, are not generally regarded as presenting pollution or health risks.

2.4. Soil analyses

The use of the best available methods for soil and plant analyses to determine the correct rates of fertiliser to apply could play an important role in minimising application rates for optimum crop production. It is necessary to standardise the methods of soil analyses used in the different EC countries. For example, it is difficult to justify the wide range of soil extractants currently used in the different countries for phosphorus analyses.

3. RECYCLING OF AGRICULTURAL WASTES

For thousands of years animal manures and other farm wastes have been recycled as the principal method of maintaining soil fertility. Almost 2000 years ago, Virgil in his agricultural advisory book of verse, the Georgics, recommended the use of organic manures for good crop production.

In recent years, with the adoption of intensive animal rearing and the ready availability of relatively inexpensive chemical fertilisers, animal manures are often regarded as a disposal problem. This is in sharp contrast to the great value placed on organic manures in the past.

3.1. Value of agricultural waste

Animal manures contain all the nutrients essential for plant growth. In the region of 80–90 % of the plant nutrients in food are excreted by the animal. These nutrients can and should be recycled to land to produce subsequent crops. This simple recycling approach helps conserve natural resources, reduce costs on the farm and at the same time minimises the risk of pollution. The rate and time of application should be related to crop needs and supplemented with chemical fertiliser to give optimum crop production.

The relative increase in fertiliser nutrient use compared to manure in Germany, over the past 100 years is illustrated in Fig. 5.

This figure shows that in recent years, and particularly since the end of World War II, more of the three major fertiliser nutrients are applied in chemical fertiliser than in manure.

The quantity of manure produced by farm animals can vary considerably. The daily values most often quoted for intensive livestock production are 45 kg for mature cattle, 4·5 kg per fattening pig and 0·1 kg per chicken [24]. There can also be a wide variation in the nutrient composition of manures [23] and this variation presents difficulty in efficient manure utilisation.

The range of variation in slurry composition from a number of pig and cattle farms is summarised in Table 2.

The wide range in composition of slurry was principally due to varying

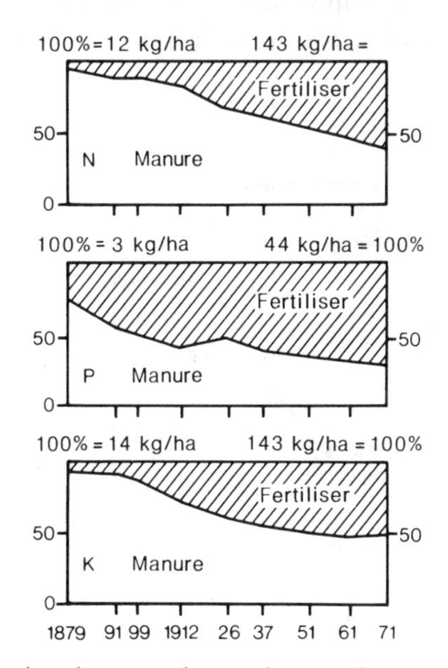

FIG. 5. Nitrogen, phosphorus, and potassium nutrients applied in fertiliser and animal manures in Germany over the past 100 years [8].

Table 2
Average and range of animal slurry composition from a study of Irish farms [25]

	% Dry matter	*kg per 10 t*		
		N	*P*	*K*
			Cattle slurry (33 farms)	
Mean	8	28	6	42
Range	1–14	8–56	1–12	8–64
			Pig slurry (25 farms)	
Mean	4	30	9	15
Range	1–13	4–70	1–34	2–33

degrees of dilution with water. In the past the conventional farmyard manure was less variable in composition and it was therefore easier to know the rates to apply to meet crop needs.

Pig manure is higher in phosphorus and lower in potassium than cattle manure, reflecting the higher phosphorus in the mainly cereal diet of the pigs and the higher potassium in the grass based diet of the cattle.

There is a positive correlation between the dry matter and fertiliser value of slurry [23]. In other words the higher the dry matter the higher the

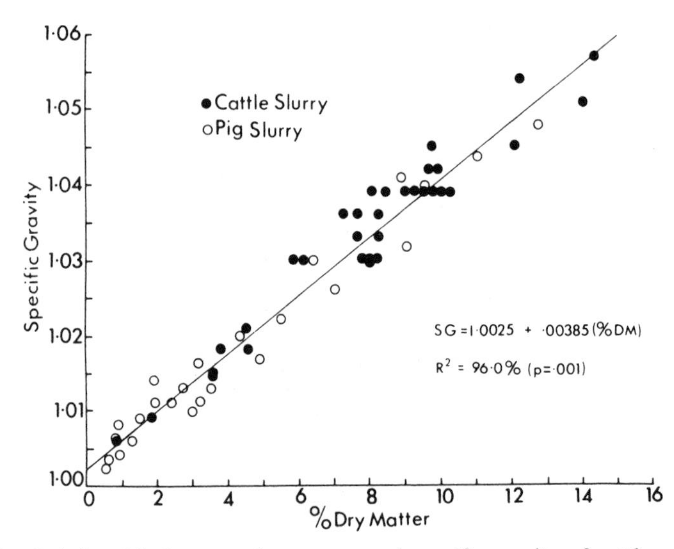

FIG. 6. Relationship between dry matter and specific gravity of cattle and pig slurry [23].

fertiliser value and vice versa. Interestingly there is a highly significant straight line relationship between dry matter and the specific gravity of slurry [23]. The relationship is illustrated in Fig. 6.

This relationship is the basis of a simple field test, developed at Johnstown Castle Research Centre, for estimating the fertiliser value of slurry and is based on a hydrometer calibrated in percentage dry matter. The dry matter is read from the hydrometer and the corresponding nutrient content of the manure is then read from an accompanying table. It is then possible to determine the correct quantity to apply or the monetary value of the slurry.

3.2. Fertiliser value

An attempt is made in Tables 3 and 4 to estimate the plant nutrients in animal manures and compare them with chemical fertiliser on a world scale and also in the EC. These estimates are based on the livestock populations and manure composition figures shown in Table 3. It should be stressed that these are approximate values and are used as a guide to obtain an overall perspective.

Table 4 shows that on a world scale there are more nutrients in animal manures than applied in chemical fertiliser. In the EC, however, more plant nutrients are applied in fertilisers.

It is clear from Table 4 that animal manures contain a large quantity of plant nutrients and are an important source for maintaining soil fertility if

Table 3

Livestock population [6] approximate waste production and composition

	1978 Livestock population (millions)		
	Cattle	*Pigs*	*Poultry*
World	1 213	732	6 468
EC	80	74	614

	Approximate manure composition			
	(tonnes/animal/year)	*(kg/10 t)*		
		N	*P*	*K*
Cattle	12·00	40	8	40
Pigs	1·20	45	16	20
Poultry	0·36	120	60	60

Table 4
Chemical fertiliser consumption [7] estimate of corresponding nutrients in animal manures for 1978 (millions of tonnes)

	Manure	*N*	*P*	*K*
			Animal manures	
World	15 667	65·0	15·0	61·0
EC	1 071	4·5	1·1	4·2
			Chemical fertilisers	
World		48·0	12·0	19·0
EC		6·4	1·9	3·4

they are recycled efficiently. There are other organic wastes such as sewage sludge, food factory waste, etc., which contain significant quantities of plant nutrients.

It should be remembered that a high proportion of cattle waste is naturally recycled to pasture by grazing animals. Approximately 90% of nutrients in manures are produced by cattle and only about 10% by pigs and poultry.

In the EC perhaps more than half the total manure production is naturally recycled by animals in the field and half is available for spreading from housed animals. We can assume that much of the manure produced at present in the EC is recycled effectively and is therefore already contributing to a lower fertiliser use. But there can be little doubt that a significant part of the manure is disposed of at high rates on limited areas of land; in these situations its fertiliser potential is not being achieved. This situation is most likely to occur with intensive pig and poultry farms. These areas of high application can usually be identified by very high soil phosphorus status.

In the Netherlands and Ireland manure banks have been established with the aim of transporting manure from areas of high pig or poultry density to areas where the manure nutrients can be used effectively for plant growth.

In general the P and K in animal manures are as effective as their chemical fertiliser equivalents. Therefore more efficient utilisation of manure could certainly lead to a saving on P and K fertiliser requirements. Perhaps the most serious situation exists on farms where the full complement of P and K fertiliser is applied even though the soil's nutrient status is already very high because of manure disposal. These excessive rates of application can lead to soil and water pollution problems and also represent an unnecessary expense for the farm.

The efficiency of manure nitrogen is generally low relative to fertiliser nitrogen. This poor efficiency can be attributed to (a) nitrate leaching, (b) denitrification and (c) volatilisation as ammonia. Approximately half the total nitrogen in manure is present as ammonia. Where manure is applied to the soil surface most of this ammonia can be lost to the atmosphere within a few days if conditions are suitable. Work carried out over 50 years ago [14] showed that 90 % of the ammonia was volatilised within four days when manure was applied to the soil surface; however, the loss was small when manure was worked into the soil immediately after application.

Research in Ireland indicates that 50–75 % of ammonia in manure can be lost within one week of application [20]. In the USA Vanderholm [27] estimates that 30–65 % of nitrogen in manure can be lost during storage and shortly after spreading by surface runoff and volatilisation of ammonia. Porter [19] estimates that half the total nitrogen in animal manures in America is lost by ammonia volatilisation. Australian work [5] shows that ammonia volatilisation can also be high from livestock wastes produced by grazing cattle on pasture.

The loss of nitrogen by denitrification may also be facilitated by livestock wastes as they can provide anaerobic conditions and readily decomposable organic matter [2].

Acidification of manure can reduce ammonia loss [17] and in the past superphosphate was added to liquid manure in Germany to reduce ammonia loss. However in the past 50 years there has been little or no practical progress made in reducing ammonia loss from manures.

Injecting manure into the soil may reduce ammonia loss but can result in losses due to denitrification. The combination of slurry injection into the soil combined with nitrification inhibitors may be worth studying.

If the loss of nitrogen from manure could be reduced it could significantly reduce the fertiliser nitrogen requirements. This could be considered as a long term possibility with considerable potential for reducing nitrogen fertiliser needs on grassland.

Guidelines have been drawn up on the use of animal manures within the EC with a view to minimising the risk of pollution and maximising their fertiliser value. In summary it is possible to say that if these guidelines were applied carefully the present level of agricultural production could be maintained with a lower input of chemical fertiliser.

4. NITROGEN FIXATION

The fact that agricultural production can be based on nitrogen fixation by legumes, as opposed to chemical nitrogen fertiliser, is illustrated by the example of New Zealand and has already been shown in Fig. 1. The extent to which this change in agricultural practice is possible within the EC is an open question. In short it is possible that a significant proportion of our present fertiliser nitrogen use could be replaced by nitrogen fixing crops. However this would involve a decrease in production. The increased use of legumes would also require higher soil phosphorus and potassium status.

There are many diverse biological nitrogen fixing systems in nature; however, the legume system is the only one of practical importance in agriculture [18].

Grain, forage and pasture legumes are capable of supplying their nitrogen needs via the fixation process and are therefore independent of mineral forms of nitrogen for growth. This is of great importance as nitrogen supply is usually limiting for plant growth in all agricultural systems. The relatively recent availability of large quantities of nitrogenous fertilisers at low cost largely replaced use of legumes in European agriculture and this resulted in higher crop yields, greater stability of yield and control over nitrogen supply.

Recently, however, the practice of modern agriculture based almost exclusively on chemical nitrogen fertiliser is coming under increasing scrutiny largely as a result of increasing costs of fertiliser production coupled with a fall off in product value. Contamination of ground water by nitrates resulting from over use of fertiliser is also a cause for worry. The fixation process uses solar energy for nitrogen production, has no distribution costs and supplies sufficient or less than the plant needs so it suffers from none of the above disadvantages.

Legume based systems are, however, characterised by lower yields, poor yield stability and there is no control over nitrogen inputs as compared with systems supplied with mineral nitrogen fertilisers. Furthermore the fixation process which doubtlessly evolved in soils low in nitrogen is at a severe disadvantage in most fertile agricultural soils where mineralisable nitrogen levels may be high. Such levels of mineral nitrogen forms depress nodulation processes and the nitrogen fixation system *per se* [1].

So the problem is how best to adapt legumes for performance in highly intensive farming systems. The euphoria of the early 1970s which considered the possibility of spreading the fixation process to non-fixers such as grasses and cereals has now been superseded by the reality of the serious difficulties involved in this approach. The genetics of the nitrogen fixing system are advancing rapidly and the first steps in the transfer of nitrogen fixing genes to non-nitrogen fixing organisms have been taken. However, the complexities involved with the legume system are daunting and much work still remains to be done before the practical benefits of genetic engineering on the nitrogen fixation process become a reality [3].

Therefore in the short to medium term the best possibilities rest with improving existing nitrogen fixing systems rather than attempting to construct new ones. A number of areas stand out as being worthy of attention:

(a) Development of legumes better able to perform in intensive systems through the use of plant selection and breeding techniques.
(b) Increasing the efficiency of symbiosis by better matching of bacteria and host plant.
(c) Improvement of inoculant technology and a better understanding of the factors affecting rhizobium performance and survival in soils.
(d) Understanding the underlying factors affecting legume crop yield stability in the field.
(e) Use of molecular genetics for the development of plant genotypes and bacterial strains with improved symbiotic performance.

Finally whether conventional or novel legume systems are involved their performance will only be as good as the management systems imposed on them. It is of great importance therefore that we know how to minimise stress and maximise output from legumes growing in intensive or semi-intensive agricultural systems.

5. LEY FERTILITY AND TILLAGE

It is well established that organic matter accumulates in the soil under grassland over many years and that it disappears rapidly after tillage. Research work in the Netherlands [13] showed that organic matter started accumulating soon after sowing and continued asymptotically for many years until the equilibrium of old grassland was attained. In this work the yield of arable crops and of grassland with a low clover content was higher, the higher the content of organic matter in the soil. This yield increase is attributed, almost entirely, to the increased nitrogen supply from the soil and the same effect could be attained by a higher nitrogen dressing. Soils high in organic matter, with over 3 % organic carbon, can contain up to 10 tonnes of nitrogen per hectare in the top 30 cm. Part of this is mineralised each year and is available for plant growth. After ploughing the rate of mineralisation is increased and relatively large quantities of nitrogen can be released and this may meet much of the needs of tillage crops in the subsequent two or three years.

Even further into the tillage rotation significant quantities are mineralised in the cool temperate regions. In Ireland, for example, the net cumulative N release from fertilised rotations of cereals and sugar beet is of the order of 3200 kg N ha^{-1} in the ten years of tillage following grazed leys [11]. At the end of this period values of mineral N of *circa* 270 kg N ha^{-1} are observed in the 0–60 cm root profile of sugar beet in June. Of more relevance to leaching losses, however, are residual levels typically of the order of 90–140 kg N ha^{-1} that remain after crop harvest [12].

The accumulation of nitrogen under grassland comes principally from applied fertiliser and manure nitrogen or fixation by legumes. Its subsequent release to tillage crops is referred to as ley fertility. Some of this mineralised in the cool temperate regions. In Ireland, for example, the net with high winter rainfall.

The northern areas of the EC tend to have soils with high organic matter relative to southern Europe. The highest organic matter contents are found in the permanent grasslands of the United Kingdom and Ireland, whereas the lowest would be found in tillage soils in southern Italy or Greece. The high accumulation of organic matter can be attributed to the higher rainfall and lower temperatures in the north. The ultimate organic matter accumulation is of course the development of peat and at the other extreme is desertification which can occur in southern areas of the EC when organic matter is depleted.

The potential of ley fertility to reduce nitrogen fertiliser use is perhaps dependent on the use of legumes or manures to build up ley fertility. Otherwise it is simply dependent on immobilising fertiliser nitrogen in soil organic matter for later use.

On grassland the plant nutrients are often concentrated in the top few centimetres of soil. Deep ploughing for reseeding of grassland can put most of these nutrients out of reach of the grass roots. Therefore in this situation shallow ploughing could decrease the fertiliser needs for establishment.

Minimum tillage or no tillage techniques for sowing cereal crops and reseeding grassland can help maintain high organic matter in the soil and therefore perhaps in the long term the need for less fertiliser nitrogen. This would also reduce the flush of nitrate leaching that can occur in the autumn on ploughed soil. Spring ploughing instead of autumn ploughing, for spring sown crops, could also reduce nitrate leaching.

6. OTHER POSSIBILITIES

In addition to the topics already covered there are undoubtedly other possibilities for improving existing fertiliser efficiency and thereby reducing the fertiliser requirements. For example on intensive dairy farms upwards of 250–300 kg N ha^{-1} is often used yet only 50–60 kg ha^{-1} is recovered in milk and animals [21]. Any approach that can improve this low efficiency would allow a reduction in fertiliser use.

The introduction of more concentrated fertiliser can sometimes lead to excessive use. For example farmers will sometimes apply urea (46 % N) at the same rate per hectare that they used for less concentrated fertiliser such as calcium ammonia nitrate (27·5 %) in the past.

Improved techniques for more accurate application of fertiliser and organic manures is an area where agricultural engineering may be able to help with the aid of microtechnology. In this area the placement of fertiliser to give maximum efficiency may also permit a reduction in fertiliser use. The development of the techniques for seedling transplanting, which was developed in Japan for sugar beet and is now being widely tested for other crops, may offer possibilities for fertiliser placement and improved efficiency.

These developing techniques would also appear to offer possibilities for reduced pesticide use.

The use of rock-phosphate instead of water soluble phosphates to maintain soil fertility could reduce phosphorus loss on soils subject to runoff. A high proportion of nutrients in fertiliser and manures may be lost from soils subject to runoff if heavy rainfall occurs shortly after application on saturated soils [21].

These losses can be greatly reduced if the nutrients are applied at the start of or during the growing season, instead of in autumn or winter. Better use of weather forecasting information to determine the timing of application could lead to reduced fertiliser need on soils subject to runoff. This would also, of course, greatly reduce the risks of eutrophication of rivers and lakes.

Phosphorus can be readily leached from peat soils so it should be applied in the growing season and should not exceed the crop needs.

Nitrate nitrogen losses due to leaching are highest on soils without crops over the winter. There will normally be low nitrate leaching from grassland. The presence of a winter cereal or other crop over the winter will also trap some nitrate that would otherwise be lost due to leaching. The use of nitrification inhibitors may also offer some possibilities to reduce nitrate leaching [22].

The type of nitrogen fertiliser can also affect the rate of loss by leaching; there is a lower loss from urea and sulphate of ammonia than from nitrate fertilisers [26].

A balanced fertiliser programme is essential for efficient fertiliser use. If one nutrient is limiting plant growth then the addition of normal rates of other nutrients can be inefficient. For example if sulphur deficiency is limiting grass yield then a farmer may put on extra nitrogen in an attempt to correct what may appear to be a shortage of nitrogen, but this extra nitrogen cannot be used by a crop that is already limited by sulphur deficiency. This is also true for any nutrient or other factor that is limiting plant growth.

Irrigation can give increased production on many soils where water is a limiting factor and several million hectares are irrigated in the EC. Water can be regarded as another essential plant nutrient and when it is limiting the efficiency of other fertiliser nutrients will be reduced. Irrigation will allow increased yield response to higher rates of nitrogen.

Irrigation is not an alternative to adequate fertiliser use but it can improve fertiliser efficiency on high value crops that can justify the capital costs involved.

Drainage will increase the availability of fertiliser nutrients by improving soil aeration and thereby facilitating the aerobically mediated energy system that transports nutrient across the membranes of root cells.

After drainage the production potential of soil is increased and in practice this will usually mean a response to higher fertiliser use.

Plant breeding to give improved cultivars has been carried out under conditions of high fertility. The breeding of plants that yield well under relatively low fertility may offer some possibilities for reducing fertiliser use.

7. CONCLUSIONS

It appears that there are several possibilities for improving fertiliser efficiency in practice and thereby maintaining present production with reduced fertiliser input. Figure 7 shows the lack of relationship between the cost of chemical inputs and cereal yields for a number of farms [16]. Some of the highest yields were obtained with the lowest inputs and vice versa.

This work suggests that high inputs are not always the answer to high yields and that good management, attention to soil analyses and other aids are necessary to minimise the inputs of agricultural chemicals required to give optimum production.

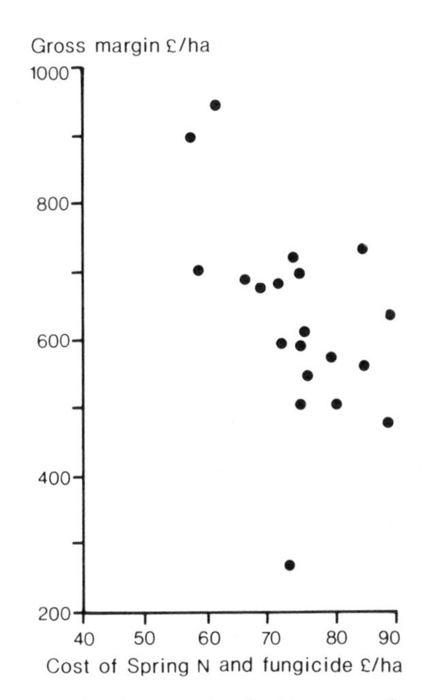

Fig. 7. There is no correlation between 'spring' inputs and gross profit margins on these 19 winter wheat crops grown on medium and heavy land in south Oxfordshire 1981/82 [16].

The adoption of legume based systems in the EC to fix atmospheric nitrogen as an alternative to fertiliser nitrogen could have an important impact on our rate of fertiliser use. A considerable body of information exists on this topic, however, it would still require considerable research and study to decide to what extent this change may be possible, the time scale required and if the reduced yield and economic consequences would be justified. Efficient recycling of organic wastes could also be of importance.

ACKNOWLEDGEMENTS

I wish to acknowledge the valuable help and suggestions from my colleagues at Johnstown Castle Research Centre.

REFERENCES

1. AMARGER, N. (1977). Nitrogen fixation by rhizobium legume associations. Proc. C.E.C. Seminar, I.N.R.A. Dijon, France.
2. BREMNER, J. M. (1977). Role of organic matter in volatilisation of sulphur and nitrogen from soils. *Soil Organic Matter Studies*, Publ. Int. Atomic Energy Agency, Vienna, no. 2, pp. 229–40.
3. BRILL, W. J. (1981). Agricultural microbiology. *Scientific American*, Sept., 146–55.
4. CHALLINOR, S. (1984). Fertiliser market review. *Fertiliser Review 1984*, Fertiliser Manufacturers Association, Cowcross Street, London, pp. 2–4.
5. DENMEAD, O. T., SIMPSON, J. R. and FRENEY, J. R. (1974). Ammonia flux into the atmosphere from a grazed pasture. *Science*, **185**, 609–10.
6. FAO (1979). *Production Yearbook 1978*, FAO, United Nations, Rome.
7. FAO (1979). *Fertiliser Yearbook 1979*, FAO, United Nations, Rome.
8. FLAIG, W., NAGAR, B., SOCHTING, J. and TIETJEN, C. (1978). Organic Materials and Soil Productivity, Soils bulletin No. 35, FAO, Rome.
9. GERVY, R. (1980). The contribution of fertilisers to food production. *Phosphorus in Agriculture*, I.S.M.A., Paris, no. 79, pp. 17–36.
10. GROVES, C. R. (1983). *An EEC Agricultural Handbook*, West of Scotland Agric. College. Res. Develop. Pub. No. 6, 3rd edn, p. 74.
11. HERLIHY, M. and O'KEEFFE, W. F. (1983). The contribution of soil reserves to the pool of available nitrogen in Irish soils. *Proc. Symposium on Nitrogen and Sugarbeet*, I.I.R.B., Brussels, pp. 145–53.
12. HERLIHY, M. (1984). An Foras Taluntais, Johnstown Castle Research Centre, Wexford, Personal communication.
13. HOOGERKAMP, M. (1973). Accumulation of organic matter under grassland and its effects on grassland and on arable crops, Agricultural Research Report 806, Pudoc, Wageningen, pp. 1–24.
14. IVERSEN, K. (1983). Loss of nitrogen by evaporation under application of liquid manure. *Fordampningstabet ved ajensundbringning 1928–33, Tidsskr.*, pl avl., **40**, 119–202.
15. LEE, J. (1984). An Foras Taluntais, Johnstown Castle Research Centre, Wexford. Personal communication.
16. MANBY, T. C. D. (1984). Selected aspects of efficient crop production. *J. Agric. Engng Res.*, **29**, 275–93.
17. MOLLOY, S. P. and TUNNEY, H. (1983). A laboratory study of ammonia volatilisation from cattle and pig slurry. *Ir. J. agric. Res.*, **22**, 37–45.
18. MURPHY, P. (1980). Biological nitrogen fixation. In: *Energy Management in Agriculture* (Ed. D. W. Robinson and R. C. Mollan), Proc. of the 1st Intern. Summer School in Agriculture, Royal Dublin Society.

19. Porter, K. S. (1975). *Nitrogen and Phosphorus*, Ann Arbor Science, Michigan.
20. Sherwood, M. T. (1980). Effect of land spreading animal manures on water quality. In: *Effluents from Livestock* (Ed. J. K. R. Gasser), Applied Science Publishers, London, pp. 379–90.
21. Sherwood, M. T. (1984). Nitrogen is too valuable to waste. *Farm and Food Research*, Agricultural Institute, Dublin, **15**(4), 114–15.
22. Slangen, J. H. G. and Kerrhoff, P. (1984). Nitrification inhibitors in agriculture and horticulture, A literature review. *Fertiliser Research*, **5**(1), 1–76.
23. Tunney, H. (1979). Dry matter, specific gravity and nutrient relationships of cattle and pig slurry. In: *Engineering Problems with Effluents from Livestock* (Ed. J. Hawkins), C.E.C. Luxembourg, pp. 430–47.
24. Tunney, H. (1980). Agricultural waste as fertilisers. In: *Handbook of Organic Waste Conversion* (Ed. M. W. M. Bewick), Van Nostrand Reinhold, New York, pp. 1–39.
25. Tunney, H. (1980). An overview of the fertiliser value of livestock waste. *Livestock Waste. A Renewable Resource*, Proc. 4th Int. Symp. Livestock Wastes. Amariollo, Texas, A.S.A.E., St. Joseph, Michigan, pp. 181–4.
26. Vahtras, K. and Wiklander, L. (1977). Leaching of plant nutrients in soils. *Acta Agric. Scand.*, **27**, 165–74.
27. Vanderholm, D. H. (1975). Nutrient losses from livestock wastes during storage treatment and handling. Proc. 3rd Int. Symp. on Livestock Wastes, Urbana, Ill., A.S.A.E., St. Joseph, Michigan, pp. 282–5.

17

The Role of Monitoring and Surveillance in Control of the Environmental Impact of Pesticides

P. J. BUNYAN

Agricultural Science Service, Ministry of Agriculture, Fisheries and Food, London, UK

SUMMARY

The nature of the effects which can be produced in soil, water and wildlife by the use of pesticides is discussed and monitoring and surveillance to assess their occurrence and significance are defined. The form and extent of background biological and chemical monitoring in the United Kingdom are described, as are the details of two schemes to examine deaths among wildlife and damage to crops thought to be the direct result of pesticide use. The recent development of improved chemical, biochemical and biological tests which can be employed in intensive field surveillance exercises to assess environmental impact during the registration and early use of pesticides is reviewed. The framework within which the various forms of monitoring and surveillance can be combined to provide a system for control of the environmental impact of pesticides is outlined and examples are given of the way in which the system has been used in practice. The benefits to be derived from the use of monitoring and surveillance and the principles to be applied to their organisation and financing are discussed. Areas requiring further attention are identified.

1. INTRODUCTION

Although chemicals have been employed to control pests and diseases of agriculture over a very long period, their use was until relatively recently on a limited scale and few environmental problems of more than a purely local nature appear to have arisen. During the first half of the twentieth century, the development of cyanide resistance in scale insects as a result of the use of cyanide for fruit treatment in California and the observation of depletion of earthworm populations in orchards following the long-term use of copper-based fungicides are notable exceptions and should perhaps have sounded a note of caution concerning the wider use of chemicals to come.

The discovery in 1940 of the insecticidal properties of DDT is usually taken to mark a watershed, when the spasmodic use of mainly inorganic or naturally occurring chemicals to combat agricultural pests was replaced by the much more widespread use of selective synthetic organic chemical biocides for pest control. In the United Kingdom, pesticide use expanded from approximately fifteen common insecticides and fungicides in 1950 to approximately two hundred chemicals in 800 formulations by 1975, with herbicides accounting for approximately 70% by weight of the active ingredients applied [1]. Since that time the rate of increase in use has probably declined, although the proportion by weight of herbicide applied has risen to nearer 80%. However the advent of newer classes of pesticides, such as the synthetic pyrethroid insecticides, often with much lower application rates has ensured that the variety of chemicals employed has continued to increase rapidly.

Anxiety has been widely expressed that because of the quantity of pesticides now being used and the broad spectrum of their biological activity, they may, either by acute or chronic action, seriously affect the distribution or survival of natural fauna and flora (including soil microflora) and pollute soil and water. Such occurrences may be designated the 'direct' effect of pesticides and can be defined as:

> Effects in environmental systems which can be specifically correlated with use of a particular pesticide or group of pesticides.

The possibility that direct effects might arise from the use of pesticides was recognised at an early stage in the development of their use [2] and well documented examples of such effects among the earlier persistent organochlorine pesticides have subsequently been demonstrated.

The development of chemical pest control has also contributed extensively to, or in some sectors stimulated an agricultural revolution in the developed countries, including those of the European Community. This has been characterised by lower labour inputs, increasing use of chemicals and machinery and higher yields of all farm produce. It has entailed changes in crop husbandry and crop distribution, which in turn have led to changes in the traditional landscape and perturbation of the environment of the fauna and flora. In Europe the landscape and the wildlife have evolved steadily over the millenia with agricultural activity as a major influence. In the most recent phase, the pace of change has greatly accelerated and developments have been so diverse and widespread that it is often difficult to comprehend their full extent or to connect changes with specific causes. Many recent agricultural changes are now seen as inimical to wildlife, but they can seldom be delineated unequivocally and even where change is proven or suspected it is often impossible to isolate a single or even the major cause, although pesticides frequently come under suspicion.

Some changes which are not directly attributable to chemicals, such as some alterations in habitat, may nevertheless have pesticides as the underlying cause, although there may be many contributing causes. Such changes may be designated the 'indirect' effect of pesticides and can be defined as:

> Effects in environmental systems which can be correlated with other changes in the system which are themselves the result of the use of pesticides.

The possibility of such indirect effects arising was also recognised early [2] but they have also proved difficult to demonstrate.

The European and other developed countries, and increasingly the less developed countries, have some form of registration procedure for new pesticides which lays down the conditions for their use. This follows consideration of their chemical and biological properties, their likely use pattern and their possible impact on the user, the consumer and the environment. Although the tests which provide data to support such registration schemes are well developed for the first two 'at risk' categories which both concern people, they are less certain for assessing risk to the environment. This is partly because they can generally only encompass direct effects, partly because they cannot encompass the variety of species and substrates which might be affected by pesticides in practice and partly because of the immediate and direct nature of many of the interactions between pesticides and the environment and the possibility that a 'cocktail' of chemicals might be involved. Thus the chemical behaviour and biological effects of new pesticides in the environment cannot be predicted with the degree of certainty with which human exposure can often be judged.

For this reason it is particularly important that there should be some system of checking for the adverse environmental impact of pesticides once

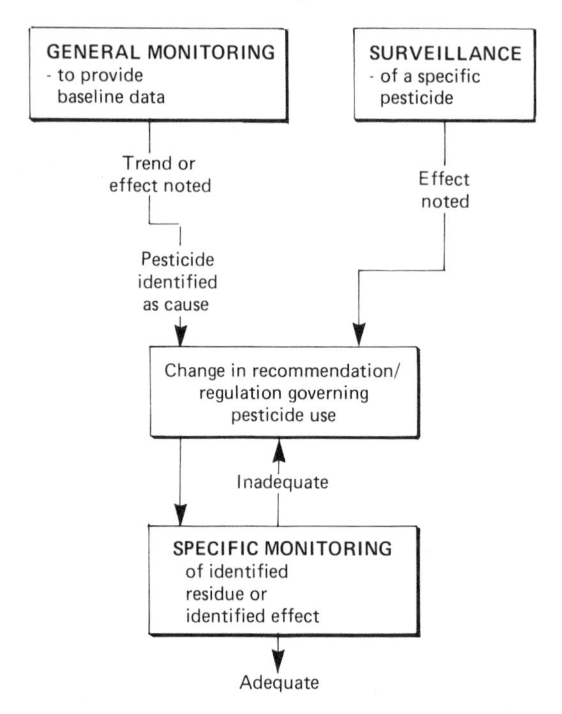

FIG. 1. Monitoring and surveillance for the environmental impact of pesticides.

in use. This can take two forms which may be termed monitoring and surveillance (see Fig. 1). Monitoring may be defined as:

> The long term and extensive measurement of chemical or biological parameters of an environmental sector or system which indicate its state.

In this type of exercise trends or effects which may be seen are often difficult to relate to specific causes, especially if the monitoring is of a very general nature. Surveillance may be defined as:

> The short term and intensive measurement of chemical and/or biological parameters of an environmental system for which the pesticide inputs are known.

In this type of exercise effects can usually be related to specific causes. Monitoring and surveillance are important complementary tools in the risk management of pesticides, especially where there is any indication from the registration data that they could have a significant environmental impact.

2. THE NATURE OF MONITORING AND SURVEILLANCE

In order to be effective, monitoring and surveillance for the environmental impact of pesticides should be as comprehensive as resources will allow. They must aim to predict or detect significant environmental effects produced by individual pesticides or by groups of pesticides and to differentiate them from the effects produced by other natural or human influences. Monitoring systems must be capable of delineating direct effects and, as far as possible, indirect effects on the environment and its flora and fauna. Indirect effects are in the long term often the more important and frequently the more difficult both to detect and to remedy.

2.1. Chemical monitoring

Monitoring for the environmental impact of pesticides has in the past relied heavily upon sampling and analysis of environmental systems and sectors for pesticides or their breakdown products. This approach has the advantage of being orientated towards particular potential problem pesticides identified on the basis of their known chemical and biological properties. It relies on the existence of a sensitive analytical method, capable of application to all the various substrates encountered in the environment. Since this is by no means universal, the approach tends to concentrate attention on certain groups of pesticides, for example, the persistent organochlorines. But many modern pesticides produce their biological effect very rapidly, residues decline quickly and they are unlikely to be detected even if very sensitive analytical methods are available. The problem is compounded by a recent trend to produce pesticides such as the synthetic pyrethroids with high biological activity. These are applied at extremely low rates and can exert a significant biological effect at levels which are below the detection limits of current analytical methods. For these reasons chemical monitoring can be biased and misleading.

Whenever residues are found, caution must be exercised to ensure that any conclusions drawn on their likely environmental impact take account of the sampling pattern utilised and of the biological significance of the levels of chemical observed. The latter may be very difficult to assess due to the paucity of experimental data on the various species and substrates encountered in practice. Widespread general environmental residue monitoring therefore tends to be unprofitable and can only be sustained as an end in itself where there may be a public health implication, particularly for instance in potable water or its sources of supply. Chemical monitoring is of the greatest use in relatively closed systems, preferably where the input is known and a knowledge of the dispersal and decline of the residues is the objective.

2.2. Biological monitoring

Although often more difficult to execute and interpret than chemical monitoring, background biological monitoring has often proved to be very effective in delineating the environmental impact of pesticides. The greatest part of the effort in this area has been directed towards regular monitoring of the distribution or population density of individual species of flora and fauna. It has several advantages. It provides essential evidence that environmental effects are observable. The effects produced are often cumulative and can be discerned after the residue has declined. Moreover, the significance of any observed effect can usually be assessed and, where appropriate, action can be taken to remedy problems. Properly targeted, biological monitoring is also capable of assessing the impact of mixtures or groups of pesticides, and of revealing the effect on whole ecosystems as well as individual species. It can detect indirect and direct effects with equal facility. However, especially in the absence of supporting chemical evidence it can be extremely difficult to unravel the real cause of any effects which may be detected by biological monitoring because of their spatial and temporal separation. This is particularly true where trends or fluctuations are noted in the long term monitoring of spatial distribution or density of individual species. Nevertheless where there appears to be a strong circumstantial connection with the use of pesticides, such observations do allow authorities to make adjustments in the conditions of pesticide use and if this is followed by more carefully targeted monitoring the outcome of such action can be checked.

2.3. Surveillance

Surveillance is particularly useful for the assessment of the environmental impact of new pesticides either where there is some doubt following the submission of data to the registration authority, or where a specific environmental effect is suspected during early commercial use. Surveillance concentrates on the suspected effect and on the substrates and species thought to be most involved. Both chemical and biological sampling are usually undertaken in order to demonstrate the chemical input, its subsequent distribution and its biological effect. Such an intensive and structured study often allows firm action to be proposed on the use of the chemicals and subsequent targeted monitoring can check on the outcome of such recommendations.

2.4. The relation between monitoring and surveillance

The ways in which chemical and biological monitoring and surveillance relate to each other in an ideal management system are summarised in Fig. 1. Where a registration authority suspects that a new pesticide or a new use for an older pesticide may have an environmental impact a surveillance exercise during early commercial use can indicate likely problems. Background monitoring may also indicate problems arising from current pesticide uses although the cause is usually difficult to pinpoint. If causes can be demonstrated, registration authorities can take appropriate remedial action. Subsequent specific monitoring can then show whether the action has been adequate to deal with the problem.

3. THE PATTERN OF MONITORING AND SURVEILLANCE IN THE UNITED KINGDOM

3.1. General monitoring

3.1.1. Biological monitoring

Biological monitoring mostly takes the form of population distribution or density measurements on individual species of wild fauna and flora. To be of use studies need to be conducted over a sufficient time span for significant fluctuations or trends to be discernible from the normal background variations in populations. If such schemes were now required *a priori* to monitor the environmental impact of pesticides they would be both difficult and extremely costly to mount. Fortunately, for many species there has been much interest in the United Kingdom over a long period, so that data on populations exist already, although the depth of knowledge varies according to the interest of naturalists and ease of data collection. Because many of the measurements are made by individuals who neither expect nor require payment, such data banks are a national resource which is highly cost-beneficial.

Birds. Reliable information exists on the numbers and distribution of many bird species. This has been collected, often over many years, by amateur and professional ornithologists. The British Trust for Ornithology (BTO) maintains continuous monitoring of the abundance and distribution of birds by the coordinated efforts of amateur members, through a number of permanent projects including the Common Bird Census (CBC), the national ringing scheme and the nest records scheme.

This is supplemented for limited periods by extensive surveys of the national distribution of particular species. The CBC was initiated in 1962 to look at farmland and woodland species in order to establish whether the use of pesticides and other agricultural practices was leading to changes in these populations. The results are published annually (for example [3]) and provide a sensitive indicator of the fate of birds breeding on farmland [4]. The nest records scheme and the national ringing scheme provide additional information on many species not covered by the CBC and sudden increases in the incidence of recoveries of ringed birds can indicate unusual mortality through pollution. A major advance was the Atlas of Breeding Birds published in 1976 [5] which summarises the results of an intensive survey of breeding birds in Britain and Ireland conducted in 1968–72.

The Wildfowl Trust regularly monitors the populations of geese, ducks and swans and the Game Conservancy maintains shooting records and carries out various censuses to monitor game bird populations. Population levels, breeding biology and pesticide residue levels in a variety of predatory bird species have been conducted for more than 20 years by the Natural Environment Research Council (NERC). Because they were originally designed to check the effects of pesticides, these last have been particularly valuable in assessing their environmental impact.

Mammals. Information on mammal populations is less comprehensive because their breeding biology, size and range make them more difficult to study. Most information relates to distribution rather than density. The Mammal Society has organised a number of surveys of small mammals and badgers [6, 7] and published an Atlas of British Mammals [8]. The otter, *Lutra lutra*, because of its conservation interest has been subjected to detailed studies in England [9], Wales [10] and Scotland [11]. Information on deer, *Cervidae*, populations has been published [12] and is also collected by the Deer Society. Various other agricultural mammalian pest species (rats, mice, coypu and mink) have also been subjected to periodic surveys by the Ministry of Agriculture, Fisheries and Food (MAFF) and a range of species including bats, *Chiroptera*, brown hare, *Lepus capensis*, stoat, *Mustela erminea*, weasel, *Mustela nivalis* and red squirrel, *Sciuries vulgaris*, are currently also being studied locally or nationally by various groups.

Amphibia, reptiles and fish. There are twelve species of amphibia and reptiles in the United Kingdom and they are of considerable conservation interest. Distribution maps have been prepared for most species by the

Biological Records Centre at the Monks Wood Experimental Station (NERC) and three rare species, the smooth snake, *Coronella austriaca*, the natterjack toad, *Bufo calamita* and the sand lizard, *Lacerta agilis*, have been the subject of special surveys by the Nature Conservancy Council. The British Herpetological Society has collected considerable data on the distribution and population densities of the more common species, including the common frog, *Rana temporaria*, common toad, *Bufo bufo* and smooth newt, *Triturus vulgaris*, which are all thought to have declined in recent years.

Freshwater fish are of considerable sporting interest and Regional Water Authorities (RWA) monitor fish stocks in most rivers. These usually rely either on netting studies or on returns from angling societies. Results are published regularly in RWA reports.

Invertebrates. Although there are many species of invertebrates in the United Kingdom, regular monitoring is limited to a few species of insects of agricultural or conservation interest. Special studies have been undertaken on a local basis on snails, crustacea and earthworms, where problems have been suspected, although such studies have been closer in concept to surveillance exercises than to monitoring.

Aphid populations are nationally monitored regularly by suction trapping [13] organised nationally through the Rothamsted Experimental Station of the Agricultural and Food Research Service (AFRS). The distribution and abundance of moths is monitored by a network of light traps manned by amateur naturalists and coordinated by Rothamsted [14] and the Biological Records Centre (NERC) has mapped the national distribution of over 100 moths [15]. The Institute of Terrestrial Ecology has established a method for monitoring butterfly populations on a number of nature reserves [16] and since 1976 this has become a national monitoring scheme which gives early warning of species decline.

Plants. Systematic mapping of the distribution of plants was initiated in 1954 by the Botanical Society of the British Isles and this formed the basis of the initial work of the Biological Records Centre (NERC). Distribution is mapped by amateur naturalists on a 10 km grid system. Information on density is not recorded except for a limited number of rare species of conservation interest which are periodically subjected to detailed surveys. Results from these surveys which have been reviewed [17], suggest that certain weeds of arable land and flora of wetlands have the highest proportion of threatened species, although only the former are primarily due to the use of herbicides. Weeds of agricultural interest have been subjected to a number of surveys (for example [18]) and the Weed Research Organisation of the AFRS has a regular programme for monitoring weed infestations on selected fields. There has not been any monitoring of the effects of pesticides applied in agriculture on wild plants and plant communities outside the target areas or from the use of herbicides in non-agricultural situations, except water. Such plants may be at hazard from misapplication or drift.

Trees, lichens and mosses have all been the subject of surveys. It seems doubtful that these species have been affected by pesticides, but rather that air pollution is the major cause of any damage seen in this area.

Ecosystems. It is clear that with a few important exceptions there is a good record of population trends for many individual species of animals and plants in the United Kingdom but information on trends in whole ecosystems is far less satisfactory. It is only with the benefit of this type of data that the indirect effects of pesticides, if they occur, are likely to appear. Long term monitoring of such systems is sparse, the methodology not well established and results are difficult to interpret. Nevertheless, two well established monitoring schemes have addressed particular facets of this problem in the United Kingdom with a good measure of success.

The Joint Cereal Ecosystem Study was set up in 1967 by the Game Conservancy to monitor grey partridge chick survival in a 62 km^2 block of cereal crops in Sussex. Monitoring included spatial and temporal observations on the populations of the partridge, the insect fauna of the cereal fields and their weed flora. There were also certain additional studies on vertebrate predators. The study elegantly demonstrated [19] how the decline of the grey partridge was related to the decreasing availability of insect food for chicks during their early development. This in turn was related to the loss of weed flora which was the habitat for important insect species following herbicide use. A further but more recent influence was shown to be the use of aphicides on cereals. Although much of this study is complete, the long term monitoring of insect fauna continues and is beginning to show trends which can be related to recent changes in agricultural practice, including changes in the use of pesticides.

Much of the methodology developed for this study is now being used in the long term monitoring of the environmental effects of three different pesticide regimes being used on the MAFF Boxworth Experimental Husbandry Farm in Cambridgeshire. In this experiment, which is closer to a surveillance exercise than to monitoring, observations are being made on the whole agroecosystem and in addition residue measurements are being made widely. It differs from the Joint Cereal Ecosystem Study since it is being conducted on a much smaller area which is itself divided into three treatment blocks, but has the advantage that pesticides, fertilisers and agronomic factors are all carefully controlled and recorded. It is too early to discern trends or inter-relationships at present.

The second regular monitoring system for total ecosystems is conducted by the Water Authorities. This scheme aims to make a biological assessment of the quality of British Rivers primarily by monitoring the freshwater invertebrate fauna. Sampling stations have been set up on all major rivers and tributaries. Aquatic fauna are sampled annually, identified and data on the presence of individuals from 85 important families are condensed to produce a 'biological score'. The methodology has been described in detail [20]. Results are published at 5-yearly intervals and include maps showing the national position. Although essentially qualitative, this scheme is an interesting approach to the problem of producing an overall national picture through comparable assessments of the wide variety of river habitats encountered in the United Kingdom. No effects ascribable to pesticides have yet been discerned. It seems likely that because of the nature of freshwater ecosystems, if there are any effects they would be masked by other forms of pollution or by changes in river management schemes.

3.1.2. Chemical monitoring

There have been many general pesticide residue monitoring exercises undertaken in the United Kingdom and elsewhere during the last 25 years. Unless carefully targeted and supported by biological monitoring they have in general not been particularly cost-effective for assessing the environmental impact of pesticides and they are likely to be even less cost-effective in future. Most of the effort has been directed towards residues of the persistent organochlorine pesticides, which with modern analytical methods can be shown to be virtually ubiquitous. Residues of a number of other pesticides can be detected at very low levels in flora, fauna, soil or water, but sampling presents a major problem. Most modern pesticides either decay too rapidly to be detected far from the point of application or are applied at such low levels that residue methods are insufficiently well developed even to identify pesticides with certainty in environmental samples, quite apart from quantifying their presence.

Biological monitoring offers by far the best chance of identifying the environmental impacts of pesticides. The picture which emerges from general chemical monitoring of the environment is often biased by the analytical methods available and by the samples examined. In addition, facile interpretation of such data can lead to the setting of legal limits for pesticide residues, particularly in soil and water, which are based solely on the limit of analytical detection. This approach takes no account of whether the residues could have a biological effect. It is a basic tenet of toxicology and ecotoxicology first stated by Paracelsus in the sixteenth century that all substances can be poisons, but it is the dose or exposure level which determines the effect they produce. To ignore this principle leads to the imposition of expensive legal controls for no good purpose.

One area where structured background residue monitoring is both necessary and worthwhile, is where public health is involved. Food and water are monitored regularly in the United Kingdom, the latter by the RWAs under the Harmonised Monitoring Scheme and results are reported regularly (for example [21] and [22]).

3.1.3. Other environmental monitoring

Most environmental monitoring for the impact of pesticides seeks to take a random sample, albeit often from carefully selected populations or environmental sectors thought to be particularly at risk. An alternative form of monitoring described below has proved to be very useful in the United Kingdom for assessment of the impact of pesticides on wild fauna and on soil. This approach only examines biological material which is alleged to have been damaged by pesticides but often leads to action which directly remedies any problems encountered.

The Wildlife Incident Monitoring Scheme. In this scheme, coordinated by the Tolworth Laboratory of MAFF, any vertebrate wildlife found dead on or near agricultural land where pesticides have recently been used can be submitted for circumstantial, veterinary, biochemical and chemical investigation designed to establish the cause of death. The scheme, which started in 1960 to investigate bird deaths caused by organochlorine pesticides, now extends to all vertebrates and to bees. The majority of incidents are reported by the farming community, the non-governmental naturalist organisations and official sources. The results are published annually (for example [23]) and have provided an invaluable picture of wildlife–pesticide interactions, allowed some of the direct effects of pesticides to be assessed and the conditions of use to be adjusted to protect wildlife.

Crop damage investigations. In this scheme, which is centred upon the MAFF Pesticide Unit at Cambridge, any farmer or agricultural adviser who suspects that crops have been damaged by pesticides, particularly herbicides, can submit samples together with the soil, for examination by plant pathologists, soil scientists and chemical analysts, in order to establish the cause of the damage. Data from the scheme, which started in 1967, have been published regularly (for example [24]). The results have provided a valuable picture of pesticide problems, most particularly the unexpected effects of herbicides under abnormal weather conditions and in the wide variety of soils encountered in the United Kingdom. The scheme has also, incidentally, provided background data on pesticide levels in soils. The investigations have allowed advice to be given to farmers in order to avoid isolated problems of a local nature or, where effects appear to be more widely spread, to alter the conditions of pesticide use to protect soils and crops.

3.2. Surveillance

If laboratory and field trials data suggest that a new pesticide could present an environmental problem, the registration authorities in the United Kingdom will often initially allow a period of limited use and call for a surveillance exercise during this period. The earliest guidelines produced for the Pesticides Safety Precautions Scheme (PSPS) included information on how such exercises might be approached. They laid stress on choosing an appropriate site with a good range of wild fauna and then making observations on the levels of activity and the diversity of species at appropriate times of the day and at appropriate periods before and after a normal commercial pesticide application. Considerable emphasis was placed on searches for dead animals, particularly birds, on the treated area or in the immediate surroundings during the post spraying period. It was suggested that if cadavers were found they should be subjected to a post mortem in order to try to establish the cause of death, although little guidance was given on the evidence which might be accepted. A number of these exercises were carried out, but they generally gave either negative results or were inconclusive.

In the course of the last 10 years this approach has been superseded by another, which is more objective. It is based on a combination of biological, biochemical and chemical measurements which are selected and coordinated in order to delineate and assess any suspected environmental impact. This approach and the techniques it employs have recently been reviewed [25]. New PSPS guidelines have been issued to help intending notifiers of pesticides to understand the requirements. Both single intensive trials (secondary testing) or a series of less intensive observations (tertiary testing) have been required from notifiers depending on the nature of the possible impact which the registration authorities wish to investigate. The surveillance exercises either concentrate on individual species thought to be particularly at risk (the variety of species chosen is steadily increasing) or in a few instances are more wide-ranging. In general the observations are designed to associate the presence of a chemical residue either with a biochemical effect in the individual animals or with changes in the population levels, where these are large enough in the area under observation to allow meaningful statistics to be applied. Residue measurements in some or all parts of the agroecosystem, sampling of local fauna for subsequent biochemical measurements and biological observations including trap, mark and release techniques all contribute to the approach. Where the latter can be coupled with blood sampling, the time course of residue decay and associated biochemical effects can also be measured.

Non-destructive testing of this type is a new development of particular value. Detailed description of the methodology used and the results obtained from a number of surveillance exercises carried out by both MAFF and some commercial notifiers have been published (for example [26–32]) and serve to highlight the variety of techniques and approaches which are possible within this conceptual framework.

4. THE APPLICATION OF MONITORING AND SURVEILLANCE IN THE UNITED KINGDOM

Monitoring and surveillance to measure the environmental impact of pesticides on soil, water and wildlife has been applied 'as necessary' rather than systematically. This is partly because the methodology has been steadily developed and refined, so that opportunities for meaningful work have been dependent on the progress in this area and partly because the registration authorities only require formal exercises where experimental results suggest that there is a real need. However any effective monitoring programme requires considerable resources.

Long term biological monitoring has mainly been undertaken by voluntary effort and has generally not required the commitment of large amounts of public money. The background built up by this effort is increasingly seen to be valuable. A modest increase in Government support for national naturalist, sporting and other appropriate monitoring organisations, could now be particularly effective in ensuring coordination, the use of standardised methodology and the complete and systematic reporting of results. Chemical monitoring is more expensive since it can only be carried out with professional input. For this reason, work is usually carried out either by national Government agencies such as MAFF, AFRS and NERC or by other publicly funded bodies including Local Authorities and RWAs. In view of the expense of such exercises the cost benefit must be properly established and this generally mitigates against large long term systems. Some work has been done from time to time in this area by independent workers, for instance in the Universities, but continuity and adequate sampling procedures have both proved difficult for such groups.

Surveillance exercises on new pesticides are usually required to be carried out by the notifying pesticide manufacturer. The form and content are specified by the regulatory authorities and further advice, together with some specialised assistance, can be provided by Government agencies. On occasions where new or more fundamental problems are perceived, or the opportunity is presented to develop better methodology, Government agencies may undertake the largest part of such exercises [27] with financial input from the notifier. Where adverse environmental effects are suspected from the use of established pesticides, any monitoring or surveillance subsequently required is generally undertaken by Government who have the responsibility for allowing the continued use of chemicals which are often by then being marketed by a number of companies.

4.1. Soils

There has been no structured long term sampling of soils in the United Kingdom in order to monitor the effects of pesticides, although much work has been undertaken for localised field trials. This is understandable given the difficulty of defining a sampling system to encompass the wide variety of soils and pesticides which may be involved. Work carried out over a long period world-wide has demonstrated that soils are robust ecosystems. Agricultural soils and their biota are constantly subjected both to natural and cultural changes, to which they are seldom incapable of adaptation. The application of pesticides, particularly herbicides, does not appear to be an exception.

The monitoring of crop yields and the diagnosis of damage which is carried out systematically (see Section 3.1.3) are the best means of uncovering any widespread impact of pesticides on soils. The problems detected have generally been limited either temporally or spatially and involve the unexpected persistence of residues on certain soil types or under abnormal weather conditions [33]. There is little evidence that soil microbial fauna is adversely affected in the long term by pesticide use [34], although there have been examples of the adaptation of fauna leading to decreased herbicide persistence [35]. In all these situations where there are no significant environmental problems, the provision of appropriate advice has proved sufficient to deal with the farming aspects. A few pesticides have been shown adversely to affect certain species of soil macrofauna (for example the effect of benomyl on earthworms) but the usage is too localised to be significant. Overall assessment of the impact of pesticides on this facet of the soil ecosystem awaits the development of better methodology (see Section 3.1.1) and its systematic application.

Some more general problems have been uncovered by crop and soil investigations. The most recent example was the observation that in certain weather conditions some ester formulations of phenoxy herbicides can cause crop damage by vapour movement rather than by spray drift [36]. Following changes in the formulation required by the regulatory authorities, the number of crop damage incidents due to this class of herbicides has declined substantially.

4.2. Water

Soils can act as a reservoir for pesticides and their breakdown products to leach into water, an aspect which causes concern. Pre-notification testing should ensure that no chemicals with considerable propensity to leach come onto the market. Nevertheless most substances will leach to some extent. In certain conditions, 'run-off' may also contribute to the pesticide burden of water. A small number of pesticides are cleared for direct application to water.

The public health involvement and the importance of water as an environmental medium have ensured that systematic chemical and biological monitoring of river and potable water is carried out. Water Authorities take samples on average about once a month, amounting nationally to approximately 2400 samples per annum under their Harmonised Programme for River Water Quality, but pesticides are only one of many types of contamination examined. Apart from occasional isolated incidents which are mainly the result of accidents, levels of pesticides do not give cause for concern.

Freshwater ecosystems tend to be more exposed to pollution than terrestrial or soil ecosystems. This has been a powerful influence in the development of the biological monitoring discussed earlier (see Section

3.1.1) for both individual freshwater species and total ecosystems. The results from the latter are difficult to assess but they suggest that, where freshwaters do not support a diverse ecosystem, river management systems or industrial chemical pollution is a more significant factor than pesticide usage.

Because of the sensitivity of freshwater ecosystems, the United Kingdom registration authority currently invariably requires a surveillance exercise on any pesticide notified for direct application to water. Surveillance on a site which involves a body of water is also frequently required on any new agricultural pesticide which is shown to be highly toxic to aquatic flora and fauna. These exercises frequently lead to changes in the label recommendations, although most have not been documented. Recent examples where accounts have been published involve the aquatic herbicide dichlobenil [31] and the synthetic pyrethroids, cypermethrin [32] and flucythrinate [37]. Dichlobenil, which is formulated as a slow release granule, was shown to give rise to water concentrations which were toxic to fish if the whole water area was treated at one time. This led to the recommendation that only 10–25 % of the water surface should be treated on any one occasion. Conversely, in laboratory tests, the synthetic pyrethroids appear to be extremely toxic to aquatic organisms, but the surveillance exercise on cypermethrin demonstrated that in practice, adsorption on suspended particulate matter considerably reduces their toxicity, allowing use near water. Flucythrinate however has a delayed toxic effect even after brief, low level exposure, so that aerial spraying is not permissible within 250 m of water.

4.3. Wildlife

Background monitoring of the population levels of individual species can point to changes which might be attributable to pesticides, but further work is generally required to confirm any suggested connection. In the United Kingdom the best known example of such a pesticide-induced effect on a particular group of species was the decline of raptorial bird populations in the period 1950/70, due to the use of organochlorine pesticides. This was demonstrated by long term studies at the Institute of Terrestrial Ecology (NERC) into the population levels, breeding biology and pesticide residues in these species [38], which contributed to phased withdrawal of organochlorine pesticides in the United Kingdom as satisfactory alternatives became available [39]. There has been a consequent recovery in the bird populations [40]. It has been suggested that the observed declines in some other vertebrate and invertebrate populations are also the result of pesticide use, but these conclusions have less certainty because of the lack of supporting residue data and the unquantifiable effects of the many other environmental influences on population levels. Similarly, monitoring of the impact of pesticide use in agroecosystems as exemplified by the Joint Cereal Ecosystem Study [19] has suggested a number of unforeseen indirect effects arising from pesticide use and has produced results which counsel caution in employing the present spectrum of pesticides. However, because of the variety of influences which affect agroecosystems this work has not so far allowed the formulation of objective proposals to improve the situation by altering pesticide use.

By contrast, the Wildlife Incident Monitoring Scheme (see Section 3.1.3) frequently relates deaths among wild vertebrates and bees directly to the use of pesticides. This has resulted in a number of changes in the conditions of use for cleared pesticides. Two striking examples are given below. The use of carbophenothion as a seed treatment was shown to be particularly hazardous to migrant geese but not to other species, feeding on newly sown cereal fields in the autumn. This led to an agreement that it should not be used for this purpose in Scotland where migrant geese regularly over-winter [41]. The use of insecticides, particularly triazophos during the flowering period of oilseed rape, was shown to be hazardous to bees. This resulted in the limitation of triazophos use only after flowering has completely finished.

Surveillance during the early commercial use of pesticides offers the most sure method for detecting their adverse impact on wildlife. A large number of exercises have been carried out by notifiers at the request of the registration authorities but the details have seldom been published. Most have been relatively simple exercises concentrating on specific aspects of concern. None have been as extensive as the work carried out by MAFF during the notification of aldicarb [27]. This used many of the techniques now available [25] and compared a single intensive trial with a series of less intensive trials on a number of sites. The intensive trial was shown to be particularly effective for demonstrating the spread of the applied pesticide and its metabolites through the local agroecosystem and linking them with biochemical effects in individual vertebrates on the site. This allowed an assessment of the biological significance of the residues observed. The application of this primary principle of ecotoxicology has been reviewed elsewhere [42]. The series of less intensive trials demonstrated that over a range of sites and conditions, the assessment of significance in the intensive trial was correct. In tandem they allowed firm recommendations to be made for aldicarb use, particularly on the nature of the formulation and the method of incorporation following soil application.

Subsequent published surveillance exercises [28–30] have developed the use of biochemical measurements in individuals for the assessment of pesticide impact, as well as sophisticated methods for assessing population changes on small sites. This approach utilises observations made during the course of the normal pre-registration medical toxicological testing for environmental impact assessment. These tests are necessary for all pesticides in order to assess their likely impact on people. Where the blood biochemistry shows significant effects, non-destructive blood sampling and testing has been developed for birds and small mammals. If this is coupled with trap, mark and release techniques, it can allow a time course to be plotted for residues and their effects. This package of methods provides a very sensitive and sophisticated approach to field trials for the early assessment of the environmental impact of pesticides which significantly augments the current approach based mainly on laboratory tests. It should now be more widely applied. It will require pesticide manufacturers to have cognisance of the environmental field testing requirements during their early toxicology testing, in order that any effects noted at high dosages can be utilised subsequently for this purpose. Unfortunately surveillance is currently very much an isolated activity carried out late in the testing regime of a new pesticide. Many of these methods have also been included in the wider environmental monitoring study at the MAFF Boxworth Experimental Husbandry Farm to which reference was made earlier (see Section 3.1.1), although this seeks to look at the impact of different pesticidal regimes rather than individual pesticides.

5. DISCUSSION AND CONCLUSIONS

In the United Kingdom, the continuing accumulation of background data from well established biological monitoring schemes, the increasing development of methodology for measuring the state of whole ecosystems and the availability of a battery of tests for the field surveillance of the early use of pesticides provide a practical framework for assessing and controlling the environmental impact of pesticides. It is questionable however, whether sufficient resources have yet been allocated to the interpretation of results obtained in the more complex biological studies and to the construction of appropriate predictive mathematical models to which the results can be input. It is also desirable that more emphasis is given to ensuring that the environmental impact of pesticides is fully explored both during the registration phase and during the ensuing usage and that adequate private and public resources are provided for this purpose. In general, national authorities have a good record for taking appropriate action when direct effects attributable to pesticides are demonstrated, either during the registration phase or later. Much of this action has never been documented because of the nature of national registration processes. Action on indirect effects has proved more difficult for technical and other reasons.

Pesticide monitoring and surveillance in the United Kingdom is organised and executed on a local or national basis by the groups undertaking the work. In order to ensure the efficient use of resources, good scientific reasons are required before work is initiated and the objectives are clearly identified before resources are committed. Furthermore, most work is designed to maximise the chances of finding biological effects or chemical residues which is only possible if there is a sound scientific basis for requiring the work to be done.

Whilst there may be over-riding reasons for supra-national organisations to provide strategic guidance on monitoring to national authorities, or for national authorities to provide similar guidance to local or specialist groups, the detail should be left to those who will carry out the work. Large externally imposed, random monitoring schemes, particularly those involving chemical analysis, are very costly of resources and seldom effective in identifying problems. Local knowledge and enthusiasm is more likely to lead to well designed and well executed schemes, giving job satisfaction to those carrying them out. In all monitoring and surveillance work, it must also be recognised that pollution of the environment can arise from many sources, of which pesticides are only one. Moreover pesticides are not used solely in agriculture and residues and their resulting biological effects can be observed in soil, water and wildlife from use in many non-agricultural situations.

Despite the advances made in this area during the last twenty years which have led to the framework of monitoring and surveillance described in this review, a number of aspects require further development work and more resources will need to be diverted to them. Among monitoring schemes, as reviewed for the United Kingdom, whilst there appears to be adequate coverage for vertebrate species, that for invertebrate species is inadequate because of the large number of species at risk. In view of the work of Potts [19] this should be remedied. Furthermore for the monitoring and surveillance of agroecosystems there is a need for the development of better methodology for detecting significant biological effects and correlating them with chemical inputs from whatever source. The development of appropriate mathematical models would be of considerable aid in this process. To improve surveillance there is a need for the development of further sophisticated, preferably non-destructive methods for assessing the sub-acute effects of pesticides on individuals of various species.

The scale of current pesticide usage and the rate of introduction of new pesticides make continued vigilance necessary to identify and remedy both the direct and indirect effects which might arise. Correctly targeted monitoring and surveillance can contribute to this process although it must be recognised that effects need to be assessed against the impact of complete farming systems on flora and fauna.

REFERENCES

1. SLY, J. M. A. (1977). Ecological effects of pesticides. *Linn. Soc. Symp. Ser.*, No. 5, 1–6.
2. ZUCKERMAN, S. (1955). Toxic chemicals in agriculture. Risks to Wildlife. Report to the Minister of Agriculture, Fisheries and Food and to the Secretary of State for Scotland, of the working party on precautionary measures against toxic chemicals used in agriculture, HMSO, London, 31 pp.
3. MARCHANT, J. and HYDE, P. A. (1980). Bird population changes for the years 1978–79. *Bird Study*, **27**, 173–8.
4. BULL, A. L., MEAD, C. J. and WILLIAMSON, K. (1976). Bird-life on a Norfolk farm in relation to agricultural changes. *Bird Study*, **23**, 163–82.
5. SHARROCK, J. T. R. (1976). *The Atlas of Breeding Birds in Britain and Ireland*, British Trust for Ornithology, Tring, Hertfordshire, 477 pp.
6. HARRIS, S. (1979). History, distribution, status and habitat requirements of the harvest mouse (*Micromys minutus*) in Britain. *Mammal Rev.*, **9**, 159–71.
7. NEAL, E. G. (1977). *Badgers*, Blandford, Poole, Dorset, 321 pp.
8. ARNOLD, H. R. (1978). *Provisional Atlas of the Mammals of the British Isles*, Biological Records Centre, The Institute of Terrestrial Ecology, Huntingdon, Cambridgeshire, 70 pp.
9. LENTON, E. J., CHANIN, P. R. F. and JEFFERIES, D. J. (1981). *Otter Survey of England 1977–1979*, Nature Conservancy Council, Banbury, Oxfordshire, 75 pp.
10. CRAWFORD, A. K., JONES, A. and MCNULTY, J. (1979). *Otter survey of Wales*, Society for the Promotion of Nature Conservation, London, 70 pp.
11. GREEN, J. and GREEN, R. (1980). *Otter survey of Scotland 1977–1979*, Vincent Wildlife Trust, London, 46 pp.
12. CHAPMAN, N. G. and CHAPMAN, D. I. (1980). The distribution of fallow deer: a worldwide review. *Mammal Rev.*, **10**, 62–138.
13. TAYLOR, L. R. (1973). Monitoring changes in the distribution and abundance of insects. Report of the Rothamsted Experimental Station for 1973, Part 2, 202–39.
14. TAYLOR, L. R., FRENCH, R. A. and WOIWOD, I. P. (1978). The Rothamsted Insect Survey and the urbanisation of land in Great Britain. In: *Perspectives in Urban Entomology*, Academic Press, New York, pp. 31–65.

15. HEATH, J. and SKELTON, M. J. (1973). *Provisional Atlas of the Insects of the British Isles Part 2, Lepidoptera*, Biological Records Centre. The Institute of Terrestrial Ecology, Huntingdon, Cambridgeshire, 105 pp.
16. POLLARD, E. (1977). A method for assessing changes in the abundance of butterflies. *Biol. Conserv.*, **12**, 115–34.
17. PERRING, F. H. and FARRELL, L. (1977). *British Red Data Books 1. Vascular Plants*, Society for the Promotion of Nature Conservation, Lincoln, Lincolnshire, 98 pp.
18. ELLIOTT, J. G., COX, T. W. and SIMONDS, J. S. W. (1968). A survey of weeds and their control in cereal crops in South East Anglia during 1967. *Proc. 9th Br. Weed Control Conf.*, pp. 200–7.
19. POTTS, G. R. (1977). Some effects of increasing the monoculture of cereals. In: *Origins of Pests, Parasite, Disease and Weed Problems*, 18th Symposium of the British Ecological Society, Blackwell, Oxford, pp. 183–202.
20. ANON. (1981). *River Quality—the 1980 Survey and Future Outlook*, National Water Council, London, pp. 38–9.
21. ANON. (1984). DOE/NWC Standing Technical Committee Report No. 37, Fourth Biennial Report, HMSO, London, 111 pp.
22. ANON. (1982). Report of the Working Party on Pesticide Residues (1977–1981). Food Surveillance Paper No. 9, HMSO, London, 38 pp.
23. FLETCHER, M. R. and HARDY, A. R. (1982). Pesticides and wildlife—a review of wildlife incidents investigated from October 1981 to September 1982. In: *Pesticide Science 1982*, Agricultural Science Service Research and Development Report, Reference Book 252 (82), HMSO, London, pp. 47–55.
24. ANON. (1982). Pesticide problems for farmers and growers. In: *Pesticide Science 1982*, Agricultural Science Service Research and Development Report, Reference Book 252 (82), HMSO, London, pp. 41–6.
25. BUNYAN, P. J. (1983). Techniques for the detection and assessment of the effects of pesticides on wildlife. In: *Pesticide Residues. Proceedings of a Conference 31 March–2 April 1980*, Reference Book 347, HMSO, London, pp. 183–93.
26. BAILEY, S., BUNYAN, P. J., JENNINGS, D. M., NORRIS, J. D., STANLEY, P. I. and WILLIAMS, J. H. (1974). Hazards to wildlife from the use of DDT in orchards: a further study. *Agro-Ecosystems*, **1**, 323–38.
27. BUNYAN, P. J., VAN DEN HEUVEL, M. J., STANLEY, P. I. and WRIGHT, E. N. (1981). An intensive field trial and a multi-site surveillance exercise on the use of aldicarb to investigate methods for the assessment of possible environmental hazards presented by new pesticides. *Agro-Ecosystems*, **7**, 239–62.
28. WESTLAKE, G. E., BUNYAN, P. J., JOHNSON, J. A., MARTIN, A. D. and STANLEY, P. I. (1982). Biochemical effects in mice following exposure to wheat treated with chlorfenvinphos and carbophenothion under laboratory and field conditions. *Pest. Biochem. Physiol.*, **18**, 49–56.
29. WESTLAKE, G. E., BROWN, P. M., BUNYAN, P. J., FELTON, C. L., FLETCHER, W. J. and STANLEY, P. I. (1982). Residues in mice after drilling wheat with carbophenothion and an organo-mercurial fungicide. *Proceedings of the Sixth International Symposium on Chemical and Toxicological Aspects of Environmental Quality*, Valbonne, France, 1980, pp. 522–7.
30. ANON. (1981). Field trials to establish the effect of agricultural chemicals on

wildlife. In: *Pesticide Science 1980*, Agricultural Science Service Research and Development Report, Reference Book 252 (80), HMSO, London, pp. 64–9.
31. TOOBY, T. E. (1978). The fate of the aquatic herbicide dichlobenil in hydrosoil, water and roach (*Rutilus rutilus* L.) following treatment of three areas of a lake. *Proc. EWRS 5th Symposium on Aquatic Weeds*, pp. 323–31.
32. CROSSLAND, N. O. (1982). Aquatic toxicology of cypermethrin. II Fate and biological effects in pond experiments. *Aquatic Toxicology*, **2**, 205–22.
33. EAGLE, D. J. (1983). Risks to succeeding crops from persistent soil acting herbicides. *Aspects Appl. Biol.*, **3**, 191–6.
34. GREAVES, M. P. (1983). Effect of pesticides on soil microfauna. In: *Pesticide Residues. Proceedings of a Conference 31 March–2 April 1980*, Reference Book 347, HMSO, London, pp. 194–200.
35. KAUFMAN, D., KATAN, Y., EDWARDS, D. F. and JORDAN, E. G. (1984). Microbial adaptation and metabolism of pesticides. In: *Agricultural Chemicals of the Future*, BARC Symposium No. 8 (Ed. James L. Hilton), Rowman and Allanheld, pp. 437–51.
36. EAGLE, D. J. (1982). Hazards to adjoining crops from vapour drift of phenoxy herbicides applied to cereals. *Aspects Appl. Biol.*, **1**, 33–41.
37. ANDERSON, R. L. and SHUBAT, P. (1984). The toxicity of Flucythrinate to *Gammarus lacustris* (Amphipoda), *Pteronarcys dorsata* (Plecoptera) and *Brachycentrus americanus* (Trichoptera): Importance of exposure duration. *Environ. Pollut.*, **35**, 353–65.
38. NEWTON, I. and HAAS, M. B. (1984). The return of the sparrowhawk. *Brit. Birds*, **77**, 47–70.
39. ANON. (1969). Further review of certain persistent organochlorine pesticides used in Great Britain, HMSO, London, 148 pp.
40. MARCHANT, J. H. (1980). Recent trends in sparrowhawk numbers in Britain. *Bird Study*, **27**, 152–4.
41. STANLEY, P. I. and BUNYAN, P. J. (1979). Hazards to wintering geese and other wildlife from the use of dieldrin, chlorfenvinphos and carbophenothion as wheat seed treatments. *Proc. Roy. Soc., London, Ser. B*, **205**, 31–45.
42. BUNYAN, P. J. and STANLEY, P. I. (1982). Toxic mechanisms in wildlife. *Regulat. Toxicol. Pharmacol.*, **2**, 106–45.

18

Pesticides et Environnement: Information et Formation des Producteurs Agricoles et des Distributeurs

J. P. BASSINO

Organisation Internationale de Lutte Biologique,
Section Régionale Ouest Paléarctique (OILB-SROP),
Paris, France

RÉSUMÉ

Pour l'agriculteur, la plante cultivée occupe la position centrale de l'agro-écosystème.

L'usage rationnel et limité des pesticides s'inscrit dans une lutte intégrée. Elle seule est capable de réduire les effets néfastes des pesticides sur l'environnement. Elle exige cependant un effort soutenu de la part des producteurs, conseillers, distributeurs....

Une importante et longue action d'information et de formation, s'appuyant sur des centres-supports (instituts, services de protection des plantes, universités...) peut faire évoluer les pratiques, en matière d'emploi des pesticides.

L'effort des équipes de scientifiques et de conseillers, poursuivi depuis une vingtaine d'années, mérite d'être encouragé, à l'échelle de l'Europe.

1. PRÉSENTATION GÉNÉRALE

Il me paraît indispensable de commencer cette présentation par une question essentielle: que peut-on faire concrètement pour concilier les exigences de la production agricole avec la préservation de la qualité de notre environnement? Plus précisément, car tel est le sujet qu'il m'a été demandé de traiter, quelle attitude peut-on avoir vis-à-vis des pesticides? Encore une fois, que peut-on faire raisonnablement?

Les 'il faut...', 'il suffit de...', 'il n'y a qu'à...' des illuminés ou des rêveurs ne changent rien à la difficulté de l'entreprise et à l'extrême complexité des agro-écosystèmes!

D'une manière générale, les pesticides fabriqués par l'Homme ou existants dans la Nature sont d'une efficacité extraordinaire—certains agissant par example sur la quasi totalité d'une population de ravageurs et... autres organismes, à la dose de quelques grammes par hectare; ils sont donc, comme des médicaments, incontestablement dangereux pour le milieu vivant.

Par ailleurs, nous connaissons bien maintenant leurs effets indésirables directs qui entraînent:

l'apparition de souches résistantes chez les arthropodes phytophages, ainsi que chez les champignons phytoparasites;
l'élévation de ravageurs secondaires au rang de ravageurs principaux;
le développement du phénomène de résurgence;
des pullulations brutales de ravageurs ou des attaques foudroyantes de parasites, à la suite de déséquilibres biologiques créés notamment par la destruction des organismes auxiliaires (parasitoïdes et prédateurs).

Ces graves inconvénients n'ont pas échappé aux scientifiques et aux techniciens de pays où, depuis une trentaine d'années, l'évolution de l'Agriculture est poussée par les deux aspirations suivantes, des producteurs et des distributeurs intéressés par ce secteur d'activitité:

augmentation des rendements (sur des surfaces agricoles limitées) et de la qualité des produits;
diminution des travaux pénibles.

En Europe, la nécessité d'une évolution vers un système de protection sanitaire des cultures, permettant un emploi aussi peu perturbant que possible des pesticides, s'est manifestée dès 1956, à la suite de pullulations d'acariens devenus résistants aux produits alors utilisés.

L'idée de lutte intégrée s'est répandue rapidement en Europe [1–15] en particulier à l'initiative de l'Organisation Internationale de Lutte Biologique (OILB).

Des équipes composées de chercheurs, ingénieurs, conseillers divers et agriculteurs ont travaillé avec ardeur et les résultats pratiques obtenus ont permis progressivement des applications en cultures, selon le schéma d'action suivant:

(1) Travaux de recherche et expérimentation.
(2) Application en diverses situations; simplification des méthodes.
(3) Information générale et formation.
(4) Réalisation par les producteurs.

Les bonnes liaisons établies entre les différents partenaires ont permis les adaptations nécessaires, à chaque étape.

La mise en oeuvre d'une lutte intégrée, ou plus modestement raisonnée, suppose des connaissances techniques et méthodologiques et aussi une attention soutenue de la part de l'agriculteur. Cela nécessite des qualités d'observation et un esprit en éveil. Conduire ses cultures en lutte intégrée, 'c'est moins simple que de traiter systématiquement'!

A ce titre, les actions d'information et de formation (qui sont complémentaires), revêtent une importance capitale. Elles concernent aussi bien les producteurs que les distributeurs et, de ce point de vue, il n'y a pas lieu d'effectuer une distinction entre ces deux groupes socio-professionnels. On peut simplement ajouter que les distributeurs doivent impérativement suivre l'évolution du monde paysan, afin de conserver la confiance de leurs interlocuteurs.

2. LA SITUATION ACTUELLE

On doit savoir que, dans l'état actuel des connaissances, les procédés de lutte dont dispose le producteur sont essentiellement chimiques. Cette situation se maintiendra encore pendant une assez longue période car le remplacement des pesticides par des moyens culturaux ou biologiques, moins perturbants pour le Mileu, ne pourra être que progressif; la maîtrise de tels moyens exigeant encore souvent d'importants travaux de recherche.

L'objectif à court et moyen terme est donc la large diffusion de méthodes de lutte chimique (parfois biologique) simples et efficaces qui prennent en compte les effects dits 'secondaires' des interventions, effets qui sont 'primordiaux' si l'on considère l'impact sur l'environnement; étant entendu que la plante cultivée occupe, dans notre approche, la position centrale de chaque agro-écosystème.

3. ACTIVITÉS DE VULGARISATION, DIFFICULTÉS, IMPORTANCE DE L'INFORMATION ET DE LA FORMATION

Nous avons vu qu'en matière d'action phytosanitaire, la seule base acceptable aujourd'hui pour les interventions (essentiellement avec des moyens chimiques) se trouve dans les principes qui débouchent sur la lutte intégrée; c'est la seule voie de progrès axée sur la protection de l'environnement.

Il faut bien avoir conscience que l'on demande alors un effort constant aux divers utilisateurs et que l'on ferme délibérément la porte à la voie de la facilité et de l'assurance, représentée par le 'calendrier des traitements'.

Tableau 1

Raisons évoquées	*Difficultés rencontrées par*	
	le vulgarisateur	*l'agriculteur*
Techniques		
méthodes trop contraignantes		●
méthodes incomplètes ou trop restreintes	●	
Psychologiques		
difficulté d'intégrer les méthodes dans l'exploitation		●
refus d'effectuer les comptages dans les champs		●
refus d'un risque de perte		●
Institutionnelles		
nombre de conseillers trop faible	●	●
chercheurs peu nombreux	●	
soutien financier de l'Etat insuffisant	●	
assistance technique insuffisante	●	●
moyens de formation technique insuffisants	●	●
Economiques		
gain espéré trop faible		●
temps nécessaire pour réaliser les contrôles trop long ou trop coûteux		●

Faire comprendre et surtout admettre le bien fondé de cette conception de la lutte—en fait, pour l'essentiel, l'utilisation bien raisonnée des pesticides—est donc au centre des préoccupations des agents de la vulgarisation, à tous les niveaux.

Une enquête effectuée récemment par nous, dans les pays de la Communauté européenne, fait apparaître un certain nombre de difficultés recontrées par les vulgarisateurs et les producteurs dans leur action de diffusion et d'application de la lutte intégrée.

Le Tableau 1 regroupe les résultats de cette enquête.

Il semblerait que, pour la vulgarisation, les entraves à la diffusion de la lutte intégrée dans le milieu agricole tiennent, indépendamment de la nature des productions et des pays, à l'insuffisance:

du nombre de chercheurs et de conseillers (les deux fonctions devant être prise au sens large);
de l'assistance technique et des moyens de formation;
du soutien financier de l'Etat.

On peut remarquer que cela concerne les institutions et non pas les individus.

Pour les agriculteurs, l'éventail des difficultés serait plus large. On retrouve, comme point commun avec les agents de la vulgarisation, des méthodes incomplètes ou trop restreintes et parfois difficiles à intégrer dans l'exploitation, mais surtout:

un nombre de conseillers trop faible;
une assistance technique et des moyens de formation insuffisants.

Enfin, si dans son principe la lutte intégrée semble bien acceptée, il n'en est pas de même pour les mesures et observations à effectuer périodiquement dans les champs. On peut penser que la réticence ainsi manifestée tient souvent à des facteurs d'ordre psychologique. En effet, ce type de travail ne correspond pas à l'image que les agriculteurs ont de leur métier.

Le refus d'un risque de perte est aussi fréquemment avancé; cette prise de position paraît plus forte que l'argument relatif à un gain espéré trop faible.

Elle pourrait être liée à une certaine crainte diffuse d'une maîtrise insuffisante des techniques et des méthodes, d'où l'importance de l'assistance technique et de la formation.

La nécessité d'une conception globale, pour le développement de la lutte intégrée, s'est rapidement imposée. Un tel sujet, qui ne correspond pas à la diffusion d'une simple technique mais à celle d'une innovation susceptible de provoquer un changement d'attitude, de pensée, de manière d'aborder la protection phytosanitaire, nécessite:

un travail d'èquipe;
une action méthodique, progressive et continue, selon un plan qui s'appuie et utilise les mentalitès et comportements locaux.

L'expérience montre que pour une culture (ou un groupe de cultures) considérée, une activité de promotion ne peut être envisagée que si les travaux de recherche et d'expérimentation sont suffisamment avancés.

Lorsque cette phase d'étude préalable est considérée comme acquise, il est possible et alors seulement, d'aborder le stade du développement; le processus général peut être le suivant:

1. Implantation de parcelles de références (ou pilotes) dans les secteurs où la lutte contre les ennemis est rèputèe difficile; la validité des données obtenues dans les laboratoires et les terrains expérimentaux est ainsi vérifiée et l'adaptation est effectuée au fur et à mesure.
2. Information aussi large et complète que possible des agents techniques chargés du développement agricole (sensibilisation).
3. Formation pratique (initiation en salle et dans les champs) de ces agents, afin qu'ils acquièrent une bonne connaissance des moyens et des méthodes à mettre en oeuvre.
4. Information (sensibilisation) des agriculteurs; utilisation de tous les médias disponibles.
5. Effort de pré-développement, dans les secteurs les plus réceptifs, par les conseillers agricoles et les agents spécialisés, avec l'appui de la Recherche et des services économiques.
6. Perfectionnement continu des techniciens par les spécialistes des instituts.
7. Extension, par approches successives, des secteurs dans lesquels se pratique une lutte plus rationnelle, première étape vers la protection intégrée.
8. Parallèlement, information générale et sensibilisation des milieux non agricoles par des séries de messages simples de vulgarisation.

L'expérimentation, l'information, la formation et le perfectionnement font partie de l'activité habituelle des animateurs locaux ou régionaux, parmi lesquels les distributeurs, qui peuvent avoir un rôle déterminant, du fait notamment des relations personnelles qu'ils entretiennent avec les producteurs.

4. ACTIONS D'INFORMATION

On peut penser intuitivement que la protection sanitaire des cultures occupe une place particulière dans l'esprit de nombreux agriculteurs. Paradoxalement, cette place est à la fois peu importante, si l'on considère seulement le coût des interventions par rapport à l'ensemble des opérations culturales, et très importante si l'on admet que la qualité de la protection conditionne en grande partie la bonne vente du produit ou détermine l'obtention de rendements intéressants.

Comme beaucoup de personnes en activité professionnelle, l'agriculteur reçoit une masse énorme de messages par des canaux très variés. Il peut arriver que le tri ne puisse plus être fait et que l'état de saturation soit atteint; l'individu se ferme, il devient, au moins temporairement, imperméable.

On peut ajouter que dans une période où l'évolution des connaissances est très rapide, il devient difficile de situer la qualité de l'information; c'est là une constatation de grande importance.

Par ailleurs, le monde agricole n'est pas homogène. Les différences de mentalité, de formation ... sont considérables. Il est de plus généralement admis que l'assimilation d'une notion nouvelle n'est obtenue que par petites touches, par étapes successives.

Il s'agit donc d'une situation générale complexe qui ne paraît pouvoir être réduite à un schéma simple et universel.

Au delà de ces difficultés, les échanges quotidiens nous font percevoir une attitude mélangée de circonspection et d'attrait pour la nouveauté, de la part du cultivateur.

Sous un autre angle, on voit qu'il peut être en état de réceptitivité:

> lorsque sa sensibilité est aiguisée par une préoccupation immédiate sur un sujet technique phytosanitaire (lutte malaisée contre un ravageur, développement épidémique d'une maladie, etc.), par la pression du milieu, de l'environnement (risque de résidus dans sa récolte, préservation de l'entomofaune ...);
>
> lorsqu'existe la nécessité impérieuse de faire des économies de trésorerie (pesticides, engrais ...) ou quand est forte l'espérance d'un gain dans un court délai.

Enfin, le producteur qui a opté pour la spécialisation manifeste fréquemment un attrait particulier pour la nouveauté, alors que celui qui pratique la polyculture a souvent une exigence moins grande et met en oeuvre des techniques moins délicates.

Un même message ne peut leur parvenir de la même manière.

Parmi les différents supports de l'information: médias, brochures, bulletins et revues spécialisées ... les fournisseurs de l'agriculture (les distributeurs) paraissent très écoutés. Bon nombre de producteurs estiment que le 'point de vente' est le meilleur moyen de découvrir les nouveaux produits ou de recueillir des informations intéressantes.

Certains agriculteurs et démarcheurs jouent enfin un rôle déterminant dans la communication de l'information; ils peuvent transmettre efficacement de l'information ou de la contre-information, notamment lorsqu'ils appartiennent à la couche de ce que les américains nomment les 'leaders d'opinion' dont on sait qu'ils sont des relais essentiels de pénétration des messages.

Tous ces systèmes plus ou moins concurrents peuvent en réalité se compléter. Nous avons le sentiment que l'ensemble des moyens actuellement disponibles n'est pas utilisé d'une manière rationnelle.

Le message n'est pas souvent bien adapté; il semble y avoir curieusement et d'une façon presque caricaturale, une oscillation généralisée entre les deux cas extrêmes suivants:

le contenu est trop technique et l'assimilation très difficile;
le renseignement est simpliste et le message prend l'allure d'une annonce publicitaire en 'coup de poing'.

Il paraît de toute façon nécessaire:

de bien adapter le message, de le simplifier au maximum, de le présenter dans le langage de l'agriculteur pour le rendre accessible à ce dernier;
d'obtenir un emploi harmonieux de l'ensemble des moyens d'information.

Nous avons essayé de résumer, dans la Figure 1, l'ensemble des éléments à prendre en compte, dans une action d'information, pour un sujet aussi difficile à cerner que celui de l'emploi des pesticides.

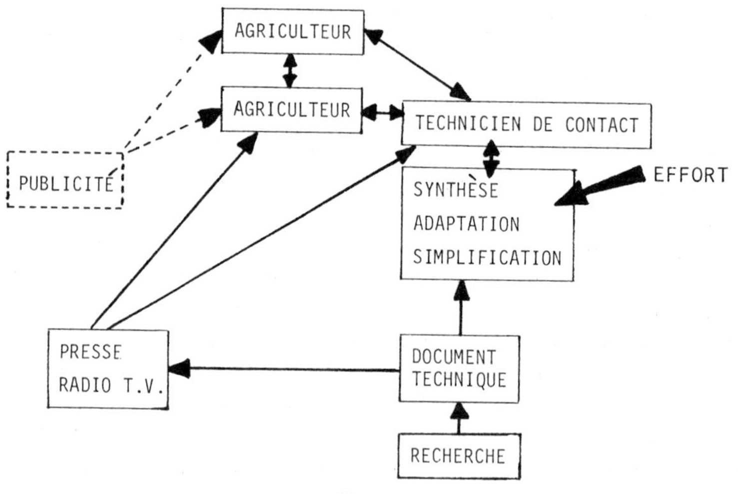

FIG. 1

D'autres moyens très performants devraient prochainement permettre un accès immédiat à des informations précises. C'est par example le cas du système télématique Iphyto développé en France par l'Association de Coordination Technique Agricole (ACTA) et les Instituts Techniques.

Enfin, on ne peut clore ce chapitre sans mentionner l'excellent effet de sensibilisation des importantes journées d'information de Valence et de Bruxelles, organisées par la C.C.E. et les instances internationales spécialisées.

5. ACTIONS DE FORMATION

La formation des agriculteurs, comme celle des vulgarisateurs (conseillers, distributeurs . . .) doit être axée sur les aspects pratiques de la mise en oeuvre de la lutte intégrée.

Elle doit par conséquent s'appuyer sur des unités d'expérimentation et de démonstration; il faut en effet montrer aux praticiens le résultat des applications en 'vraie grandeur'.

A l'échelon local ou régional, la formation des conseillers agricoles vulgarisateurs et des agriculteurs-techniciens peut être conçue, en fonction de contraintes diverses, suivant deux schémas très différents:

une action progressive;
un cours intensif.

5.1. Formation progressive, selon trois étapes

Les trois étapes sont:

la sensibilisation;
l'initiation pratique;
le perfectionnement.

Le stage de sensibilisation

Au cours de deux journées de travail en salle, et grâce aux apports d'exposés magistraux, suivis de discussions, on décrit la protection sanitaire et on propose une autre démarche: la lutte intégrée.

Le stage d'initiation pratique

Il peut comprendre deux parties principales réparties sur trois journées d'étude:

les généralités;
le mode opératoire, soit au plan général, soit rapporté à une culture ou à un groupe de cultures.

Le stage de perfectionnement technique et/ou méthodologique
Il peut prendre des formes différentes, selon la nature des besoins exprimés par les futurs stagiaires (une consultation préalable s'impose donc).

5.2. Le cours intensif
Il ne peut s'adresser qu'à des conseillers, à des distributeurs ou, peut-être, à des producteurs ayant une excellente formation technique et une bonne expérience professionnelle.

Si l'on se rapporte à des situations vécues, il apparaît que le cours intensif a pour principale caractéristique de provoquer un 'choc' qui répond à un objectif de sensibilisation d'un groupe de stagiaires non encore bien informé sur l'esprit qui anime l'action de lutte intégrée et les techniques qu'elle fait mettre en oeuvre.

Il existe toutefois un risque de voir les personnels non chevronnés perdre pied face à la masse d'informations nouvelles assénées en un laps de temps relativement court.

Ainsi, pour des raisons d'efficacité manifestes, il semble souhaitable de ne retenir pour de tels stages que des agents ayant deux ou trois années d'expérience professionnelle.

La durée optimale d'une telle session paraît être de quatre ou cinq semaines.

La formation des producteurs et des distributeurs est réalisée, de manière fort différente, selon les pays de la Communauté européenne et, semble-t-il, de façon sectorielle, essentiellement à l'initiative d'organismes techniques spécialisés (Instituts, Universités, Services de la Protection des Végétaux ...) ou d'institutions à vocation économique.

L'accroissement de cette activité de formation, dans des centres servant de support technique et matériel, mériterait d'être encouragée et développée à l'échelle de l'Europe.

Des propositions pourraient alors être faites pour la réalisation d'une opération concertée de bonne dimension, étalée sur cinq années, ce qui semble correspondre à une période optimale.

6. EN CONCLUSION

Il est important pour les agriculteurs de mettre en oeuvre des moyens de production rationnels et économes.

Cet aspect économique très important ne doit cependant pas faire perdre de vue les incidences d'ordres qualificatif, écologique et toxicologique de la protection phytosanitaire, qui correspondent généralement aux préoccupations des particuliers et des collectivités. En ce domaine, la seule base aujourd'hui acceptable pour les interventions de l'agriculteur se trouve dans les principes qui débouchent sur la lutte intégrée.

Les études écologiques approfondies réalisées au cours des vingt dernières années ont permis d'acquérir une meilleure connaissance des agro-écosystèmes et de mettre au point des méthodes de lutte, contre les déprédateurs des cultures, incontestablement plus satisfaisantes que celles axées sur des interventions de nature systématique.

Bien que la recherche et l'expérimentation doivent être poursuivies, encouragées et développées, ce sont la formation et l'information qui méritent la plus grande attention et, sans doute, un effort prioritaire; l'accord semble quasiment général sur ce point.

Il est essentiel que cette activité s'adresse aux producteurs et aussi aux distributeurs qui jouent un rôle incontestable de véhicule des innovations et qui influencent souvent la prise de décision des agriculteurs.

S'il n'en est pas ainsi, les effets de l'amont—c'est-à-dire des organismes de recherche et d'appui technique et méthodologique—ne peuvent réellement porter leurs fruits.

Nous l'avons vu, l'évolution vers une meilleure protection de l'environnement passe par un emploi rationnel et limité des pesticides. Elle nécessite une information permanente et une activité de formation considérable.

La qualité de notre cadre de vie mérite bien l'effort de tous !

RÉFÉRENCES

1. ACTA (1984). Les auxiliaires—ennemis naturels des ravageurs des cultures. Brochure, 4 p.
2. ACTA (1984). Index des produits phytosanitaires—France—Afrique, 20ème édition, 616 p.
3. AMARO, P. *et al.* (1982). Introduçao a protecçao integrada. FAO—DGPPA—Lisboa, Vol. 1, 276 p.
4. BAILLY, R. (1982). Instituts techniques et protection des plantes-systèmes actuels d'information et perspectives. *Journée sur la communication en protection des plantes—présent et futur*—GPAIH, Paris, 7 décembre.
5. BASSINO, J. P. (1981). L'information technique: sa communication vers l'agriculteur. Bulletin technique d'information, 360, pp. 393–402.
6. BASSINO, J. P. (1983). Evolution vers des systèmes de production intégrée en arboriculture fruitière. Colloque AIDEC Dijon 19–21 avril.
7. BASSINO, J. P. et MOUCHART, A. (1983). Formation et promotion de la lutte intégrée dans les pays de la Communauté européenne. Analyse descriptive ACTA-CCE, 45 p.

8. Cauderon, A. (1981). Sur les approches écologiques de l'Agriculture. *Agronomie*, **1**(8), 611–16.
9. Cavalloro, R. et Piavaux, A. (1983). CEC programme on integrated and biological control—Agriculture—Progress report 1979/1981. EUR 8273, 344 p.
10. C.C.E.—Graffin, Ph. (1982). Integrated crop protection. Proceedings of a symposium held at Valence, France, 18–19 June 1980, AA. Balkema Pub., 407 p.
11. Deroo, M., Fougeroux, A. et Waksman, G. (1983). I. Phyto—Système d'information sur les produits phytosanitaires—INFODIAL—Recueil des conférences du 2ème congrès et exposition internationale sur les bases et banques de données—Paris, 24–27 mai, pp. 244–6.
12. Miller, A. (1983). Integrated pest management: psychosocial constraints. *Protection Ecology*, **5**, 253–67.
13. O.E.P.P. (1982). Recherches en cours en matière de lutte intégrée dans les pays de l'O.E.P.P. (no. 1). Publications O.E.P.P., série B., 85, 82 p.
14. OILB/SROP (1974). Les organismes auxiliaires en verger de pommiers. Brochure no. 3. 1ère édition, 242 p.
15. O.I.L.B. (1980). Conference on future trends on Integrated Pest Management. Bellagio 30 May–4 June, 75 p.
16. Pelerents, C. (1979). La contribution de la Communauté européenne dans le domaine de la lutte intégrée. C.R. Symposium international OILB/SROP sur la lutte intégrée en agriculture et en forêt, Vienne, 8–12 octobre, pp. 383–6.
17. Steiner, H. (1979). Recherches pour le développement d'une production agricole intégrée. Cahiers des Entretiens Ecologiques de Dijon, 3–4, p. 64.

SESSION IV

Conclusions and Recommendations

Chairman: V. O'Gorman
(*National Board for Science and Technology, Dublin, Ireland*)

19

Report on Session I

FINN BRO-RASMUSSEN

Laboratory of Environmental Science and Ecology, Technical University of Denmark, Lyngby, Denmark and Chairman of the Scientific Advisory Committee on Ecotoxicity of Chemical Compounds of the European Communities' Commission (CSTE/Ecotox), Brussels, Belgium

1. INTRODUCTION

There are three fundamental principles to be considered when we talk about proper and safe use of chemicals in our societies. These principles are (cf. Fig. 1):

The principle of good practice, which for the purpose of this Symposium should be Good Agricultural Practice—often abbreviated to: GAP.

The principle of human health protection, which is often practised through reference to Acceptable Daily Intakes, or the ADI-values for individual chemicals. For obvious reasons this principle should always be kept in mind, although it is not directly the object of our discussions here to-day, and

The principle of environmental quality evaluation (EQE), which has been the focus of Session I of the Symposium, and of this report.

The three principles are not mentioned in an order of priority. The sequence is rather reflecting the order in which they have emerged historically as identified concepts. I shall take advantage of them as a framework for my review of the lectures and discussions of Session I.

In doing so, I shall advocate that the three principles are important individually, but also that they are of equal importance. They should be respected in their entirety as a coordinated set of conditions and rulings to which we shall adhere when we deal with pesticides in practical, administrative or scientific professions.

Among the principles, the third one, viz. on Environmental Quality

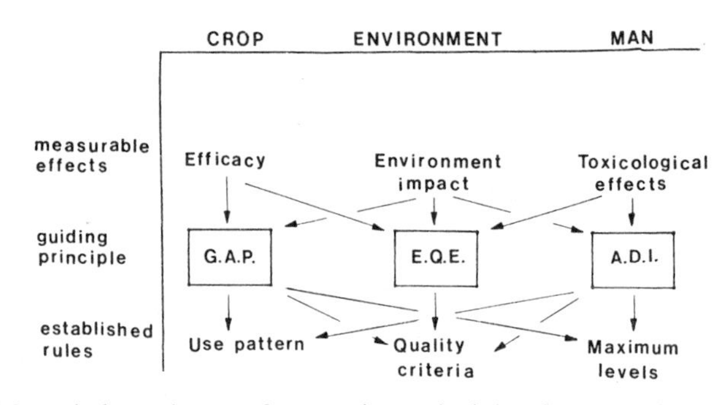

FIG. 1. Mutual dependency of governing principles for use of chemicals in agriculture.

Evaluations (EQE), has developed gradually during the latest years. In my recollection it originates from discussions of the Codex Alimentarius Commission/Committee of Pesticide Residues in the 1960s from which an official concept of 'Lowest possible contamination' emerged. It was adopted in 1971 by a FAO/WHO Working Party of Government Experts on Pesticides (Copenhagen, 1971) as one guiding principle out of three.

I would like to express the hope for the present Symposium that this third principle may now develop fully into a principle of Environmental Quality. This might permit us later to refer to a full set of 'Dublin principles' for the evaluation and the proper use of chemicals in agriculture.

2. PESTICIDES

2.1. Environmental quality and good agriculture practice

I find good reasons for expressing such hope when I turn to the lectures of Session I, which deal with pesticide problems. In summarising the experiences and observations from the session, I have identified at least four themes, all of which have been discussed thoroughly and which all gave significant contributions to a further development of our safeguarding principles. I shall comment on these themes one by one:

(1) We have been presented with the need for full environmental/ ecotoxicological evaluations of pesticide chemicals individually.

To a certain extent such evaluations take place already today in the EEC countries as part of pesticide regulations and/or precautionary schemes for new pesticides. It seems evident, however, that many well-known pesticide chemicals in practical and regular use for a number of years, still need to be regulated, and possibly re-registered on the basis of the latest scientific information.

Re-evaluation of 'old' pesticide chemicals has been initiated in recent years in some countries, but it might well be encouraged and supported by further common initiatives within the EEC. In comparison, or rather in contrast to other chemicals and chemical products, it should be borne in mind that pesticides are developed and intended for use *in* the environment. For the prediction of their fate and effect therefore, we should require that a full environmental profile should be established. In the words of Directive 79/831 (i.e. 6. amendment to Directive nr. 67/548) this requires more than just the level 0 (base set) information known from notification of other chemicals.

As far as the fate is concerned we heard (Dr Stanley) that a Predicted Environmental Concentration (the PEC) is central for the description of the environmental behaviour of pesticides or, in other words, for the exposure analysis as part of an environmental hazard assessment. We were reminded of the great number of data and information which are needed to perform such fate description.

Just as important is the need for ecological effect evaluation of pesticides. This was the subject of Professor Lebrun's and Professor Koeman's lectures, and of several contributions to the debate from the auditorium. On that basis, it is suggested that we try to initiate and to coordinate our efforts in the EEC countries to define those ecological effects which should be included as essential for a full environmental hazard assessment of the individual pesticide chemicals.

From Professor Lebrun's presentation we may note that effects on natural communities as well as on ecosystems are relevant. His schemes might well serve as the starting point for our definition of such structural and functional disturbances which might be important for the evaluation process.

(2) There is a need for the evaluation of multiple or concerted actions of several pesticides and of effects from continued use of pesticides on the same fields.

This is an important question on which several speakers have already commented during the Symposium. The background was presented *inter alia* by Professor Koeman in some of his examples, especially when he dealt with normal and disrupted regulatory functions of natural ecosystems. In one example he reported on 25–30 applications of insecticides in Nicaraguan cotton fields per season. This is of course an extreme example—though not outside our experiences—but it is pointing towards the ultimate result, namely a so-called 'environmental hit back', and it did actually result in severe reductions of the productivity of the cultivated fields.

Disturbances at the community and ecosystem level were also demonstrated in the examples of Professor Lebrun in his report from Belgian sugar beet growing areas. On these fields, pesticides have been used regularly throughout 20 years with the result that both structural and functional disturbances are noted in the soil ecosystem, including effects on microbial life, insect fauna, arthropods, etc.

Similarly we should expect that reductions in the avifauna, e.g. partridges (*Perdix perdix*), which were first studied by Potts and coworkers in the United Kingdom, are the result of toxic pressure from continued uses of pesticides. Parallel observations are made in other countries, e.g. in Denmark as we have heard during the discussions.

The ecological effects described in various examples may result from pesticide uses in a direct or an indirect way. We may draw a line from the early observations 10–20 years ago of, for instance, bird poisonings caused by the eating of mercury-dressed seeds, or via uptakes of foodchain transported DDT or related compounds. At that time we changed the agricultural practices by banning the DDT and the mercury-containing fungicides.

2.2. Needs for the future

Today it seems as if we cannot rest on earlier achievements in our safeguarding of nature. We must establish a continued monitoring activity and surveillance of pesticide effects, and it is suggested that Professor Lebrun's observations from sugar beet fields are not only prolonged, but also extended to cover other situations of heavily treated agro-ecosystems. A coordinated network of European reference areas and observation fields of known history of pesticide management might be a suitable approach to meet this requirement, e.g. under the auspices of an EEC-sponsored COST-project.

This proposal has been put forward by Professor Koeman, and it has been seconded from various sides of the auditorium. In a very specific manner it is suggested that studies should concentrate on a number of community and ecosystem parameters among which the following would be of immediate significance:

changes in abundance and diversity of soil microflora and -fauna, as well as terrestrial and aquatic invertebrates, and
experimental field studies aiming at cause-effect analysis of multiple pesticide patterns.

(1) It seems to be appropriate to call for a re-definition of the concept of Good Agricultural Practice.

The presently accepted concept of GAP rests on a general understanding that pesticide usage—when referring to environmental issues—should not cause any 'undue hazards' to elements of fauna and flora, and no 'undue contamination' of the environment should be presented.

It is evident from the discussions during the Symposium that such definition needs to be better structured and—above all—to be expressed in more specified, quantitative terms. It is necessary to identify minimised pesticide loads, possibly correlated to minimum uses of pesticides. Research projects such as the current plans for establishing pesticide-free protection zones between agricultural areas and the open land—so-called pesticide exclusion strips—may be utilised in that direction. By specific studies of these 'conflict-zones' (see below) we may achieve a better understanding of and a rational approach to a minimising concept as a basis for future pesticide usage.

Included in the redefined code of practices should also be the aspects of improved application techniques. The example given by one of the lecturers (Koeman), namely concerning endosulfan uses in tropical areas, could be quoted—together with other examples—in support of this. In the specific case, a successful endosulfan spraying programme was performed on a land area neighbouring to an aquatic system without damaging the fish life, which is contrary to the experiences of the other spraying practices.

The explanation for this success was for a major part connected to the fact that a very precise technique was used which permitted application at low dosage levels.

There is a need for established criteria and for quality objectives for specific target areas (agro-ecosystems) as well as non-target areas likely to be exposed, either intentionally or not-intentionally, to pesticides or their residues.

This brings me to a final observation concerned with the environmental effects of pesticides, namely the establishment of a clear definition of the boundaries between target areas and the non-target areas, i.e. what I have called the 'conflict-zone' between the arable, agricultural land, and the neighbouring environmental regions. In short, it is essential that environmental quality criteria are defined separately for the two spheres, i.e. for the pesticide treated agro-ecosystem, and for the untreated or non-target areas.

For the untreated land, the principles of environmental quality protection should be applied within the framework of recognised nature conservation codes and practices. However, it is not yet clear which criteria should apply for the arable land, or how they should be established. Dr Bassino (France) pointed to that difficulty during the discussions of Session I by his simply phrased, though intricate question: How do we work with the ecosystems?

I shall not attempt to answer Dr Bassino's question here, but I am tempted to respond by reversing it with the words, that we can only solve the problem of establishing quality criteria for the agro-ecosystems by actually working with the systems! It will serve the purpose if we create a clear delimitation of the individual ecosystems, and similarly, if we define the lines of demarcation between them. Some of the above-mentioned projects, e.g. on pesticide exclusion strips, and on ecological monitoring of agro-ecosystems may be important examples in this respect.

3. FERTILISERS

3.1. Environmental quality and a need for improved practices

In my summary of the three key-note lectures on fertilisers and the environment, reference will be made to the same three principles as above, viz. the principles of Good Agricultural Practice, of Human Health Protection, and of Environmental Quality Evaluation. It may be convenient, however, to start with an overall conclusion from the three lectures. This conclusion is presented as a personal statement, which I suggest should be given full attention by the EEC-Commission. The statement is supported by the facts and evidence which have been presented during the Symposium, and it may be sounded as follows:

> Evidence has been presented that nitrate (NO_3^-) in excessive amounts should be considered a contamination, which causes detrimental effects in the aquatic environment, and which may develop into hazard for man.
>
> Nitrate contamination is of major concern in a number of European countries—possibly all—and it should presently be counted among our most serious environmental problems.

The problem of nitrate contamination is emerging
in vegetable foods, and possible feed,
in drinking water supplies, and
in the aquatic environment,
and the present agricultural practices have been identified as a major contributor to the emergence and the development of the contamination.

3.1.1. Nitrates in food, feed and drinking water
In his presentation, Dr Selenka reported on the human intake of nitrates based on information from the Federal Republic of Germany. His experience corresponds closely to similar, earlier reports from other countries, e.g. Denmark, the Netherlands, Switzerland, and (in some regions) USA. The major proportion of human intake of nitrate is to-day supplied by a selection of vegetable foods, mainly leafy vegetables and some root crops, and in amounts which seem to have increased parallel to the increased use of fertilisers in modern cultivations. Quantitatively, this contribution is recorded to vary for Europe as a whole from 20 % to 60 % of the established acceptable daily intake values.

This situation, however, has become vulnerable because the remainder of our total nitrate intake is supplied by drinking water, which is a highly variable source of nitrate. From various countries the disturbing fact is presented, that nitrate levels in drinking water are increasing to concentrations above the permissible levels of (25 or) 50 mg litre^{-1}.

As typical examples we are informed that 10–25 % of individual water supplies in some European regions (e.g. Denmark, parts of FRG, eastern part of Britain) are becoming contaminated above the health protection limits. This suggests the equivalent of some 10 % of the nationwide (or regional) supplies of water, and it leads us to conclude that up to the order of 10 % of the populations in these regions are at risk of receiving water of unsafe quality. The Danish figures from recent surveys amount to 8 % above 50 mg litre^{-1}, and 19 % above 25 mg litre^{-1}.

From some countries it is further reported that increases in groundwater nitrate levels have been determined during the last 2–3 decades. Rates of yearly increments up to 2–4 mg NO_3^- per litre have been observed on a regular basis.

There seems to be a consensus on the causes of this situation, namely the continuous leaching of nitrates from the root zone of top soils. This is a process which in colloquial common terminology is often referred to by the unpleasant name of 'nitrate bomb' as a reflection of the threatening nature of the development. The leaching is characterised as a free movement of dissolved nitrate with soil water, and it is dominantly influenced by such factors as soil type, hydraulic conditions, geology of the underground, etc.

Evidently, it is certainly a situation calling for timely countermeasures.

3.1.2. Nitrate in aquatic ecosystems

The situation which has been described concerning the nitrate contamination in the aquatic environment is also serious, but with different implications than for those relating to drinking water supplies. In Professor Vighi's exposé on the river Po situation, it was disturbing to hear of the degree of eutrophication of the Adriatic coastal waters, and of its connection to agricultural diffuse sources, and to urban point sources of nitrogen- (and phosphorus-) containing nutrients.

In many of its details, that situation is similar to the deterioration of water quality in the marine environment which is increasingly experienced around the Danish peninsula and islands, and in the Baltic Sea region. Dr Kofoed presented comprehensive information on this subject in his lecture. A total of 150 000 tons of nitrogen is estimated to be leaching (or discharged) per year into the Danish inland waters, and it is scientifically accepted that this is the most important cause for the ecological imbalance which is developing in these marine ecosystems.

In relation to the Baltic Sea area increasing areas of deserted benthic communities are reported. The number of episodes of oxygen depletion (including fish kills) in fjords and coastal regions also seem to be increasing, with close correlation with the increase in nitrogen load and with the degree of eutrophication.

Thorough studies of possible sources indicate that this sombre picture of undesirable changes can be related *inter alia* to a number of important sources of nitrogen. According to present estimates, 20 % of the input is contributed from atmospheric fall-out, while 65 % of the remainder is related to agricultural sources. These are non-point sources (or so-called area contributions) and more localised, though scattered discharges from farm areas (dunghills, etc.).

These reports provide for an estimate of the load as a result of leaching from agricultural areas. Present estimates indicate an amount of 40–65 kg N ha^{-1} yr^{-1}. That figure, however, should be compared with the further mass balance studies described in the recent NPO-report (Nitrogen-Phosphorus-Organic Matter) presented at this symposium by the Danish National Agency for Environmental Protection. The agency claims that the leaching more likely is of the order of 70–90 kg N ha^{-1} yr^{-1}.

The difference between the figures is of course important, as it has been reflected by some of the discussions during the Symposium. I shall take advantage, however, of not having involved myself in the details of that discussion. I prefer to express, as an opinion, that more important than the disagreement on figures is the common agreement on the existence of an unwanted development, namely that considerable losses of nitrogen take place from agricultural areas, that there is evidence of increasing losses due to developed practices, and that the losses are considered to be a major—though not the sole—source of the threatening consequences, which have been one of the subjects of Session I of the Symposium.

I shall take these observations as background for my proposals and suggestions, which I find important as elements of a recommendation for the future activities concerning the proper use of fertilisers, and for the improvement of our environmental quality.

3.1.3. The need for reduction of nitrate contamination

A code of agricultural practices for fertilising is suggested as a means of reducing the agricultural contribution of nitrate to groundwater and to surface waters. It should take into account that good practices shall include full consideration to be given to the quality aspects of our environment. The leaching of nitrates from agricultural areas is correlated to the total load of nitrogen-supply, but it is realised also that the greatest problems are presently connected to the unsatisfactory management of farmyard manure (FYM). Standards should be set, therefore, for the conditions and practices concerned with:

storage and handling;
distribution, possibly including commercial exchange of FYM; and
utilisation, including questions of amounts, techniques and timing for liquid as well as solid farmyard manures and for silage.

As part of this it may prove necessary to establish maximum load values for the fertilising with various forms of nitrogen-supplies.

Educational programmes should be considered important, even indispensable parts of any action plan connected to proper uses of fertilisers, including farmyard manures and silages. Such programmes should turn towards individual farmers, and they should be concerned with questions of crop selection, time of fertilising, amounts of fertilisers, etc.

The agricultural extension services and advisory activities are especially important in a period of changing practices. It is important that the requirements from environmental aspects are strengthened and brought into the educational activity as early as possible. The specific and immediate aim of the educational efforts should be the minimising of losses and pollution from fertilising chemicals and products.

It may be part of the educational obligations that the attention of the general public is also attracted to the problems, in so far as general costs, directly or indirectly, may have unavoidable, wider effects, eventually to the degree that costs should be shared by the community as a whole.

3.2. Research needs

Appropriate research will, clearly, be an essential prerequisite for improvements in agricultural practices and for further evaluation of environmental quality. The assignment of research priorities may become possible on the basis of this Symposium.

There is an urgent need for increased knowledge and experimental studies of the leaching processes which are resulting from various forms of cultivation techniques, e.g. crop rotation patterns, introduction of second crops, various winter growing crops, minimal soil treatment techniques, etc. Included in these programmes, continued attention should be given to the distinction between industrial fertilisers and farmyard manures, and long-term experiments with fertilisers as well as with FYM should be encouraged.

Recent research and experiences have shown that the losses of nitrogen as ammonia (NH_3) from farmyard manures may be more intensive than earlier realised. This point-source of aerial nitrogen transport and deposition poses problems which deserve further investigation possibly aiming towards establishment of limiting application rates, or similar precautions.

One question which has attracted special attention during the Symposium is the possible reduction of nitrate in/and below the root zone, and in groundwater. The question is of significance for the understanding of the fate of nitrates during the leaching processes, and it is recommended that it should be the object for further research. This is also the case for the chemical and microbiological processes which govern the mineralisation process of organic matter from solid and liquid farmyard manures.

4. FINAL REMARKS

This Symposium has been devoted to the integrated evaluation of problems of chemicals in agriculture and in the environment. It is understood that these problems must also involve economic consequences as well as the efforts needed for their solution.

Investment in improved management and techniques of farmyard manure usage may reduce future costs of nutrient supplies to agriculture considerably, due to improved efficiency in utilisation of natural resources. It is unavoidable, however, that other costs, e.g. coverage of research needs, measures to purify contaminated waters, etc., may not be recoverable in the short term.

The paying of such costs from other funds is of course a matter of greatest concern for all of us. It is to be expected that we shall all have to share the burden of suitable and adequate water supplies in the future and of the protection of our environment. In some countries such burdens are discussed in the form of duties or point-taxes (especially on water), while in others (e.g. in Sweden) taxes have already been placed on fertilisers.

The logic of our present situation might even suggest the imposition of taxes on individual farm animals as well, if we are following the PP (polluter pay) principles to which we have earlier committed ourselves for the protection of our environment. Unpopular as it may be to all parties, such questions must be dealt with when we are meeting the challenges of the present situation.

20

Report on Session II

F. P. W. WINTERINGHAM*
Formerly Head, Joint FAO/IAEA Chemical Residues Programme, Vienna, Austria

1. INTRODUCTION

As indicated in the Background Paper, if present levels of agricultural productivity are to be sustained within the EC the needs for agrochemical usage are clear. In terms of abundance, variety and quality of food there are, likewise, clear benefits to the consumer. Moreover, if current world trends in terms of arable land resources, population, and demands *per capita* continue, then agrochemical usage must extend and even intensify during the immediate decades ahead. This takes fully into account current research and development. But it also takes into account the time needed to bring research application to the point of effective impact.

This is not to imply that all agrochemicals are equally necessary or equally beneficial. Indeed, some, if perhaps not all, agrochemical practices can and should be questioned from time to time, for example, the use of antibiotics as a nutritional aid in meat production [8], about which questions are currently being prompted by the possible hazard to human health as a result of spreading microbial resistance to antibiotics [9].

Time has, inevitably, precluded the discussion of many aspects of needs and benefits in addition to those so ably presented during Session II. Therefore this report will briefly refer to certain additional aspects before drawing conclusions and making any recommendations. In particular, since the characteristic theme of this Symposium has been the simultaneous review of both sides of the agrochemical usage coin, some reference will be made to needs and benefits in relation to costs and risks.

* Present address: Darbod, Harlech, Gwynedd LL46 2RA, UK.

2. COMMENTS

2.1. The time element

The papers and discussions have concerned needs and benefits on the one hand; risks and costs on the other hand. However, these two pictures are by no means static and can change with time. Firstly, as a result of continuing agroecosystem–chemical interactions [14]. The development of pest resistance to agricultural pesticides is an example. Secondly, because new information becomes available in an otherwise unchanged situation. The relatively recent recognition that undesirable or unacceptable levels of nitrate in some water bodies are due to local or area agricultural practices is an example. Thirdly, because research and development may themselves suggest alternative agricultural practices and needs. Thus, minimum tillage may require increased herbicide usage. Integrated pest management may obviate the need for some pesticides. In short, the cost or risk/need or benefit ratio changes with time and underlines the importance of constantly looking at both sides of the coin.

2.2. Agrochemical energy aspects

The rise in energy costs of the last decade has prompted a number of studies of the energy equivalents and costs of agricultural inputs and the corresponding equivalents of their outputs [2, 3, 10, 12], especially those relating to fertiliser [13] and pesticide [11] inputs. The declining ratio: energy equivalent of harvested product/energy equivalent of all inputs (including or excluding that of human labour) has provided a useful integrative index of agricultural development since the times of the primitive hunter–gatherer [10, 15]. Of particular interest here is that despite the generally-declining ratio with time, the use of fertilisers or pesticides *per se* can actually result in an increased harvest yield of energy equivalent greater than that of the agrochemical input. Thus, data on the use of fertiliser nitrogen on winter wheat yields in the UK [7] implied that when nitrogen was applied at the optimum level (80 kg N ha^{-1}) the ratio increased from approximately 2·1 without fertiliser to 2·4 with fertiliser [15]. This effect was also illustrated by Dr Marshall during this Symposium. Other data [13] indicate a similar gain as a result of fertiliser use on other crops such as rice, maize, potatoes, and in forestry timber production. The energy equivalents of crop loss reversal or yield increase suggest an even greater return for pesticide use than for fertiliser; here, largely because of the very much smaller weights of pesticides and their energy equivalents needed per hectare [10, 11, 12].

It appears, therefore, that the mainly fossil carbon-based input of agrochemicals is more than balanced by the increased solar energy thereby harvested. This surely represents an unusual situation in the context of energy use generally.

The importance of quantifying cost/benefit ratios in terms of energy seems likely to increase in future, especially in relation to the supply and use of agrochemicals in developing countries. Here, the energy equivalents of agrochemical inputs will represent a far bigger proportion of national energy budgets than those of industrialised countries [3] unless there is a dramatic change in the current world disparities in energy use *per capita* [15].

2.3. Countermeasures and costs

Critical questions arise when the cost or risk/benefit ratio becomes unacceptably high on the basis of expert and impartial judgement. What can be done about it? How long will it take to introduce alternative practices or to apply countermeasures? What will it cost? Who will pay for it? At international level even more difficult questions arise. For whom is the ratio judged to be unacceptable? For the inhabitants of the African Sahel or for the citizens of Los Angeles? In the first place rational and impartial judgement of acceptability implies the ability to quantify the risk/benefit ratio.

A monetary approach to this problem was suggested more than a decade ago in connection with the benefits of atomic energy and consequent risks to the population of radiation exposure [5]. It was assumed that the *net* 'benefit' B of some product or operation is given by the equation:

$$B = V - P - S - D$$

where V is the *gross* value, P the production and maintenance cost, S is the cost of limiting exposure (E) to a given level of safety, and D the cost equivalent of the population 'detriment'. 'Detriment' has been defined as the 'mathematical concept of the harm incurred from a radiation dose, taking into account not only the probabilities of each type of deleterious effects but the severity of the effects as well'. That is, account is taken of stochastic and non-stochastic effects [6].

To simplify the argument, V and P are assumed to remain constant. The problem, then, is to minimise $S + D$, subject always to B remaining positive (i.e. of definite benefit to the community), and exposure being kept within some acceptable limit. The concept is illustrated in Fig. 1.

Clearly, $S + D$ will be minimal when the differential, $\mathrm{d}(S + D)/\mathrm{d}E$ is zero,

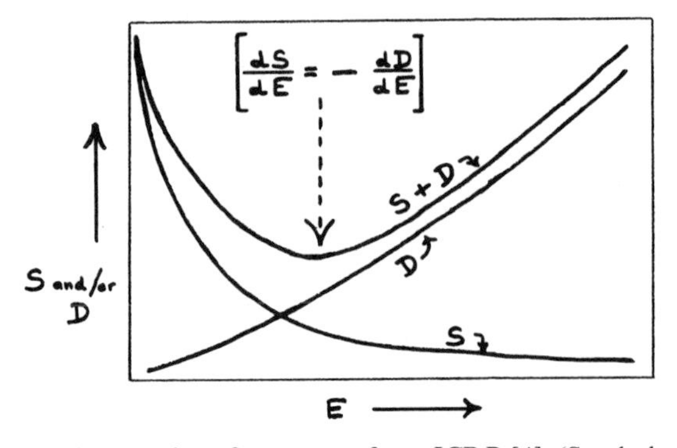

FIG. 1. Differential cost–benefit concept, from ICRP [4]. (Symbols as defined in text.)

i.e. when $dS/dE = -dD/dE$ or 'when the cost of achieving the next element of reduction of E is larger than the value of the resulting reduction in detriment'.

This concept was an attempt to quantify the cost–benefit equation in the relatively simple context of radiation hazards to man which involve largely one toxic mechanism. Moreover, for various reasons (e.g., the nature and behaviour of radionuclides in the biosphere) when human health is adequately protected it can reasonably be assumed that associated environmental hazards can be neglected *per se*. The range of agrochemicals, on the other hand, inevitably involves a wide range of different effects and quite different potentially toxic or ecotoxic mechanisms. There are also the possibilities of harm to the environment and its resources without posing any direct threat to human health. This is reflected in the need for, and in the development of, laboratory 'microcosm' testing procedures, e.g. for aquatic ecosystems [1].

The sensitive and difficult question of who should bear the costs of any constraints or necessary countermeasure can be illustrated. There is a growing problem within the EC as well as in some other countries [16] of continuing losses of soil and fertiliser nitrogen and the evidently associated and rising levels of dissolved mineral nitrogen in some derived water bodies as discussed by Dr Selenka and others during this Symposium. However, there is also evidence that the agricultural contribution to the problem can be contained without impairing essential agricultural production although changes and costs would surely be involved [16]. For example, nitrogen movement from the soil can be reduced by improved soil, fertiliser and irrigation management, and in some cases by the introduction of new or different crop rotation practices. The technology also exists for directly controlling or reducing existing nitrate levels in water [4] but its large scale application would also involve considerable cost.

Should the burden of indicated changes at farm level be left on the farmer who could not possibly have foreseen the problem, and who has simply discharged his task of maximising yields in the face of growing demands? Should the costs of water treatment be borne by the equally blameless water authority or by the state? In any event, as also mentioned by Dr Fairclough at this Symposium, who is the polluter? The manufacturer, importer, or distributor? The agricultural adviser? The farmer? Or even ourselves who demand the cheapest and abundant food?

Finally, when limits to nitrate levels in water are agreed by toxicologists in the context of human health, or by ecotoxicologists in the context of aquatic ecosystem quality how can these limits be enforced either locally, nationally, or internationally? Clearly, there is no simple answer to these complex socio-economic questions. What is clear, however, is that the need for, and extension of, nitrogen fertiliser use will surely continue into the immediate future. These questions certainly have important implications for the CEC and for the EC-based agrochemical industry.

2.4. Highlights of Session II

Before drawing conclusions some important points made by speakers during the session on 'Needs and benefits' might usefully be reiterated.

(1) In their respective papers Drs Gebhard and Ahrens and their colleagues brought out two important points. Firstly, the complementary nature of fertilisers, pesticides, and the use of high-yielding or selected crop varieties. Wise fertiliser application not only provided for increased yield but also for more effective chemical protection and improved quality of the product. Secondly, they drew attention to a much underestimated and often ignored aspect of agrochemical use, that is that the agrochemically based intensive agriculture of today effectively protects large areas of marginal land and forest which would otherwise have to be cleared and cultivated to meet even present day food demands.

(2) Dr Cooke provided salutary and timely data about soil capacity to supply crop nutrients and the need to supplement this capacity if acceptable yields were to be attained and sustained. His data also illustrated clearly the scope for improved soil and fertiliser management within the EC. He also drew attention to the complementary nature of different kinds of chemical application.

(3) Dr Marshall provided interesting and useful data on fertiliser

production and use, not only in monetary terms but also in terms of fossil carbon energy inputs and the greater solar energy harvested to which some reference has already been made.

(4) Dr Calmajane provided an unusual overview on agrochemical usage and trends over the last four decades within the EC. Moreover, he illustrated the effects not only in terms of productivity per hecare, but in terms of economics and employment. He incidentally illustrated the progress of agrochemical research and development by comparing the kg per hectare of copper fungicide once needed to the mere grams per hectare now required with some of the newer formulations.

3. CONCLUSIONS AND RECOMMENDATIONS

3.1. Conclusions

(1) In retrospect two omissions can be identified. The first is of a technical nature. The critical and increasingly important role of irrigation management for minimising the losses of mineral nitrogen from the root zone by leaching and denitrification has not been adequately discussed. Irrigation is, after all, especially complementary to fertiliser nitrogen behaviour and the consequences of these losses were certainly recognised.

The second omission is, perhaps, more philosophical in nature. Too little attention has been paid to the difference between 'maximal' yields in the sole context of agricultural production and 'optimal' yields in the wider context of both production and environmental quality protection. Professor Lieth drew timely attention to this omission in discussion. This again illustrates the need for constant appraisal of the changing risk/need ratios—here in relation to the growing demands within the EC for improved environmental quality control as well as for agricultural productivity.

(2) Both needs and benefits of agrochemical usage can be clearly recognised and often quantified.

(3) Risks and undesirable side-effects can also be recognised. However, there is good evidence that many of these problems can be contained by modified production and protection practices or by the introduction of countermeasures on the basis of existing technology; but cost and change at farm and national levels would be involved.

(4) Critical study of the cost or risk/benefit ratio is essential to any realistic assessment of priorities, of needs for changed agricultural practices or for the introduction of countermeasures. The ratio tends to change with

time for various reasons. This calls for ongoing monitoring and appraisal. It is sometimes useful to consider cost/benefit ratios in terms of resources such as energy. Such ratios become important under the conditions of limited resources as obtain in some developing countries.

(5) The prospects for an extension of agrochemical usage world-wide and even for intensification within the EC for the immediate future cogently argue for improved international harmonisation of recommended limits for agrochemical residues in the environment as well as in food; also for improved guidelines for use at farmer and advisory service level for not exceeding these limits. Such harmonisation not only helps to minimise disparities between nations in terms of human health and environmental quality; it also facilitates international trade in agrochemicals and agricultural products.

3.2. Recommendations

The proceedings of international conferences are burdened with long lists of recommendations for new research, new projects, etc. Most remain unimplemented. This is, perhaps, neither surprising nor even undesirable because such recommendations often reflect more delegates' research interests than real problem-solving needs and priorities. Therefore, one general recommendation only is outlined here on the basis of the preceding conclusions: Priority be given to the subjects identified below (3.2.1) for attention in shorter-term CEC-activities which are aimed at optimising agrochemical usage in the EC by taking simultaneously into account both the needs of agriculture and of environmental quality protection. A single project is suggested for longer-term R and D activities of the CEC (see 3.2.2 below).

3.2.1. Short-term priorities

(a) Guideline methodology for monitoring changes in environmental quality, especially soil fertility, agroecosystem-derived water quality, pest and wildlife populations and their response to agrochemical applications and residues.

(b) Simple methodology for assessing local or on-site needs for agrochemical usage at farm level; e.g. for fertiliser-N application as a function of location, climate, crop, initial soil nutrient status, irrigation, etc.

(c) Guideline limits for pesticide residues in environmental media and biota as distinct from food and feed, and simple guidelines for use at farm level for not exceeding the guideline limits.

(d) Needs and priorities for harmonising criteria of agrochemical use, acceptability and methodology for predicting environmental residues and effects.

(e) Identify research and development priorities and needs in relation to alternative agricultural practices or countermeasures with minimal impairment of agricultural production and farmer welfare.

(f) To make available to the media responsible, balanced, and easily understood reports on agrochemical risk/benefit ratios, this not only in relation to the EC but also in relation to world needs because of their very important implications for the EC in terms of trade and responsibility. In particular to draw greater attention to the fact that current intensified agricultural practices in the EC provide for a considerable protection of undisturbed land amenities which otherwise would have to be brought under cultivation.

In dealing with these subjects the importance of close liaison and cooperation with other organisations concerned cannot be overstressed, for example with the UN agencies such as FAO, UNEP and WHO, with national agricultural, forestry and fisheries advisory services, and with recognised non-governmental and voluntary organisations (see Bassino and Bunyan).

3.2.2. Suggested longer-term R and D projects

The papers and discussions during this Symposium surely suggest that the growing problem of fertiliser nitrogen residues should command the highest priority for its activities under items (b) and (e) above.

Overall, the 'nitrogen problem' of the EC, which involves the huge waste of harvested nitrogen in urban communities, can be seen, at least in part, as a consequence of amending agricultural soils, either by too concentrated animal wastes or by mineral fertilisers with too high N/C ratios—see especially, pp. 25 and 70 or reference 16. Finally, therefore, in relation to R and D of the CEC for the future [17] a timely project would be one to study the possibilities of modifying existing and future Community infrastructures to provide for the acceptable, economic and discrete collection of all human, animal and plant waste for processing and non-intensive recycling to agriculture.

REFERENCES

1. Cairns, J. (1979). Hazard evaluation with microcosms. *Int. J. Environmental Studies*, **13**(2), 95–9.

2. CREMER, H. D. (1979). Energy input and food production. In: *Science and Technology for Development. Proc. of U.N. Conference, Vienna*, Laupp and Göbel, Tübingen, pp. 9–15.
3. FAO (1979). *Energy for World Agriculture*, FAO, Rome, Agriculture Series No. 7, pp. i–xxvii and 1–286.
4. FRANCIS, C. W. and CALLAHAN, M. W. (1975). Biological denitrification and its application in treatment of high-nitrate waste water. *J. Environ. Qual.*, **4**(2), 153–63.
5. ICRP (1973). Implication of Commission recommendations that doses be kept as low as readily achievable. International Commission on Radiological Protection (ICRP). Publication No. 22, Pergamon Press, Oxford, pp. 1–17.
6. ICRP (1977). Recommendations of the International Commission on Radiological Protection (IRCP). Publication No. 26. Pergamon Press, Oxford, pp. i–v and 1–42.
7. LEWIS, D. A. and TATCHELL, J. A. (1979). Energy in UK Agriculture. *J. Sci. Fd Agric.*, **30**(5), 449–57.
8. MULLER, Z. O. (1983). Recent developments in feed additives. In *Chemistry and World Food Supplies* (Ed. L. W. Shemilt), Proc. Int. Conference, Manila, 1982, Pergamon Press, Oxford, pp. 249–67.
9. PEARCE, F. and CHERFAS, J. (1984). Antibiotics breed lethal food poisons. *New Scientist*, **103** (1421), 3–4.
10. PIMENTEL, D. and HALL, C. W. (Eds) (1984). *Food and Energy Resources*, Academic Press, New York, pp. i–xiv and 1–268.
11. PRICE JONES, D. (1975). The energy relations of pesticides. *Span*, **18**(1), 20–2.
12. ROBINSON, D. W. and MOLLAN, R. C. (Eds) (1982). *Energy Management in Agriculture*, Royal Dublin Society, Dublin, pp. i–viii and 1–441.
13. SCHUFFELEN, A. C. (1975). Energy balance in the use of fertilizers. *Span*, **18**(1), 18–20.
14. WINTERINGHAM, F. P. W. (1979). Agroecosystem–chemical interactions and trends. *Ecotoxicology & Environ. Safety*, **3**(3), 219–35.
15. WINTERINGHAM, F. P. W. (1983). Biomass cultivation and harvest: Global trends and prospects. *Outlook on Agric.*, **12**(1), 21–7.
16. WINTERINGHAM, F. P. W. (1984). Soil and fertilizer nitrogen: Review of studies worldwide with particular reference to plant nutrition and environmental protection. FAO/IAEA/GSF. Vienna. IAEA Technical Report Series No. 244, pp. i–x and 1–107.
17. CEC (1984). Eurofutures—the challenges of innovation, CEC–FAST report (Forecasting and Assessment in Science and Technology), Butterworths, London, pp. 1–199.

21

Report on Session III

A. F. H. Besemer
Agricultural University, Wageningen, The Netherlands

1. THE OBJECTIVE

The aim of Session III was to analyse the instruments and procedures available for optimising pesticide and fertiliser uses with particular reference to their environmental impact.

It is obvious that the use of pesticides and chemical fertilisers continues to be indispensable in modern agriculture in EC countries, and elsewhere, and also in the public health field. The main reason for the still extensive use of pesticides is the lack of a sufficient range of adequate alternative methods for the protection of crops in any or perhaps most crop/pest or crop/weed situations. However, it has to be recognised that in recent years in many parts of the world, including the EC countries, successful alternative crop protection methods and procedures are being developed. Some are already in operation or at a stage of practice in which pesticide chemicals play only a modest role (see 2(2) and Section 3.2 below).

2. WAYS AND MEANS FOR THE OPTIMISATION OF THE USE OF PESTICIDES AND FERTILISERS

The topics covered in Session III included the following:

(1) Pesticide legislation and regulatory measures including pesticide labelling.
(2) Integrated Pest Management (IPM).
(3) Integrated Fertiliser Management.

(4) Optimisation of application techniques and improved formulations directed on minimising the environmental impact.
(5) Monitoring of the fate and biological effects of pesticides in the environment.
(6) Information and training, including the ongoing education of extension officers, retailers and farmers with regard to the 'Good agricultural practice of pesticides', which also takes into account the environmental impact. The presentation and discussion of these topics are reviewed *seriatim* as follows.

3. COMMENTS

3.1. Pesticide legislation and pesticide authorisation or registration procedures including label clearance relevant to environmental protection (M. Hascoet)

In all or nearly all EC countries, the protection of the environment is increasingly being taken into account in pesticide authorisation or registration procedures.

In several countries a similar trend can be seen in relation to the regulatory procedures for fertilisers.

The 'Guidance on environmental phenomena and wildlife data' as developed in the booklet: 'Pesticides' of the Council of Europe (5th edn) and the Report of the FAO Second Expert Consultation on Environmental Criteria for Registration of Pesticides (FAO Plant Production and Protection paper nr. 28-1981) form a basis of the environmental risk assessments in the framework of the pesticide registration procedures in several EC countries.

The author stressed the fact that several assay methods are now available to assess and evaluate details of possible undesirable environmental effects such as, fish toxicity, effects on 'beneficial' parasites and predators of noxious insects and mites, bee toxicity. Also methods are now available for the evaluation of the fate of the pesticide in soil, ground- or surface water, etc., and the effects on soil and water fauna or flora, e.g., soil chromatography, leaching experiments in soil columns, lysimeter studies and computer models based on physical and chemical properties of the pesticides. Some of these methods were developed by international bodies like OECD and the EC Directive on dangerous substances.

It has to be recognised that many of these methods provide information on the *potential* environmental hazards but they are often not adequate for predicting actual environmental risks. In the light of the latter it is obvious that there is a need to develop assay methods simulating the complex environmental conditions met in practice if the pesticide is used according to the instructions or recommendations. The development of such methods is under way in several EC countries but no action is undertaken at the moment to harmonise these within EC.

In order to share the laborious task involved with the development of these methods and to avoid duplication, the harmonisation and choice of a set of most-suitable test methods in this field is necessary.

The EC commission could play an important role as initiator of the desirable harmonisation of test methodology and procedures enabling a reliable prediction of the possible environmental impact as a consequence of the (future) use of the pesticide. The availability of harmonised requirements and test methods for all EC countries may also be advantageous for manufacturers concerned with national registrations in various EC countries since it may result in lower costs for the registrations.

Although the concept of EC approved pesticides seems to be remote and not practicable at the moment, the manufacturers may be satisfied if they have the possibility to provide one and the same technical dossier to the national authorities concerned when they are seeking an authorisation in various EC countries. In this way unnecessary expenses for the agrochemical industry will be avoided.

A few years ago the EC commission was active in this field and on the right road to establish harmonised requirements, including also requirements with regard to environmental impact of pesticides in the framework of the proposed Directive on EC approved Pesticides. Also a start was made on harmonisation of relevant test methodology. It is regrettable that the harmonisation activity in this field has slowed down, nearly to zero, due to hesitations and lack of consensus between EC member countries at Council level.

The original design of the proposed Directive may have been too ambitious at that time. Moreover, there was general consensus at the Dublin meeting on the desirability of recommending that the EC Commission should undertake again and now as a very urgent and, preferably as a separate matter, the harmonisation of requirements and matching test methods at Community level including methods simulating the environmental conditions which will be met when a pesticide will be used in practice.

In the process of developing these methods, especially those which enable a reliable prediction on the fate and effects of a pesticide in the agricultural ecosystem and in other compartments of the environment, the type of surveillance described by Bunyan (see Section 3.5) should be considered. The experience already available in the UK in this field may be a good starting point. The term surveillance has to be understood as defined by Bunyan in his contribution at the Dublin symposium: 'Surveillance to identify the environmental impact of pesticides': 'The short term and intensive measurement of chemical and/or biological parameters of an environmental system for which the pesticide inputs are known. Effects can usually be related to specific causes'.

One should be aware that developing harmonised requirements for the evaluation of environmental effects and test methods complying with these is only a first, but however, important step. Consensus should also be reached between responsible experts within the EC on the interpretation of the data in terms of the likelihood of actual environmental (and other) hazards and on the criteria which should be observed in this matter.

3.2. Integrated Pest Management (IPM) (R. Diercks)

The author described the state of art of the Integrated Pest Control in EC countries.

IPM according to the definition developed by the FAO Panel of Experts on Integrated Pest Control is a pest management system that, in the context of the associated environment and the population dynamics of the pest species, utilises all suitable techniques and methods in as compatible a manner as possible and maintains the pest population at levels below those causing economic injury.

It is generally recognised that IPM is one of the best instruments to delay the risks of development of a high level of pesticide resistance and to decrease the pesticide load on the environment.

Successful IPM schemes were developed and are operative in fruit-culture for the control of insect and spider mite pests, in glasshouse cultures of fruiting vegetables (control of red spider mite with the predatory mite *Phytoseiulus* and White fly with its parasite *Encarsia formosa*) and for weed control in various field crops.

Two main elements were elucidated:

(1) The importance of the concept of the economic (damage) threshold level.
Control measures should only be carried out if there is a risk of exceeding this threshold level.
With regard to the threshold level mentioned above, it has to be

noted that this level is not of one population level, but varies according to the external and internal conditions of the crop, the pest population, and the population of beneficial organisms.

(2) The need of a sufficient number of available and suitable pesticides with a selective mode of action. The pesticide must be effective against the target pest organism but exert no severe influence on populations of non-target organisms, including 'beneficial' organisms such as parasites, predators or antagonists of pest organisms in the agricultural ecosystem.

In order to promote further development of IPM schemes there will be an increasing demand of highly selective, moderately persistent pest control agents. There will also be a continuous demand of new products to replace older ones that have become ineffective due to the development of resistance. It has, however, to be recognised that the success of IPM on crops on which several pests occur in the same period is not guaranteed with a selective pesticide for one pest, but the availability of a selective pesticide also for the 'last' pest on the crop for which such pesticide was not available before.

Severe constraints on the development of new pesticides, in particular of selective pesticides, are the high costs of research and development and of the requirements for registration or authorisation. The development of a pesticide in the laboratory and the field, from the first screening up till the introduction on the market with full scale production, requires nowadays about 20–30 million US dollars equivalent. At the time of introduction $\frac{1}{3}$ or more of the patent life of the pesticide is already past. From an economic point of view, the development of a new compound will in general be only worthwhile for the industry concerned, if there are prospects of a broad market for a reasonably long period to ensure a satisfactory return of investment. With a broad spectrum pesticide it will take another $\frac{1}{3}$ of the patent life before the returns are in balance with the investments up till that time.

For the desired selective pesticide it will take a longer period; often the returns will not equal the investments at all.

A selective pesticide for a specific pest/crop situation on a crop of limited economic importance will only become easily available if the product can also be used on a reasonable scale on one of the world's key crops, such as maize, cotton or citrus.

Further development of IPM for other specific crop/pest situations often requires highly specialised and time consuming research before introduction in practice is possible and there may also be the need for intensive instruction and training of the growers concerned.

In the light of the particulars mentioned above and of the theme of the Dublin symposium it has to be stressed that IPM may be an important tool to attain a restrictive and judicious use of crop protection chemicals.

In order to speed up the development and effective introduction of IPM for new crop/pest situations it is recommended that the EC initiate and assist further necessary research in this field.

The success of IPM schemes is not only dependent on the outcome of research programmes but also on training and permanent education in this field. Therefore, the EC could also play an important and fruitful role with regard to the latter aspect preferably in cooperation with other bodies in this field, such as the OILB (Organisation Internationale de la Lutte Biologique). Instructions and training could take the form of demonstrations in practice, workshops, etc., in which the outcome of the research is 'translated' into the growers' knowledge and language.

3.3. Integrated fertiliser management (J. C. Remy and H. Tunney)

With regard to the risks of undue pollution of ground and surface water resulting from the extensive use of chemical fertilisers both authors shared the opinion that nitrogen (nitrate) leaching from agricultural soils has given rise to alarm not only in many EC countries, but also elsewhere in the world.

There has been a considerable increase in the use of nitrogen fertilisers in most countries in Europe since World War II. The amounts applied per ha in EC countries show, however, large differences. In The Netherlands the average rate of nitrogen fertiliser per ha is the highest, over 240 kg ha^{-1}, whereas Ireland and Greece are at the other end of the range with averages of about 43 and 33 kg ha^{-1} respectively.

Phosphorus and potassium fertilisers cause less concern. The environmental pollution due to leaching of phosphatic compounds from agricultural soils and the eutrophication resulting from it is in general more localised and of minor importance compared with other sources of 'Phosphorus' pollution in ground and surface water (e.g. industrial, detergents, washing powder, etc.).

Thus the main attention has to be focused on nitrogen (nitrate). The conclusion was drawn that there are several possibilities for improving fertiliser efficiency in practice and still maintaining current production levels with a reduced fertiliser input. The latter could result in a decreased risk of water pollution especially with nitrate.

Both authors drew attention to the possibilities of, and also the importance of, growing an inter-crop during the autumn and early winter period when the soil humidity in general has reached the 'saturation' level and additional rainfall may easily cause leaching of N (nitrate) to ground and surface water.

The water quality criteria foreseen in the EC in the near future of 50 mg litre^{-1} nitrate may prove difficult to meet in several regions and action has to be taken to decrease the nitrate leaching either by a lower input of nitrogen fertiliser or by changing the soil fertiliser management practice.

In the light of the already occurring rather dramatic situation in many regions with regard to environmental hazards arising from nitrate there is an urgent need to intensify the research on this matter in order to improve the situation without reducing agricultural production too much.

Tunney was also of the opinion that the possibilities of a legume-crop based crop rotation system should not be overlooked. The fixation of atmospheric nitrogen by the legume-crop could replace to some extent the fertiliser nitrogen and thus have a considerable impact on the rate of chemical N-fertiliser needed and used for a succeeding crop. It was recognised that such crop rotation has been largely discontinued in most regions of the EC for many years.

Extensive information exists on this topic. It would, however, still require considerable research and considerations to decide to what extent a return to such crop rotation could be of value and practicable in EC countries and whether reduced overall yields and their economic consequences would be justified.

In discussing the topics Ms C. Klitsie, Ministry of Agriculture, The Netherlands, confirmed the high average input of nitrogen fertilisers. These high amounts are, however, not the main cause of nitrate contamination of groundwater. In areas with intensive dairy farming about 400 kg N ha^{-1} is applied. The nitrate content in the ground water in these areas is lower than the proposed EC maximum level of 50 mg litre^{-1}, i.e. about 35 mg litre^{-1}. The amount of N-fertiliser on arable land, 150–200 kg N ha^{-1}, also does not result in nitrate levels in ground water exceeding the EC maximum level. In The Netherlands the nitrate ground and surface water problem is related to the huge amounts of natural manure often in liquid form produced in the intensive pig and poultry farming, which is not cultivated-land related. This type of farming is concentrated in The Netherlands on light sandy soils.

The ground water monitoring network covering 350 sampling points has shown that especially in the sandy areas with intensive dairy and poultry farming the ground water pollution with nitrate gives rise to concern. Solving this manure problem is a great challenge for The Netherlands and gets high priority.

Professor Amberger, München, suggested that:

(1) Intensifying the use of chemical fertilisers is the only way to reach economic agricultural production, but the amounts of fertilisers used should be in accordance with the nutrition balance (input–output in balance). It is highly important to educate the farmers on this subject.

(2) Chemical fertilisers provide the highest return together with the lowest burden on the environment (see also Kofoed, Session I, and Cooke, Session II). The reason for this is the period of application, i.e. shortly before sowing or early in the growing period. During the growing period in general no leaching occurs.

(3) The critical issue with the application of manure is the time between manuring and mineralisation. Mineralisation proceeds during late autumn and early winter when crops are harvested and in general the soil is not covered by a vegetation. Winter rain on the saturated soil will cause nitrate leaching. Chemical fertilising should be carried out in such a way that immediately after harvest of, for example, a cereal crop, not more than 10 kg N ha^{-1} remains in the upper soil. In a similar soil about 50–60 kg N ha^{-1} will be found in October/November following autumn applications of natural manure, due to mineralisation. This amount is ready for leaching through the winter rain.

(4) Possibilities: N-levels in the soil include: crop rotation with an inter-crop covering the soil during late autumn and a part of the winter period. 'Fertilising' the soil with straw and, perhaps, the application of nitrification inhibitors.

(5) Dressing with liquid manure provides no solution for the 'nitrate-problem'. Usually liquid manure is applied mid-October or shortly thereafter, soon after harvest of main field crops. The total ammonium-nitrogen is mineralised in about 2–3 weeks and the nitrate developed during this process may be leached to the ground water by winter rains.

Professor Amberger expressed the doubt as to whether the maximum level of 50 mg $litre^{-1}$ nitrate as water quality criterion foreseen in the EC is really necessary from a public health standpoint. To adhere to this level will give rise to problems in many areas and high expenses will be involved.

3.4. Optimisation of application techniques and improved formulations directed on minimising the environmental impact (D. Seaman)

At present most pesticides are applied using high- or intermediate-volume aqueous sprays having a broad spectrum of droplet sizes. Other modes of application are seed dressing and applications of granular formulations of pesticides in the furrow or broadcast on the soil surface.

Seed dressing is an effective method, where the seed requires protection or where insects or fungi attack the young plant. Granular formulations are applied for control of soil borne nematodes and/or insects and for volatile herbicides. Availability of the pesticides in the soil is not immediate and relies on the volatility or water solubility of the active ingredient.

In Europe the aerial application of pesticides is still limited, especially to field crops which cover a large acreage, such as cereals, potatoes, rape and sugar beets.

Application is at higher concentrations of pesticide in the spray liquid than for ground based applications, using 20–50 litre ha^{-1}.

ULV (Ultra Low Volume) application, which applies 2·5–5 litre ha^{-1} of a concentrated spray, without water as carrier, is used rather rarely in Europe with a small number of active ingredients and special formulations.

Although the conventional spray application techniques are in general effective, they are often not fully efficient, since there is always a loss of spray liquid not reaching the target crop.

Depending on the density of the crop, application techniques and apparatus, distance between the nozzles or other devices and the crop canopy, wind velocity and thermal air currents, a part of the spray liquid, sometimes a considerable part, is not deposited on the crop but reaches the soil near (under) or at some distance from the target crop. Very small droplets may float away in the air, sometimes up to large distances. The already small droplet is getting still smaller due to evaporation and may cover longer distances.

A new development to reduce spray drift and thus environmental impact, to improve spray retention and to reduce spray volumes is the so called controlled droplet application (CDA), produced by spinning-disc based systems.

Spinning discs, either with horizontal or vertical discs, generate if properly used a droplet spectrum much closer to 'monosize' than conventional sprayers. The droplet size depends on the disc speed, e.g. for the application of herbicides to soil or arable crops a CDA technique is used in which 80% of the droplets are within the range of 200–300 μm. The absence of very small droplets largely eliminates spray drift whereas the absence of both small and large droplets provides for a reduction in the amount of spray liquid whilst maintaining good crop coverage.

The introduction of CDA in practice although promising is still limited. Part of the reason for this is that each and every formulation needs to be biologically tested to demonstrate an equivalent or better effect than those of conventional sprayers. The biological effect may show differences because:

(a) spray coverage and distribution of the pesticide on the crop is different and
(b) the spray liquid is about ten times more concentrated, in relation to the lower volume used.

Furthermore pesticide formulations may in many cases require optimising for CDA.

There is a need for further research and training in this promising technique.

Another promising new application technique to improve droplet deposition and reduce the amount of spray liquid which does not reach the crop is the electrically charged spray. Several slightly different methods are under study; one of these, for example highly charged low conductivity organic liquids, may also generate 'monosize' droplets of 50–100 μm.

The concept of electrical charging of sprays to improve performance and to decrease the environmental impact due to minimising drift is promising but still in its infancy. It must be recognised that the distribution on the target crop is different from conventional applications which can affect the overall biological effect.

There is a huge task of designing and testing a whole range of special formulated products to establish this technique.

Much attention has been given in recent years to optimise pesticide formulations resulting in an improved biological effectiveness, improved safety for the operator and less undesirable side-effects on the environment.

With regard to the latter the controlled release formulations represent a promising new development. Controlled release may provide for the active ingredient becoming available during or after a certain time period.

This, in turn, can provide various advantages, e.g. improved persistence of otherwise too volatile compounds, improved retention so that a granular systemic insecticide, for example, allowed top sediment in flooded rice paddy, does not lead to toxicity of fish in the overlying water.

A very elegant way to obtain controlled release is by microencapsulation.

Tiny particles or droplets of a pesticide of a few μm diameter are coated with a plastic film.

As well as increasing persistence, microencapsulation can also reduce mammalian toxicity (thus improving operator safety) and plant phytotoxicity. Despite all these potential benefits only very few products have reached the market or are near to introduction in practice.

The addition of polymers to the spray liquid in order to reduce drift requires further testing to justify their uses. For a small number of pesticides the addition of oil is beneficial; it may hamper the evaporation from small spray droplets.

3.5. Monitoring the fate and biological effects of pesticides in the environment (P. J. Bunyan)

Monitoring the fate and biological effects is an important instrument to check whether the authorised or recommended uses are safe and not causing undue hazards to the environment. Such monitoring programmes should always be carried out for newly introduced pesticides, especially in cases where extensive use may be expected. Such monitoring programmes should also be carried out when a pesticide which had initially a limited use is to become highly extended.

The pro's and con's of the now available monitoring methods were clearly illustrated. The so called surveillance programmes as referred to and defined were recognised as the method of choice. However, there is still a need for the development of more sophisticated methods for assessing subacute effects of pesticides on non-target organisms in actual field conditions. There is also a need for the development of better methodology for detecting significant biological effects, including behavioural changes in total ecosystems and correlating them with total input of pesticide chemicals and their toxic metabolites from whatever source.

In addition to the related conclusions of Section 3.1 (above) the meeting also concluded that the EC could play an important role in initiating the necessary research on the subjects of monitoring and surveillance especially with the aim of harmonising methodology and programmes.

3.6. Information and training, including continuing education of Extension officers, retailers and farmers with regard to 'good agricultural practice' taking into account the environmental impact of pesticide use (J. P. Bassino)

The main parties involved are the pesticide users, including the pest operators and the retailers.

3.6.1. Extension to and training of users and pest operators

The Extension Services may play a very important role in the improvement of the pesticide situation in many countries, also in the EC. The information to the farmer and pest operators provided by the Extension Services should not only include the agricultural aspects, such as a good and justified choice of a pesticide, the timing of the application and the application methods, but also the safety aspect. Farmers still have to learn how to use the pesticides safely, i.e. safe for the operator, safe for third parties, including the consumer of the treated agricultural crop, and safe from the point of view of the environment. One of the main tasks of the Extension is to encourage the pesticide user to apply the authorised pesticide only when such use is indispensable, and to avoid any unnecessary or unjustified use. Instruction and training on Integrated Pest Management should be provided on a permanent basis.

3.6.2. Extension and guidance to retailers

The retail trade of pesticides still gives rise to concern in several countries. It has to be recognised that the retailer plays an important role in conveying information to the farmer. This information may not always be in accordance with the information intended by the Extension Services.

In the light of their important role it is essential that the Extension Services try to get a grip on the retail trade and provide the retailers with the necessary instructions for the satisfactory storage and handling of pesticides.

Both parties involved should be instructed and continuously updated on the concept of 'good agricultural practice' of pesticides as defined by the Codex Committee on Pesticide Residues (October 1972) with an addition by the author of the report of Session III.

'Good agricultural practice' in the use of pesticides is defined as the officially recommended or authorised usage necessary for the control of pests under practical conditions at any stage of production, storage, transport, distribution and processing bearing in mind the variations within and between regions and which takes into account the minimum quantities necessary to achieve adequate control, applied in a manner so as to leave a residue which is the smallest amount possible and which is toxicologically acceptable . . . 'and in such a manner that undue hazards to fauna and flora elements the preservation of which is desired will be avoided and undue contamination of the environment in general will be prevented'.

22

Overall Conclusions of the Symposium

UMBERTO COLOMBO

President, Comitato Nazionale per la Ricerca e per lo Sviluppo dell'Energia Nucleare e delle Energie Alternative (ENEA), Rome, Italy and Chairman of the European Communities' Committee for the European Development of Science and Technology (CODEST), Brussels, Belgium

It is not easy to present, in the short time available for this report, the main conclusions and recommendations stemming from this highly interesting and stimulating Symposium. Fortunately, we have just listened to three excellent summaries of the sessions into which the Symposium has been articulated.

Let me first of all express the view that it has been timely for the EC to hold a symposium in which both the environmental effects of agrochemicals and their economic importance have been discussed, in an attempt to find adequate solutions through the integration of ecological considerations in the planning and operation of agriculture.

In the present situation we must urgently define tools to allow us to reconcile the economic imperatives with the absolute need to take into proper account the quality of our environment and to protect the health of populations.

It should not be forgotten, in fact, that Western Europe represents about 7% of world arable land, and because of the intensive character of its agriculture uses about 25% of all the pesticides, and about 17% of the fertilisers consumed on a world basis.

The governments and policy makers of the European countries (indeed of all countries) are often pressed by short-term problems, and have been, in general, unable to mould their agricultural policies with strategies that sufficiently take into account the needs for environmental protection.

This Symposium has brought, in my opinion, some useful elements to form this link-up, in order to lay the foundations for Community initiatives and for cooperation on a European level in the analysis of some key problems, from which action should follow at different levels.

It is first of all necessary to develop criteria for the use of fertilisers and pesticides in the full awareness of their environmental consequences. The Commission needs the advice of scientific experts in order to establish these criteria.

It is necessary that an adequate regulatory legislation be set up in relation to the use of agrochemicals. In particular, I should like to underline what Commissioner Narjes told us in his excellent opening speech, about the insufficient information available on the toxic and mutagenic effects of some of the pesticides which are already on the market.

In recent years this problem has been tackled by several organisations at both national and international levels. For example, the International Commission for Protection against Environmental Mutagenes and Carcinogenes has made a critical review of the criteria for bio-assay assessment, selection and test performance.

On the basis of this and of similar work, it should be possible to assess the environmental effects of new pesticides before their introduction in the market. This assessment could then form an integral and obligatory part of the official approval procedure for pesticides, in order to select those active products which offer the best relative advantages.

Whilst laboratory tests provide a useful first selection to eliminate those new products that exhibit toxic, mutagenic or carcinogenic properties, these tests alone are not sufficient, and field trials are necessary in order to assess the ultimate effects on the basis of persistence, transformation, diffusion and bio-accumulation of potentially dangerous products.

Uniform guidelines for the conduct and evaluation of these experiments are clearly desirable not only with the aim of ensuring a homogeneously high level of environmental protection throughout the Community, but also to eliminate internal barriers to trade and distortions of competition which are two important objectives of the Treaty of Rome.

In this connection, it should be recalled that the Commission as early as 1976 submitted to the EEC Council of Ministers a proposal on the approval of plant protection products, seeking to establish the conditions under which pesticides may circulate freely within the Community, whilst ensuring safety for users, consumers, wildlife and the environment in general. The proposed measure provided especially for the harmonisation of requirements for the determination of environmental effects, and a Community procedure for their uniform assessment.

For a variety of reasons, some political, others due to the complexity of the subject, the Council has hitherto been unable to reach agreement on the proposal. This Symposium, I believe, has reiterated the importance of environmental concern in the use of agrochemicals, expressing the desire that the 1976 proposal, after a possible updating to take into account the progress of scientific knowledge in the field and the improvement in assessment methodologies, be considered by the Council for early approval.

As far as the use of fertilisers is concerned, regulatory actions should aim at achieving in practice respect of the EEC limit for the concentration of nitrates in drinking water (50 mg litre^{-1}). The Symposium has in fact shown that nitrogen (in the form of nitrate ion) is one of the main environmental problems at the Community level. The image of the 'nitrate bomb' is self-explanatory.

Regulation could include norms for the storage and distribution of animal manure and, where necessary and appropriate, indicate possible alternative uses for this manure (for example in the production of biogas).

The problem of applying the 'Polluter Pays Principle' has been raised in connection with the need to prevent an excessive use of nitrates. This matter is quite delicate and ought to be further analysed before making specific proposals, such as for example a rise in fertiliser price or an award system to those farmers that manage to achieve high productivity levels through a more prudent and environment-conscious use of nitrogen fertilisers.

Attention to avoid an excessive use of nitrogen fertilisers, as well as of pesticides, could stimulate policy measures with wider implications. For example, why should farmers insist in producing more of the crops which Europe already produces in substantial excess, such as sugar beet, wheat, barley, grapes, that require massive inputs of nitrogen fertilisers and pesticides and, once produced, create problems for the Community? This issue has been considered in the recently undertaken review of Common Agricultural Policy, and appropriate measures are being introduced. The process is slow, however, because decisions on production methods are taken often at the individual farm level.

The Community should establish criteria for the assessment of the costs and benefits of the use of agrochemicals. This assessment must, however, be based on a comprehensive approach rather than on strictly agricultural considerations. It should in particular include an evaluation of the full implications on water supply. It should also help to decide, in each region and case, whether to continue with the trend of an ever increasing intensity of agricultural practices, or shift towards less input-intensive cultivation, using greater amounts of land. This has obvious implications on the structure of farms, on employment, on regional policies. While on narrow economic grounds there seems to be little doubt of the convenience of intensive agricultural practices, the inclusion in the evaluation of wider socio-economic implications may well give surprising results also in Europe. One could almost reverse the approach to the problem of agricultural production, and say that the basic principle to observe should be how to minimise the use of agrochemicals, and yet fulfil the targets established for agricultural production.

The Symposium identified important directions for scientific and technological research to be carried out within the Community, with the objectives of improving the assessment of the undesirable consequences of fertiliser and pesticide use, and developing methods to optimise the use of agrochemicals also through the development of new 'bio-rational' products and pest control techniques.

Amongst the research lines suggested, let me stress the importance of basic research on the appraisal of the loss of genetic diversity due to the use of agrochemicals. An affluent region such as Europe cannot in fact escape the obligation to preserve as much as possible genetic diversity.

The fate of pesticides in terms of biotic and abiotic degradation ought to be attentively studied. Regional monitoring networks should be established, wherever this is possible, keeping in mind that we are dealing with somewhat random events in which the critical periods are often short.

It is also important to develop mathematical models for the assessment of agrochemical risk to the environment, on the basis of data on the chemical properties of the products and on farming practices, with the appropriate description of hydrologic cycles. I am fully aware of the difficulties and limitations of mathematical modelling, but believe that their introduction in this case and, above all, a careful interpretation of their results in the light of empirical evidence, might lead to the definition of preventive policies which would reduce at the outset the possibility of damage to health, to the environment, to property, and so obviate the high cost of rehabilitation and restoration.

Research should include assessment of the process of build-up of post resistance to agricultural pesticides. We were exposed to some quite interesting and alarming data on this process, that involves a significant proportion of insect pest species.

This leads to the need to develop more specific, selective, narrow-band pesticides, which are important because they reduce the direct danger to human health. Obviously, narrow-band products have economic implications given the high cost of research and development needed to discover them, and the often limited size of their market.

A most promising area of research concerns the development of an integrated approach, mainly chemical and biological, but including also physical techniques, to the control of pests. Biotechnologies will no doubt play a primary role in the development of integrated pest control methods. Let me quote the possibility of developing new strains of bacteria and viruses which are pathogenic for the parasite insects. An even more ambitious goal would be that of endowing, through modern genetics, the plant systems with pest-resistant traits. The Commission is now launching a programme of research on biotechnologies, which will no doubt take into account the potential applications of agricultural interest, including the problems discussed in this Symposium.

But more research is also needed on fertilisers and on fertiliser practices themselves.

Biotechnologies give us the hope of substantial reduction in the future use of nitrogen fertilisers, for example, by the introduction of the nitrogen-fixing gene into the genomes of non-fixing crops. Even without thinking of this very long term objective, one could aim at the attainment of symbiotic conditions between nitrogen fixing micro-organisms such as Rhizobium and the roots of cereal crops, thus extending a property exhibited by legumes.

The European Communities' research and development activities, which will be conducted as part of the Environmental Research Programme on 'Air Pollution Effects on Terrestrial and Aquatic Ecosystems' are also of great potential significance. The Programme includes research on the effects of the interaction between chemical pollutants and biotic pathogenes. One area of potential application of this research concerns the study of the interactions between inorganic pollutants (including SO_X, NO_X, ozone, heavy metals) and agrochemicals or their degradation products.

The Symposium stressed the importance of information and training in modern agricultural practices, with due consideration to environmental implications. Agriculture Extension Services should be developed and widely diffused in the Community. Given the great climatic, agricultural and technological diversity among Community countries, it is clear that information services and technical assistance to the farmers must be tailor-made to the specific requirements of the various countries and regions.

Rather than aiming at sophisticated objectives, these services should strive to provide assistance, at the farm level, in assessing in quantitative terms fertiliser and pesticide requirements of crops, and the appropriate times of their application. Quick, simple and reliable methods to achieve this must be developed.

The gap now existing in most countries between the information and agricultural extension functions, however rudimentary they may be, and the scientific and technological research, must be reduced, if we want on the one hand to orient research in order to increase its relevance to the solution of agricultural problems, and on the other hand to improve and widen the quality of the information and assistance supplied to farmers.

Mr Chairman, agriculture is becoming, in Europe as it has already become in the United States, a highly capital-intensive industry. It will become increasingly also a science and technology-intensive industry. A cluster of new technologies will strongly interact with agriculture in the future. Among these are the biotechnologies; information technologies at all levels; remote sensing from space (in relation to climate and weather forecasting, soil conditions, agricultural problems); new materials (including plastics for mulching and greenhouses); renewable energies (biogas, biomass, . . .).

We need to attract young men and women to this traditional and yet rejuvenating sector, in which up to recently the average age of workers (both farmers and employees) has gone on increasing to worrying values. Environmental concern, which is at the top of the value-system in which the new generations recognise themselves, may well help to attract an increasing number of young persons into agriculture. This is particularly important for Europe, a region where unemployment has reached the alarming level of 19 million. In my own country we are particularly concerned with the extremely high rate of unemployment in the population age range between 18 and 25.

Policies to develop and modernise the agricultural system, with emphasis on environmental protection, and the particular issue of marginal lands should be essential components of overall economic and social policies.

A final consideration: in the year 2000 the population of the planet will have reached the level of 6 billion, that is double the world population in 1958, the year the Treaty of Rome was signed.

Arable land will not have doubled, but mankind will live in conditions of nutrition and health almost certainly better than those prevailing in 1958. This is due, among other things, to the use of agrochemicals. What we need, therefore, is not to encourage attitudes hostile to technology, but to learn how to optimise its application, thus contributing to the improvement of the quality of life for all.

European scientists and experts in both agricultural practices and environmental protection can provide a valuable contribution to achieving the objectives identified in our Symposium.

ANNEX I

List of Participants

Commission of the European Communities
Kommission der Europäischen Gemeinschaften
Commission des Communautés Européennes
Rue de la Loi 200, B-1049 Brussels

Cabinet Narjes
J. Schüler
O. von Schwerin

Cabinet Dalsager
H. Fluger

Directorate-General for Environment Consumer Protection and Nuclear Safety
A. J. Fairclough
M. Cornaert
G. Del Bino
B. Lefevre
N. Mergan
G. Morrison
P. Testori-Coggi

Directorate-General for Science Research and Development—Joint Research Centre
Ph. Bourdeau
H. Ott
H. Barth
P. L'Hermite
G. Weidenbach
K. Kristensen

Directorate-General for Agriculture
G. Hudson
J. Lougheed

Economic and Social Committee
Wirtschafts- und Sozialausschuss
Comité Économique et Social
Rue Ravenstein 2, B-1000 Brussels
N. Mooney
T. Roseingrave

Irish Steering Committee
An Foras Forbartha
St Martin's House
Waterloo Road
Dublin 4
W. McCumiskey (*Chief Executive*)
E. Mulkeen
B. Clarke
F. Curran
P. Flanagan
L. Griffin
E. Lacey
C. Malone

National Board for Science and Technology
Shelbourne House
Shelbourne Road
Dublin 4
V. O'Gorman

Symposium Staff
An Foras Forbartha
E. Armonici
P. Bailey
M. Casey
G. Egan
P. Kinsella
L. O'Reilly
M. Ryan
S. Stone
R. Walpole

National Board for Science and Technology
D. O'Gorman
L. Sheehan

Belgium/Belgien/Belgique

Cosse, Mr J.
GIFAP
12 Avenue Harmoir
Brussels

David, Mr H.
European Environmental Bureau
29 Rue Vautier
1040 Brussels

Evans, Mr D.
Monsanto Europe
270–272 Avenue de Tervueren
1150 Brussels

Firmin, Mr R.
Dow Corning Europe
154 Chaussee de la Hulpe
1170 Brussels

Klatte, Mr E.
European Environmental Bureau
29 Rue Vautier
1040 Brussels

Koole, Mr D.
Norsk Hydro Belgium SA
149 Avenue Louise, Bte 11
1050 Brussels

Lebrun, Mr Ph.
Université Catholique de Louvain
Place Croix du Sud, 5
1348 Louvain-la-Neuve

Meert, Mr E.
CMg-ENGRAIS
148 Chaussee de Charleroi, Bte 5
1060 Brussels

Stehr, Mr H.
73 Rue d'Arlon
1040 Brussels

Van Haecke, Mr P.
Belgian Science Policy Office
Wetenschapsstraat 8
1040 Brussels

Van Quaquebeke, Mr R.
European Parliament Liberal Group
3 Boulevard de l'Empereur
1000 Brussels

Vlassak, Mr K.
K. U. Leuven
Kard Mercierlaan 92
B-3030 Heverlee-Leuven

Denmark/Dänemark/Danemark

Andreasen, Mr A.
Superfos a/s
Frydenlundsvej 30
DK-2950 Vedbaek

Bro-Rasmussen, Mr F.
Technical University of Denmark
Laboratory of Environmental Science
and Ecology
DK-2800 Lyngby

Christensen, Mr H
Agency of Environmental Protection
Strandgade 29
DK-1401 Copenhagen, K

Dam Kofoed, Mr A.
State Experimental Station Askov
Vejenvej 55
DK-6600 Vejen

Hansen, Mr J.
Ministry of Agriculture
Bureau of Land Data
Enghavevej 2
DK-7100 Vejle

Hansen, Mr K.
Game Biology Station
Kalø
DK-8410 Lunde

Henningsen, Mr J.
Agency of Environmental Protection
Strandgade 29
DK-1401 Copenhagen, K

Hunding, Mr C.
NAEP Freshwater Laboratory
52 Lysbrogade
DK-8600 Silkeborg

Jensen, Mr E.
Statens Plantevaernscenter
Lottenborgvej 2
DK-2800 Lyngby

Løkke, Mr H.
Agency of Environmental Protection
Centre of Terrestrial Ecology
Gyden 2
DK-2860 Søborg

Morgensen, Ms E.
Agency of Environmental Protection
Strandgade 29
DK-1401 Copenhagen, K

Mortensen, Mr K.
Ministry of Agriculture
Slotsholmsgade 10
DK-1216 Copenhagen, K

Olsen, Ms H.
Agency of Environmental Protection
Strandgade 29
DK-1401 Copenhagen, K

Skriver, Mr K.
Danish Agricultural Advisory Centre
Kongsgaardsvej 28
DK-8260 Viby J

Sommer, Mr S.
Agency of Environmental Protection
Strandgade 29
DK-1401 Copenhagen, K

Wichmann, Mr H.
Agency of Environmental Protection
Strandgade 29
DK-1401 Copenhagen, K

Finland/Finnland/Finlande

Pahkala, Mr O.
Ministry of the Environment
PI 306
SF-00531 Helsinki 53

France/Frankreich/France

Balland, Mr D.
Compagnie Française de l'Azote
46 Rue Jacques Dulud
92200 Neuilly-sur-Seine

Bassino, Mr J. P.
Association de Coordination Technique Agricole
149 Rue de Bercy MNE
75595 Paris, Cedex 12

Calmejane, Mr F.
Rohm & Haas France
185 Rue de Bercy
75579 Paris, Cedex 12

Chabason, Mr L.
Ministère de l'Environment
14 Bvd du General LeClerc
92524 Neuilly-sur-Seine

Hascoet, Mr M.
INRA—Station de Phytopharmacie
Route de St-Cyr
78000 Versailles

Ignazi, Mr J.
CdF Chimie AZF
Tour Aurore, Cedex 5
92080 Paris, Defense 2

Martinière, Mr G. de la
Ministère de la Recherche et de la Technologie
1 Rue Descartes
75005 Paris

Peirani, Mr R.
Ministère de l'Industrie et Commerce
30/32 Rue Guersant
75840 Paris, Cedex 17

Remy, Mr J.
INRA
Départment d'Agronomie
Rue F. Christ, BP 101
02000 Laon, Cedex

Scorraille, Mr G. de
Federation Nationale de l'Industrie des Engrais
58 Avenue Kleber
75794 Paris, Cedex 16

Severin, Mr F.
ACTA
4 Place Gensoul
69287 Laon, Cedex 1

Thiault, Mr J.
Ministère de l'Agriculture
175 Rue du Chevaleret
75646 Paris, Cedex 13

Federal Republic of Germany/ Bundesrepublik Deutschland/ République Fédérale d'Allemagne

Ahrens, Mr C.
Bayer AG
Pflanzenschutzzentrum Monheim
Ressort Anwendungstechnik
5090 Leverkusen-Bayerwerk

Amberger, Mr A.
Technische Universität München-Weihenstephan
Fachgruppe Bodenkunde
8050 Freising 1

Blaschke, Mr M.
Aktionskreis Umwelt und Tierschutz Ostwestfalen
Auf dem Tie 3
4800 Bielefeld 15

Böttcher, Mr O.
Industrieverband Pflanzenschutz e.V.
Karlstr. 21
6000 Frankfurt/Main

Carneim, Ms C.
Aktionskreis Umwelt und Tierschutz Ostwestfalen
Tempelhofer Weg 82
4800 Bielefeld 1

Diercks, Mr R.
Petristrasse 7
7000 München 90

Dieterich, Mr F.
BELF
5300 Bonn 1

Gebhard, Mr H. F.
Universität Hohenheim
Institut für landwirtschaftliche Betriebslehre
Postfach 700
5620 Stuttgart 70
Stuttgart-Hohenheim

Jürgens, Ms S.
BASF AG
Carl-Bosch-Strasse
PO Box 220
6803 Limburgerhof

Kohlemeyer, Mr M.
Schell Strasse 5
D-463 Bochum, 1

Kördel, Mr W.
Fraunhofer-Institut für Toxikologie
Grafschaft/Hochsauerland
5948 Schmallenberg

Leber, Mr G.
Industrieverband Pflanzenschutz e.V.
Karlstrasse 21
6000 Frankfurt/Main

Lieth, Mr H.
Universität Osnabrück
F.B. 5 Biologie/Chemie
Postfach 4469
4500 Osnabrück

Nieder, Mr H.
Verband der Chemischen Industrie
Fachverband Stickstoffindustrie
Karlstrasse 19/21
6000 Frankfurt 1

Schulze-Weslarn, Mr K.
Bundesministerium für Ernährung, Landwirtschaft und Forsten
Rochusstr. 1
Bonn-Duisdorf

Selenka, Mr F.
Ruhr-Universität Bochum
Institut für Hygiene
Universitätsstr. 150
4630 Bochum

Thormann, Mr A.
Federal Environmental Agency (Umweltbundesamt)
1 Bismarckplatz
1000 Berlin 33

von Weizsäcker, Mr E. U.
Institute for European Environmental Policy
Aloys Schultz Strasse 6
53 Bonn

Waitz, Mr G.
Hoechst AG
Hofackerstr. 14
6200 Wiesbaden-Naurod

Weber, Ms B.
European Parliament Environment Committee
Sickingenstr. 1
6900 Heidelberg

Greece/Griechenland/Grèce

Antonakov, Mrs M.
Ministry of Agriculture
3–5 Hippokratous St
10679 Athens

Balayannis, Mr P.
Agricultural College of Athens
Iera Odos 75
GR-118 55 Athens

Stavros, Mr R.
Ministry of Agriculture
3–5 Hippokratous St
10679 Athens

Ireland/Irland/Irlande

Benson, Mr F.
An Bord Pleanala
Irish Life Centre
Lower Abbey Street
Dublin, 1

Brogan, Mr J.
An Foras Taluntais
Dunsinea Research Centre
Castleknock
Co. Dublin

Caffrey, Mr J.
ICI (Ireland) Ltd
5 South Frederick Street
Dublin, 2

Carey, Mr M.
Department of Fisheries & Forestry
Forest & Wildlife Service
1–3 Sidmonton Place
Bray
Co. Wicklow

Cassidy, Mr E.
An Bord Pleanala
Irish Life Centre
Lower Abbey Street
Dublin, 1

Cathcart, Mr R.
Federation of Irish Chemicals Industries
13 Fitzwilliam Square
Dublin, 2

Chambers, Mr P.
Department of Pharmacology
Trinity College
Dublin, 2

Champ, Mr T.
Central Fisheries Board
Mobhi Boreen
Glasnevin
Dublin, 9

Convery, Mr F.
REPC
University College Dublin
Richview
Clonskeagh Drive
Dublin, 6

Couchman, Mr J.
FACE
Johnstown House
Carlow

Coyle, Mr J.
School of Science
Athlone Regional Technical College
Athlone
Co. Westmeath

Cunningham, Mr P.
An Foras Taluntais
Sandymount Avenue
Dublin, 4

Daly, Mr D.
Geological Survey of Ireland
Beggars Bush
Haddington Road
Dublin, 4

Daly, Mr O.
An Foras Taluntais
Sandymount Avenue
Dublin, 4

Deigan, Mr M.
Laois County Council
County Hall
Portlaoise
Co. Laois

Desmond, Mr T.
Cavan County Council
Courthouse
Cavan

Devlin, Mr B.
Department of Agriculture
6C Agriculture House
Kildare Street
Dublin, 2

Dowding, Mr P.
Department of Botany
Trinity College
Dublin, 2

Eades, Mr J.
An Foras Taluntais
Oak Park Research Centre
Carlow

Flanagan, Mr P.
An Foras Forbartha
Pottery Road
Deansgrange
Co. Dublin

Gately, Mr T.
An Foras Taluntais
Johnstown Castle
Co. Wexford

Good, Mr J.
University College Cork
Lee Maltings
Prospect Row
Cork

Gray, Mr N.
Department of Botany
Trinity College
Dublin, 2

Hassett, Mr M.
'Sheena'
Charleville Road
Tullamore
Co. Offaly

Howell, Mr R.
Department of Agriculture
3E Agriculture House
Kildare Street
Dublin, 2

Kenny, Mr P.
An Foras Forbartha
St Martin's House
Waterloo Road
Dublin, 4

Kiernan, Mr P.
Goulding Fertilisers
Fitzwilton House
Wilton Place
Dublin, 2

Lawless, Ms K.
BASF Ireland Ltd
Clonee
Co. Meath

Leahy, Mr E.
Irish Distillers Ltd
Midleton Distillery
Midleton
Co. Cork

Lynch, Mr M.
Department of Agriculture
Pesticide Control Unit
24 Upr Merrion Street
Dublin, 2

Lynch, Mr M.
IIRS
Ballymun Road
Dublin, 9

McAleese, Mr D.
University College Dublin
Belfield
Dublin, 4

McAteer, Mr W.
Department of Agriculture
Kildare Street
Dublin, 2

McCullen, Mr R.
Irish Farmers Association
Farm Centre
Bluebell
Dublin, 12

McCumiskey, Mr W.
An Foras Forbartha
St Martin's House
Waterloo Road
Dublin, 4

McGee, Mr R.
Department of Environment
Custom House
Dublin, 1

McLoughlin, Mr J.
Department of Agriculture
Abbotstown
Castleknock
Co. Dublin

Moore, Mr D.
Department of Environment
Custom House
Dublin, 1

Murphy, Mr J.
Irish Farmers Association
Farm Centre
Bluebell
Dublin, 12

O'Doherty, Mr G.
Department of Environment
Custom House
Dublin, 1

O'Donnell, Mr C.
An Foras Forbartha
Pottery Road
Deansgrange
Co. Dublin

O'Donohue, Mr S.
Department of Fisheries & Forestry
Leeson Lane
Dublin, 2

O'Dowd, Mr P.
Fisheries Conservation Manager
Electricity Supply Board
27 Fitzwilliam Street
Dublin, 2

O'Gorman, Mr F.
An Foras Forbartha
St Martin's House
Waterloo Road
Dublin, 4

O'Gorman, Mr V.
National Board for Science & Technology
Shelbourne House
Shelbourne Road
Dublin, 4

O'Hare, Mr P.
An Foras Taluntais
Oak Park Research Centre
Carlow

O hOgain, Mr S.
College of Technology
Bolton Street
Dublin, 1

O'Leary, Mr S.
Hoechst (Ireland) Ltd
Cookstown Industrial Estate
Tallaght
Co. Dublin

O'Regan, Mr J.
Office of Public Works
Newtown
Trim
Co. Meath

O'Reilly, Mr C.
Irish Agrochemicals Association
Temple Hall
Blackrock
Co. Dublin

O'Sullivan, Ms M.
Department of Fisheries & Forestry
Abbotstown
Castleknock
Co. Dublin

Prendeville, Mr G.
Department of Botany
University College
Cork

Pugh, Mr D.
Veterinary College
Ballsbridge
Dublin, 4

Quigley, Mr J.
State Laboratory
Abbotstown
Castleknock
Co. Dublin

Regan, Mr S.
ACOT North West Region
Ballyhaise
Co. Cavan

Robinson, Mr D.
An Foras Taluntais
Kinsealy Research Centre
Malahide
Co. Dublin

Rochford, Mr J.
Bradley, Kavanagh, Rochford & Associates
7 Mulgrave Street
Dun Laoghaire
Co. Dublin

Ryan, Mr E.
ACOT National Office
Frascati Road
Blackrock
Co. Dublin

Ryan, Mr E.
An Foras Taluntais
Kinsealy Research Centre
Malahide
Co. Dublin

Ryder, Mr C.
Office of Public Works
17–19 Lower Hatch Street
Dublin, 2

Scannell, Ms Y.
c/o Law School
Trinity College
Dublin, 2

Sherwood, Ms M.
An Foras Taluntais
Johnstown Castle
Co. Wexford

Stafford, Mr L.
N.E.T.
60 Northumberland Road
Dublin, 4

Stanley, Mr M.
COPA
c/o Irish Farm Centre
Bluebell
Dublin, 12

Timpson, Mr J.
Sligo Regional Technical College
Ballinode
Sligo

Toner, Mr P.
An Foras Forbartha
Pottery Road
Deansgrange
Co. Dublin

Tracey, Mr T.
ACOT Regional Office
Lyons Estate
Newcastle
Co. Dublin

Tunney, Mr H.
An Foras Taluntais
Johnstown Castle
Co. Wexford

Ni Uid, Ms G.
National Board for Science & Technology
Shelbourne House
Shelbourne Road
Dublin, 4

Vial, Mr V.
An Foras Taluntais
19 Sandymount Avenue
Dublin, 4

Walsh, Ms M.
State Laboratory
Abbotstown
Castleknock
Co. Dublin

Whyte, Mr M.
ACOT Agricultural Education Centre
Corduff
Lusk
Co. Dublin

Young, Mr T.
Grassland Fertilisers
75 Merrion Square
Dublin, 2

Italy/Italien/Italie

Azzolina, Ms C.
Anic Agricoltura SpA
Via Medici Del Vascello 26
20138 Milan

Bianchi, Mr A.
Anic Agricoltura SpA
Via Medici Del Vascello 26
20138 Milan

Braun, Mr H.
Fertiliser & Plant Nutrition Service
FAO UN
Via Delle Terme di Caracalla
1-00100 Rome

Colombo, Mr U.
ENEA
Viale Regina Margherita 125
00198 Rome

Guarrera, Mr L.
CNCD
43 Via XXIV Maggio
00187 Rome

Ravenna, Mr R.
Fertimont Sfor
Piazza Repubblica 14/16
Milan

Sequi, Mr P.
University of Udine
Piazzale M Kilbe 4
30100 Udine

Triolo, Mr L.
FARE—ENEA
Cre Casaccia
Rome

Vighi, Mr M.
Università delgi Studi
Facultà di Agraria
Via Celoria 2
20133 Milan

Vita, Mr G.
ENEA CRE
Casaccia CP 2400
Rome

Luxembourg

Mirgain, Mr G
Centrale Paysanne Luxembourg
16 Bvd d'Avranches
L-2980 Luxembourg

Netherlands/Die Niederlände/Pays-Bas

Besemer, Mr A.
Agricultural University Wageningen
Hartense Weg 30
6705 BJ Wageningen

Don, Ms A.
Ministry of Agriculture & Fisheries
PO Box 20401
2500 EK Den Haag

Gerritsen, Mr H.
Agricultural Board
Prinsevinkenpark 19
2585 HK Den Haag

Klitsie, Ms C.
Ministry of Agriculture & Fisheries
PO Box 20401
2500 EK Den Haag

Koeman, Mr J.
Department of Toxicology
Agricultural University Wageningen
De Treyen 12
6705 BJ Wageningen

Prinz, Mr W.
Institute for Soil Fertility
PO Box 30003
Haren

van Burg, Mr P.
Netherlands Research & Advisory
Institute for Fertilisers
Thorbeckelaan 360
2564 BZ Den Haag

van der Es, Mr W.
SSC & M
Koepelstr. 60
Bergen op Zoom

van Haasteren, Mr J.
Ministry of Housing, Physical Planning
and Environment
PO Box 450
2260 MB Leidschendam

Norway/Norwegen/Norvège

Marstrander, Mr R.
Norsk Hydro A/S
PO Box 2594
Solli

Sweden/Schweden/Suède

Gunnarsson, Mr O.
Supra AB
Box 516
S-261 24 Landskrona

Switzerland/die Schweiz/Suisse

Carmichael, Mr N.
DOW Chemical Europe
Bachtoblestrasse 3
CH 8810 Horgen

Keefer, Mr C.
SRNDO Ltd
PO Box 4002
Basle

Zehler, Mr E.
International Potash Institute
PO Box 121
CH-3048 Worblaufen

United Kingdom/
Vereinigtes Königreich/
Royaume-Uni

Adams, Mr N.
Norsk Hydro Fertilisers Ltd
Levington Research Station
Ipswich IP10 0LU

Bainbridge, Mr A.
Ministry of Agriculture Fisheries &
Food
Horseferry Road
London SW1P 2AE

Baldock, Mr D.
Institute for European Environmental
Policy
10 Percy Street
London W1P 0DR

Bibbings, Mr R.
Trades Union Congress
Congress House
Great Russell Street
London WC1

Brady, Mr J.
Imperial College
Silwood Park
Ascot
Berks SL5 7PY

Broadbent, Mr L.
BASIS
Junipers
Manorial Road
Parkgate
South Wirral L64 6QW

Brownlie, Mr T.
Department of Agriculture & Fisheries
Room 311 Chesser House
500 Gorgie Road
Edinburgh EH11 3AW

Bryan, Ms K.
Severn Trent Water Authority
Abelson House
2297 Coventry Road
Sheldon
Birmingham

Bunyan, Mr P.
Ministry of Agriculture, Fisheries & Food
Great Westminster House
Horseferry Road
London SW1P 2AE

Carson, Mr M.
Ulster Farmers Union
475 Antrim Road
Belfast BT15 3DA

Clark, Mr L.
Water Research Centre
Medmenham Laboratory
PO Box 16 Marlow
Bucks SL7 2HD

Conway, Mr G.
Centre for Environmental Technology
Imperial College of Science & Technology
48 Princes Gate
London SW7 1LU

Cooke, Mr G.
Rothamsted Experimental Station
Harpenden
Herts AL5 2JQ

Croll, Mr B.
Anglian Water
Ambury Road
Huntingdon
Cambs PE18 6NZ

Crozier, Mr J.
ICI plc
PO Box 1
Billingham
Cleveland TS23 1LE

Cutler, Mr J.
Department of Agriculture & Fisheries
DAFS
East Craigs
Edinburgh EH12 8NJ

Davies, Mr P.
ICI plc
Medical Department
PO Box 8
Billingham
Cleveland TS23 1LE

Frame, Mr J.
West of Scotland Agricultural College
Auchincruive
Ayr KA6 5HW

Gaunt, Mr I.
British Industrial Biological Research
Woodmansterne Road
Carshalton
Surrey SM5 4DS

Geake, Mrs A.
British Geological Survey
Maclean Building
Crowmarsh Gifford
Wallingford
OX10 8BB

Gracey, Mr H.
Department of Agriculture for Northern Ireland
Greenmount Agricultural & Horticultural College
Antrim

Green, Mr R.
Agricultural & Allied Workers
National Trade Group
Headland House
308 Gray's Inn Road
London WC1X 8DS

Hardwick, Mr D.
Ministry of Agriculture, Fisheries & Food
Great Westminster House
Horseferry Road
London SW1P 2AE

Hydes, Mr O.
Department of the Environment
Room A4—34 Romney House
43 Marsham Street
London SW1P 3PY

King, Mr N.
Department of the Environment
Room A33—34 Romney House
43 Marsham Street
London SW1P 3PY

Le Grice, Mr S.
Agricultural Development and Advisory Service—Welsh Office
Trawsgoed
Aberystwyth
Dyfed SY23 4HT

Marshall, Mr J.
ICI plc
Agricultural Division
PO Box 1
Billingham
Cleveland TS23 1LJ

Marston, Ms M.
Royal Society for Protection of Birds
The Lodge
Sandy
Bedfordshire SG19 2DL

Mathieson, Mr A.
The East of Scotland College of Agriculture
Veterinary Division
Bush Estate
Penicuik
Midlothian EH26 0QE

Matthiessen, Mr P.
Tropical Development Research Institute
College House
Wright's Lane
London W8

Moss, Mr C.
NW Water Division
PO Box 12—Newtown House
Buttermarket Street
Warrington
Cheshire

Mottram, Mr J.
Fertiliser Manufacturer's Association
Greenhill House
90–93 Cowcross Street
London EC1M 2AE

Needham, Mr P.
Ministry of Agriculture, Fisheries & Food
Great Westminster House,
Horseferry Road
London SW1P 2AE

O'Keeffe, Mr M.
Monsanto Europe
Thames Tower
Burley's Way
Leicester

Osborn, Mr D.
NERC
Monkswood Experimental Station
Huntingdon PE17 2LS

Owen, Mr T.
ICI plc
Jealotts Hill Research Station
Bracknell
Berks RG12 6EY

Park, Mr J.
Ministry of Agriculture, Fisheries & Food
Great Westminster House
Horseferry Road
London SW1P 2AE

Payne, Mr M.
National Farmers Union
Agriculture House
Knightsbridge
London SW1X 7NJ

Perfect, Mr J.
Tropical Development Research Institute
College House
Wright's Lane
London W8

Peters, Mr. J.
Department of the Environment
Room 1116—Tollgate House
Houlton Street
Bristol BS2 9DJ

Power, Mr A.
Ministry of Agriculture, Fisheries & Food
Room 29A
55 Whitehall
London SW1A 2EY

Richards, Mr I.
Norsk Hydro Fertilisers Ltd
Levington Research Station
Levington
Ipswich IP10 0LU

Roberts, Mr G.
Institute of Hydrology
McLean Building
Crowmarsh Gifford
Oxfordshire

Robinson, Mr R.
Wessex Water Authority
Wessex House
Passage Street
Bristol BS2 0JQ

Rowe, Mr R.
Dow Chemical Company
King's Lynn
Norfolk

Seaman, Mr D.
ICI plc
Plant Protection Division
Jealotts Hill Research Station
Bracknell
Berks

Seebohm, Mr R.
Department of Trade & Industry
Ashdown House
123 Victoria Street
London SW1E

Simms, Mr D.
Department of the Environment
34 Romney House
43 Marsham Street
Westminster
London SW1

Solbe, Mr J.
Medmenham Laboratory
PO Box 16, Marlow
Bucks SL7 2HD

Stanley, Mr P.
Agricultural Science Service
Ministry of Agriculture, Fisheries & Food
Slough Laboratory
London Road
Slough SL3 7HJ

Vowles, Mr J.
Norsk Hydro Fertilisers
Levington Research Station
Levington
Ipswich IP10

Wastle, Mr J.
Scottish Development Department
Pentland House
47 Robb's Loan
Edinburgh EH14 1TY

Williams, Mr A.
ICI plc
PO Box No. 1
Billingham
Cleveland TS23 1LB

Winteringham, Mr P.
Darbod
Harlech
Gwynedd LL46 2RA
Wales

International Scientific Journalists

Bock, Mr E.
Commission of the European Communities
DGXII—Rue de la Loi 200
Brussels
Belgium

Caulfield, Ms K.
22 Earls Terrace, 1
London, W8
United Kingdom

Deliggeorges, Ms S.
29 Rue du Louvre
75002 Paris
France

Dienel, Mr R.
c/o APA Gunoldstr. 14
1199 Wien
Austria

Ford, Mr B.
Mill Park House
57 Westville Road
Cardiff CF2 5DF
Wales

Langley, Ms P.
20 Rue du Commandant Mouchotte
75014 Paris, France

Noldechen, Mr A.
Rothschildallee 28
6000 Frankfurt-am-Main
FR Germany

Parman, Mr G.
Kirkeveien, Rute 15
1450 Nesoddtangen
Norway

Riccio, Ms A.
Via Pietro Cossa 5
20122 Milano
Italy

Rothery, Mr B.
c/o IIRS
Ballymun Road
Dublin, 9
Ireland

Saetter-Lassen, Mr C.
Kirseskoven
4640 Fakse
Denmark

Soederlund, Mr S.
Moraenvaegen, 2
13651 Handen
Sweden

van Kasteren, Mr J.
Otterlaan 8
2623 CX Delft
Netherlands

Weber, Mr R.
Im Chapf 141
5225 Oberbozberg
Switzerland

Wentein, Mr G.
Denderstraat 23
9322 Gijzegem
Belgium

Wright, Mr P.
200 Grays Inn Road
London WC1 8EZ
United Kingdom

Index